Health Risk and Exposure Assessment for Ozone

Second External Review Draft

Chapter 5 Appendices

DISCLAIMER

This draft document has been prepared by staff from the Risk and Benefits Group, Health and Environmental Impacts Division, Office of Air Quality Planning and Standards, U.S. Environmental Protection Agency. Any findings and conclusions are those of the authors and do not necessarily reflect the views of the Agency. This draft document is being circulated to facilitate discussion with the Clean Air Scientific Advisory Committee to inform the EPA's consideration of the ozone National Ambient Air Quality Standards.

This information is distributed for the purposes of pre-dissemination peer review under applicable information quality guidelines. It has not been formally disseminated by EPA. It does not represent and should not be construed to represent any Agency determination or policy.

Questions related to this preliminary draft document should be addressed to Dr. Bryan Hubbell, U.S. Environmental Protection Agency, Office of Air Quality Planning and Standards, C539-07, Research Triangle Park, North Carolina 27711 (email: hubbell.bryan@epa.gov).

EPA-452/P-14-004c
February 2014

Health Risk and Exposure Assessment for Ozone

Second External Review Draft

Chapter 5 Appendices

U.S. Environmental Protection Agency
Office of Air and Radiation
Office of Air Quality Planning and Standards
Health and Environmental Impacts Division
Risk and Benefits Group
Research Triangle Park, North Carolina 27711

Appendix 5-A

Description of the Air Pollutants Exposure Model (APEX)

Table of Contents

5A-1. OVERVIEW **2**

5A-2. MODEL INPUTS **4**

5A-3. DEMOGRAPHIC CHARACTERISTICS **5**

5A-4. ATTRIBUTES OF INDIVIDUALS **5**

5A-5. CONSTRUCTION OF LONGITUDINAL DIARY SEQUENCE **6**

5A-6. KEY PHYSIOLOGICAL PROCESSES MODELED **8**

5A-7. ESTIMATING MICROENVIRONMENTAL CONCENTRATIONS **10**

5A-7.1. MASS BALANCE MODEL **10**

5A-7.2. FACTORS MODEL **16**

5A-8. EXPOSURE AND DOSE TIME SERIES CALCULATIONS **17**

5A-9. MODEL OUTPUT **19**

5A-10. REFERENCES **21**

List of Tables

Table 5A-1. Ventilation coefficient parameter estimates (b_i) and residuals distributions (e_i) from Graham and McCurdy (2005). 10

List of Figures

Figure 5A-1. Illustration of the mass balance model used by APEX. 11

Figure 5A-2. Example of microenvironmental and exposure concentrations for a simulated individual over a 48 hours simulation. (H: home, A: automobile, S: school, P: playground, O: outdoors at home). 18

Figure 5A-3. The percent of simulated children (ages 5-18) at or above 8-hour average O_3 exposures while at moderate or greater exertion. 20

This Appendix briefly describes the EPAs Air Pollutants Exposure (APEX) model.

5A-1. OVERVIEW

APEX is the human inhalation exposure model within the Total Risk Integrated Methodology (TRIM) framework (US EPA 2012a,b). APEX is conceptually based on the probabilistic NAAQS Exposure Model (pNEM) that was used to estimate population exposures for the 1996 O_3 NAAQS review (Johnson et al., 1996a; 1996b; 1996c). Since that time the model has been restructured, improved, and expanded to reflect conceptual advances in the science of exposure modeling and newer input data available for the model. Key improvements to algorithms include replacement of the cohort approach with a probabilistic sampling approach focused on individuals, accounting for fatigue and oxygen debt after exercise in the calculation of ventilation rates (Isaacs et al., 2008), and new approaches for construction of longitudinal activity patterns for simulated persons (Glen et al. 2008; Rosenbaum et al., 2008). Major improvements to data input to the model include updated air exchange rates (AERs), population census and commuting data, and the daily time-location-activities database. These improvements are described later in this chapter.

APEX estimates human exposure to criteria and toxic air pollutants at local, urban, or regional scales using a stochastic, microenvironmental approach. That is, the model randomly selects data on a sample of hypothetical individuals in an actual population database and simulates each individual's movements through time and space (e.g., at home, in vehicles) to estimate their exposure to the pollutant. APEX can assume people live and work in the same general area (i.e., that the ambient air quality is the same at home and at work) or optionally can model commuting and thus exposure at the work location for individuals who work.

The APEX model is a microenvironmental, longitudinal human exposure model for airborne pollutants. It is applied to a specified study area, which is typically a metropolitan area. The time period of the simulation is typically one year, but can easily be made either longer or shorter. APEX uses census data, such as gender and age, to generate the demographic characteristics of simulated individuals. It then assembles a composite activity diary to represent the sequence of activities and microenvironments that the individual experiences. Each microenvironment has a user-specified method for determining air quality. The inhalation exposure in each microenvironment is simply equal to the air concentration in that microenvironment. When coupled with breathing rate information and a physiological model, various measures of dose can also be calculated.

The term *microenvironment* is intended to represent the immediate surroundings of an individual, in which the pollutant of interest is assumed to be well-mixed. Time is modeled as a sequence of discrete time steps called *events*. In APEX, the concentration in a microenvironment may change between events. For each microenvironment, the user specifies the method of concentration calculation (either mass balance or regression factors, described later in this paper), the relationship of the microenvironment to the ambient air, and the strength of any pollutant sources specific to that microenvironment. Because the microenvironments that are relevant to exposure depend on the nature of the target chemical and APEX is designed to be applied to a wide range of chemicals, both the total number of microenvironments and the properties of each are free to be specified by the user.

The ambient air data are provided as input to the model in the form of time series at a list of specified locations. Typically, hourly air concentrations are used, although temporal resolutions as small as one minute may be used. The spatial range of applicability of a given ambient location is called an air district. Any number of air districts can be accommodated in a model run, subject only to computer hardware limitations. In principle, any microenvironment could be found within a given air district. Therefore, to estimate exposures as an individual engages in activities throughout the period it is necessary to determine both the microenvironment and the air district that apply for each event.

An *exposure event* is determined by the time reported in the activity diary; during any event the district, microenvironment, ambient air quality, and breathing rate are assumed to remain fixed. Since the ambient air data change every hour, the maximum duration of an event is limited to one hour. The event duration may be less than this (as short as one minute) if the activity diary indicates that the individual changes microenvironments or activities performed within the hour.

An APEX simulation includes the following steps:

1. <u>Characterize the study area</u> - APEX selects sectors (e.g., census tracts) within a study area based on user-defined criteria and thus identifies the potentially exposed population and defines the air quality and weather input data required for the area.
2. <u>Generate simulated individuals</u> - APEX stochastically generates a sample of simulated individuals based on the census data for the study area and human profile distribution data (such as age-specific employment probabilities). The user must specify the size of the sample. The larger the sample, the more representative it is of the population in the study area and the more stable the model results are (but also the longer the computing time).
3. <u>Construct a long-term sequence of activity events and determine breathing rates</u> - APEX constructs an event sequence (activity pattern) spanning the period of simulation for each simulated person. The model then stochastically assigns breathing rates to each event, based on the type of activity and the physical characteristics of the simulated person.
4. <u>Calculate pollutant concentrations in microenvironments</u> - APEX enables the user to define any microenvironment that individuals in a study area would visit. The model then calculates concentrations of each pollutant in each of the microenvironments.

5. Calculate pollutant exposures for each simulated individual - Microenvironmental concentrations are time weighted based on individuals' events (i.e., time spent in the microenvironment) to produce a sequence of time-averaged exposures (or minute by minute time series) spanning the simulation period.

6. Estimate dose - APEX can also calculate the dose time series for each of the simulated individuals based on the exposures and breathing rates for each event. For O_3, the adverse health metric of interest is decrement in forced expiratory volume occurring in one second (FEV_1). This algorithm responsible for combining the time series of APEX estimated exposure and breathing rates for individuals is discussed in greater detail in the main body of the REA, Chapter 6.

The model simulation continues until exposures are determined for the user-specified number of simulated individuals. APEX then calculates population exposure statistics (such as the number of exposures exceeding user-specified levels) for the entire simulation and writes out tables of distributions of these statistics.

5A-2. MODEL INPUTS

APEX requires certain inputs from the user. The user specifies the geographic area and the range of ages and age groups to be used for the simulation. Hourly (or shorter) ambient air quality and hourly temperature data must be furnished for the entire simulation period. Other hourly meteorological data (humidity, wind speed, wind direction, precipitation) can be used by the model to estimate microenvironmental concentrations, but are optional.

In addition, most variables used in the model algorithms are represented by user-specified probability distributions which capture population variability. APEX provides great flexibility in defining model inputs and parameters, including options for the frequency of selecting new values from the probability distributions. The model also allows different distributions to be used at different times of day or on different days, and the distribution can depend conditionally on values of other parameters. The probability distributions available in APEX include beta, binary, Cauchy, discrete, exponential, extreme value, gamma, logistic, lognormal, loguniform, normal, off/on, Pareto, point (constant), triangle, uniform, Weibull, and nonparametric distributions. Minimum and maximum bounds can be specified for each distribution if a truncated distribution is appropriate. There are two options for handling truncation. The generated samples outside the truncation points can be set to the truncation limit; in this case, samples "stack up" at the truncation points. Alternatively, new random values can be selected, in which case the probability outside the limits is spread over the specified range, and thus the probabilities inside the truncation limits will be higher than the theoretical untruncated distribution.

5A-3. DEMOGRAPHIC CHARACTERISTICS

The starting point for constructing a simulated individual is the population census database; this contains population counts for each combination of age, gender, race, and *sector*. The user may decide what spatial area is represented by a sector, but the default input file defines a sector as a *census tract*. Census tracts are variable in both geographic size and population number, though usually have between 1,500 and 8,000 persons. Currently, the default file contains population counts from the 2000 census for every census tract in the United States, thus the default file should be sufficient for most exposure modeling purposes. The combination of age, gender, race, and sector are selected first. The sector becomes the *home sector* for the individual, and the corresponding air district becomes the *home district*. The probabilistic selection of individuals is based on the sector population and demographic composition, and taken collectively, the set of simulated individuals constitutes a random sample from the study area.

The second step in constructing a simulated individual is to determine their employment status. This is determined by a probability which is a function of age, gender, and home sector. An input file is provided which contains employment probabilities from the 2000 census for every combination of age (16 and over), gender, and census tract. APEX assumes that persons under age 16 do not commute. For persons who are determined to be workers, APEX then randomly selects a *work sector*, based on probabilities determined from the commuting matrix. The work sector is used to assign a *work district* for the individual that may differ from the home district, and thus different ambient air quality may be used when the individual is at work.

The commuting matrix contains data on flows (number of individuals) traveling from a given home sector to a given work sector. Based on commuting data from the 2000 census, a commuting data base for the entire United States has been prepared. This permits the entire list of non-zero flows to be specified on one input file. Given a home sector, the number of destinations to which people commute varies anywhere from one to several hundred other tracts.

5A-4. ATTRIBUTES OF INDIVIDUALS

In addition to the above demographic information, each individual is assigned status and physiological attributes. The status variables are factors deemed important in estimating microenvironmental concentrations, and are specified by the user. Status variables can include, but are not limited to, people's housing type, whether their home has air conditioning, whether they use a gas stove at home, whether the stove has a gas pilot light, and whether their car has air conditioning. Physiological variables are important when estimating pollutant specific dose. These variables could include height, weight, blood volume, pulmonary diffusion rate, resting metabolic rate, energy conversion factor (liters of oxygen per kilocalorie energy expended), hemoglobin density in blood, maximum limit on metabolic equivalents of work (MET) ratios (see below), and endogenous CO production rate. All of these variables are treated probabilistically taking into account interdependencies where possible, and reflecting variability in the population.

Two key personal attributes determined for each individual in this assessment are body mass (BM) and body surface area (BSA). Each simulated individual's body mass was randomly sampled from age- and gender-specific body mass distributions generated from National Health and Nutrition Examination Survey (NHANES) data for the years 1999-2004.[1] Details in their development and the parameter values are provided by Isaacs and Smith (2005). Then age- and gender-specific body surface area can be estimated for each simulated individual. Briefly, the BSA calculation is based on logarithmic relationships developed by Burmaster (1998) that use body mass as an independent variable as follows:

$$BSA = e^{-2\,2781} BM^{0\,6821} \qquad (5A\text{-}1)$$

where,

BSA = body surface area (m^2)

BM = body mass (kg)

5A-5. CONSTRUCTION OF LONGITUDINAL DIARY SEQUENCE

The activity diary determines the sequence of microenvironments visited by the simulated person. A longitudinal sequence of daily diaries must be constructed for each simulated individual to cover the entire simulation period. The default activity diaries in APEX are derived from those in the EPA's Consolidated Human Activity Database (CHAD) (US EPA, 2000; 2002), although the user could provide area specific diaries if available. There are over 53,000 CHAD diaries, each covering a 24 hour period, that have been compiled from several studies. CHAD is essentially a cross-sectional database that, for the most part, only has one diary per person. Therefore, APEX must assemble each longitudinal diary sequence for a simulated individual from many single-day diaries selected from a pool of similar people.

APEX selects diaries from CHAD by matching gender and employment status, and by requiring that age falls within a user-specified range on either side of the age of the simulated individual. For example, if the user specifies plus or minus 20%, then for a 40 year old simulated individual, the available CHAD diaries are those from persons aged 32 to 48. Each simulated individual therefore has an age window of acceptable diaries; these windows can

[1] Demographic (Demo) and Body Measurement (BMX) datasets for each of the NHANES studies were obtained from http://www.cdc.gov/nchs/nhanes/nhanes_questionnaires htm.

partially overlap those for other simulated individuals. This differs from a cohort-based approach, where the age windows are fixed and non-overlapping. The user may optionally request that APEX allow a decreased probability for selecting diaries from ages outside the primary age window, and also for selecting diaries from persons of missing gender, age, or employment status. These options allow the model to continue the simulation when diaries are not available within the primary window.

The available CHAD diaries are classified into *diary pools*, based on the temperature and day of the week. The model will select diaries from the appropriate pool for days in the simulation having matching temperature and day type characteristics. The rules for defining these pools are specified by the user. For example, the user could request that all diaries from Monday to Friday be classified together, and Saturday and Sunday diaries in another class. Alternatively, the user could instead create more than two classes of weekdays, combine all seven days into one class, or split all seven days into separate classes.

The temperature classification can be based either on daily maximum temperature, daily average temperature, or both. The user specifies both the ranges and numbers of temperatures classes. For example, the user might wish to create four temperature classes and set their ranges to below 50 °F, 50-69 °F, 70-84 °F, and above a daily maximum of 84 °F. Then day type and temperature classes are combined to create the diary pools. For example, if there are four temperature classes and two day type classes, then there will be eight diary pools.

APEX then determines the day-type and the applicable temperature for each person's simulated day. APEX allows multiple temperature stations to be used; the sectors are automatically mapped to the nearest temperature station. This may be important for study areas such as the greater Los Angeles area, where the inland desert sectors may have very different temperatures from the coastal sectors. For selected diaries, the temperature in the home sector of the simulated person is used. For each day of the simulation, the appropriate diary pool is identified and a CHAD dairy is randomly drawn. When a diary for every day in the simulation period has been selected, they are concatenated into a single longitudinal diary covering the entire simulation for that individual. APEX contains three algorithms for stochastically selecting diaries from the pools to create the longitudinal diary. The first method selects diaries at random after stratification by age, gender, and diary pool; the second method selects diaries based on metrics related to exposure (e.g., time spent outdoors) with the goal of creating longitudinal diaries with variance properties designated by the user (Glen et al., 2008); and the third method uses a clustering algorithm to obtain more realistic recurring behavioral patterns (Rosenbaum 2008).

The final step in processing the activity diary is to map the CHAD location codes into the set of APEX microenvironments, supplied by the user as an input file. The user may define the number of microenvironments, from one up to the number of different CHAD location codes (which is currently 115).

5A-6. KEY PHYSIOLOGICAL PROCESSES MODELED

Ventilation is a general term describing the movement of air into and out of the lungs. The rate of ventilation is determined by the type of activity an individual performs which in turn is related to the amount of oxygen required to perform the activity. Minute or total ventilation rate is used to describe the volume of air moved in or out of the lungs per minute. Quantitatively, the volume of air breathed in per minute ($\dot{V}_I$) is slightly greater than the volume expired per minute ($\dot{V}_E$). Clinically, however, this difference is not important, and by convention, the ventilation rate is always measured on an expired sample or $\dot{V}_E$.

The rate of oxygen consumption ($\dot{V}_{O2}$) is related to the rate of energy usage in performing activities as follows:

$$\dot{V}_{O2} = EE \times ECF \quad (5A\text{-}2)$$

where,

$\dot{V}_{O2}$ = Oxygen consumption rate (liters O_2/minute)
EE = Energy expenditure (kcal/minute)
ECF = Energy conversion factor (liters O_2/kcal).

The ECF shows little variation and typically, commonly a value between 0.20 and 0.21 is used to represent the conversion from energy units to oxygen consumption. APEX can randomly sample from a uniform distribution defined by these lower and upper bounds to estimate an ECF for each simulated individual. The activity-specific energy expenditure is highly variable and can be estimated using metabolic equivalents (METs), or the ratios of the rate of energy consumption for non-rest activities to the resting rate of energy consumption, as follows

$$EE = MET \times RMR \quad (5A\text{-}3)$$

where,

EE = Energy expenditure (kcal/minute)
MET = Metabolic equivalent of work (unitless)
RMR = Resting metabolic rate (kcal/minute)

APEX contains distributions of METs for all activities that might be performed by simulated individuals. APEX randomly samples from the various METs distributions to obtain values for every activity performed by each individual. Age- and gender-specific RMR are estimated once for each simulated individual using a linear regression model (see Johnson et al., 2002) [2] as follows

$$RMR = [b_0 + b_1\,(BM) + \varepsilon]F \qquad \text{(5A-4)}$$

where,

RMR	= Resting metabolic rate (kcal/min)
b_o	= Regression intercept (MJ/day)
b_1	= Regression slope (MJ/day/kg)
BM	= body mass (kg)
ε	= randomly sampled error term, N{0, se)[3] (MJ/day)
F	= Factor for converting MJ/day to kcal/min (0.166)

Finally, Graham and McCurdy (2005) describe an approach to estimate $\dot{V}_E$ using $\dot{V}_{O2}$. In that report, a series of age- and gender-specific multiple linear regression equations were derived from data generated in 32 clinical exercise studies. The algorithm accounts for variability in ventilation rate due to variation in oxygen consumption, the variability within age groups, and both inter- and intra-personal and variability. The basic algorithm is

$$\ln(\dot{V}_E / BM) = b_0 + b_1 \ln(\dot{V}_{O2} / BM) + b_2 \ln(1 + age) + b_3\ gender + e_b + e_w \qquad \text{(5A-5)}$$

where,

ln	= natural logarithm of variable
$\dot{V}_E / BM$	= activity specific ventilation rate, body mass normalized (liter air/kg)
b_i	= see below
$\dot{V}_{O2} / BM$	= activity specific oxygen consumption rate, body mass normalized (liter/O_2/kg)
age	= the age of the individual (years)
$gender$	= gender value (-1 for males and +1 for females)
e_b	= randomly sampled error term for between persons N{0, se), (liter air/kg)
e_w	= randomly sampled error term for within persons N{0, se), (liter air/kg)

[2] The regression equations were adapted by Johnson (2002) using data reported by Schofield (1985). The regression coefficients and error terms used by APEX are provided in the APEX physiology input file.

[3] The value used for each individual is sampled from a normal distribution (N) having a mean of zero (0) and variability described by the standard error (se)

As indicated above, the random error (ε) is allocated to two variance components used to estimate the between-person (inter-individual variability) residuals distribution (e_b) and within-person (intra-individual variability) residuals distribution (e_w). The regression parameters b_0, b_1, b_2, and b_3 are assumed to be constant over time for all simulated persons, e_b is sampled once per person, while whereas e_w varies from event to event. Point estimates of the regression coefficients and standard errors of the residuals distributions are given in Table 5A-1.

Table 5A-1. Ventilation coefficient parameter estimates (b_i) and residuals distributions (e_i) from Graham and McCurdy (2005).

Age group	Regression Coefficients[1]				Random Error[1]	
	b_0	b_1	b_2	b_3	e_b	e_w
<20	4.3675	1.0751	-0.2714	0.0479	0.0955	0.1117
20-<34	3.7603	1.2491	0.1416	0.0533	0.1217	0.1296
34-<61	3.2440	1.1464	0.1856	0.0380	0.1260	0.1152
61+	2.5826	1.0840	0.2766	-0.0208	0.1064	0.0676

[1] The values of the coefficients and residuals distributions described by equation (5A-5).

5A-7. ESTIMATING MICROENVIRONMENTAL CONCENTRATIONS

The user provides rules for determining the pollutant concentration in each microenvironment. There are two available models for calculating microenvironmental concentrations: mass balance and regression factors. Any indoor microenvironment may use either model; for each microenvironment, the user specifies whether the mass balance or factors model will be used.

5A-7.1. Mass Balance Model

The mass balance method assumes that an enclosed microenvironment (e.g., a room within a home) is a single well-mixed volume in which the air concentration is approximately spatially uniform. The concentration of an air pollutant in such a microenvironment is estimated using the following four processes (and illustrated in Figure 5A-1):

- Inflow of air into the microenvironment;
- Outflow of air from the microenvironment;
- Removal of a pollutant from the microenvironment due to deposition, filtration, and chemical degradation; and
- Emissions from sources of a pollutant inside the microenvironment.

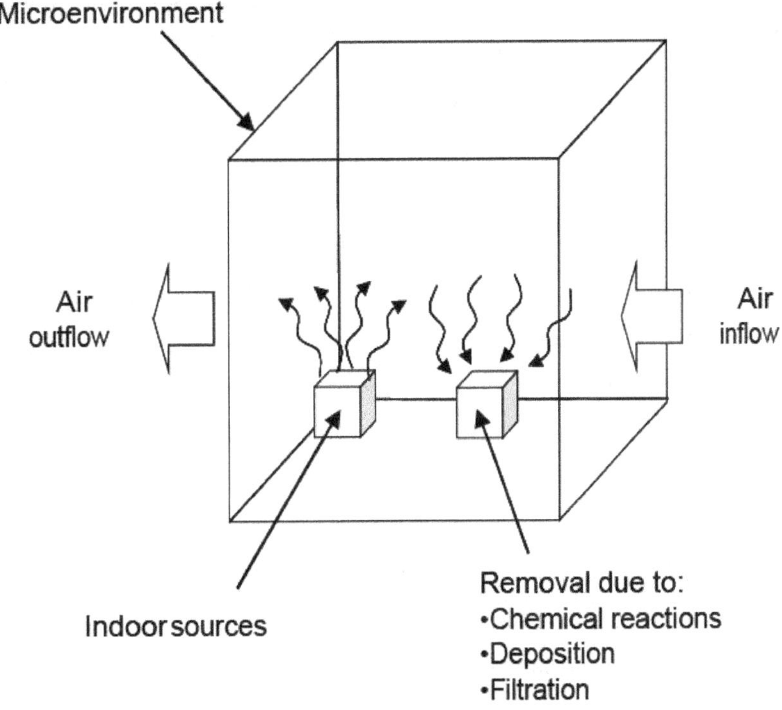

Figure 5A-1. Illustration of the mass balance model used by APEX.

Considering the microenvironment as a well-mixed fixed volume of air, the mass balance equation for a pollutant in the microenvironment can be written in terms of concentration:

$$\frac{dC(t)}{dt} = \dot{C}_{in} - \dot{C}_{out} - \dot{C}_{removal} + \dot{C}_{source} \tag{5A-6}$$

where,

$C(t)$	=	Concentration in the microenvironment at time t
$\dot{C}_{in}$	=	Rate of change in $C(t)$ due to air entering the microenvironment
$\dot{C}_{out}$	=	Rate of change in $C(t)$ due to air leaving the microenvironment
$\dot{C}_{removal}$	=	Rate of change in $C(t)$ due to all internal removal processes
$\dot{C}_{source}$	=	Rate of change in $C(t)$ due to all internal source terms

Concentrations are calculated in the same units as the ambient air quality data, e.g., ppm, ppb, ppt, or $\mu g/m^3$. In the following equations concentration is shown only in $\mu g/m^3$ for brevity. The change in microenvironmental concentration due to influx of air, $\dot{C}_{in}$, is given by:

$$\dot{C}_{in} = C_{outdoor} \times f_{penetration} \times R_{air\,exchange} \tag{5A-7}$$

where,

$C_{outdoor}$ = Ambient concentration at an outdoor microenvironment or outside an indoor microenvironment ($\mu g/m^3$)

$f_{penetration}$ = Penetration factor (unitless)

$R_{air\ exchange}$ = Air exchange rate (hr^{-1})

Since the air pressure is approximately constant in microenvironments that are modeled in practice, the flow of outside air into the microenvironment is equal to that flowing out of the microenvironment, and this flow rate is given by the air exchange rate. The air exchange rate (hr^{-1}) can be loosely interpreted as the number of times per hour the entire volume of air in the microenvironment is replaced. For some pollutants (especially particulate matter), the process of infiltration may remove a fraction of the pollutant from the outside air. The fraction that is retained in the air is given by the penetration factor $f_{penetration}$.

A proximity factor ($f_{proximity}$) and a local outdoor source term are used to account for differences in ambient concentrations between the geographic location represented by the ambient air quality data (e.g., a regional fixed-site monitor) and the geographic location of the microenvironment. That is, the outdoor air at a particular location may differ systematically from the concentration input to the model representing the air quality district. For example, a playground or house might be located next to a busy road in which case the air at the playground or outside the house would have elevated levels for mobile source pollutants such as carbon monoxide and benzene. The concentration in the air at an outdoor location or directly outside an indoor microenvironment ($C_{outdoor}$) is calculated as:

$$C_{outdoor} = f_{proximity} C_{ambient} + C_{LocalOutdoorSources} \quad (5A\text{-}8)$$

where,

$C_{ambient}$ = Ambient air district concentration ($\mu g/m^3$)

$f_{proximity}$ = Proximity factor (unitless)

$C_{LocalOutdoorSources}$ = The contribution to the concentration at this location from local sources not represented by the ambient air district concentration ($\mu g/m^3$)

During exploratory analyses, the user may examine how a microenvironment affects overall exposure by setting the microenvironment's proximity or penetration factor to zero, thus effectively eliminating the specified microenvironment.

Change in microenvironmental concentration due to outflux of air is calculated as the concentration in the microenvironment $C(t)$ multiplied by the air exchange rate:

$$\dot{C}_{out} = R_{air\,exchange} \times C(t) \tag{5A-9}$$

The third term ($\dot{C}_{removal}$) in the mass balance calculation (equation 5A-6) represents removal processes within the microenvironment. There are three such processes in general: chemical reaction, deposition, and filtration. Chemical reactions are significant for O_3, for example, but not for carbon monoxide. The amount lost to chemical reactions will generally be proportional to the amount present, which in the absence of any other factors would result in an exponential decay in the concentration with time. Similarly, deposition rates are usually given by the product of a (constant) deposition velocity and a (time-varying) concentration, also resulting in an exponential decay. The third removal process is filtration, usually as part of a forced air circulation or HVAC system. Filtration will normally be more effective at removing particles than gases. In any case, filtration rates are also approximately proportional to concentration. Change in concentration due to deposition, filtration, and chemical degradation in a microenvironment is simulated based on the first-order equation:

$$\begin{aligned}\dot{C}_{removal} &= \left(R_{deposition} + R_{filtration} + R_{chemical}\right) \times C(t) \\ &= R_{removal} \times C(t)\end{aligned} \tag{5A-10}$$

where,

$\dot{C}_{removal}$	=	Change in microenvironmental concentration due to removal processes (µg/m³/hr)
$R_{deposition}$	=	Removal rate of a pollutant from a microenvironment due to deposition (hr^{-1})
$R_{filtration}$	=	Removal rate of a pollutant from a microenvironment due to filtration (hr^{-1})
$R_{chemical}$	=	Removal rate of a pollutant from a microenvironment due to chemical degradation (hr^{-1})
$R_{removal}$	=	Removal rate of a pollutant from a microenvironment due to the combined effects of deposition, filtration, and chemical degradation (hr^{-1})

The fourth term in the mass balance calculation represents pollutant sources within the microenvironment. This is the most complicated term, in part because several sources may be present. APEX allows two methods of specifying source strengths: emission sources and concentration sources. Either may be used for mass balance microenvironments, and both can be used within the same microenvironment. The source strength values are used to calculate the term $\dot{C}_{source}$ (µg/m³/hr).

Emission sources are expressed as emission rates in units of µg/hr, irrespective of the units of concentration. To determine the rate of change of concentration associated with an emission source S_E, it is divided by the volume of the microenvironment:

$$\dot{C}_{source,SE} = \frac{S_E}{V} \qquad (5A\text{-}11)$$

where,

$\dot{C}_{source,SE}$	=	Rate of change in $C(t)$ due to the emission source S_E (µg/m³/hr)
S_E	=	The emission rate (µg/hr)
V	=	The volume of the microenvironment (m³)

Concentration sources (S_C) however, are expressed in units of concentration. These must be the same units as used for the ambient concentration (e.g., µg/m³). Concentration sources are normally used as additive terms for microenvironments using the factors model. Strictly speaking, they are somewhat inconsistent with the mass balance method, since concentrations should not be inputs but should be consequences of the dynamics of the system. Nevertheless, a suitable meaning can be found by determining the rate of change of concentration ($\dot{C}_{source}$) that would result in a mean increase of S_C in the concentration, given constant parameters and equilibrium conditions, in this way:

Assume that a microenvironment is always in contact with clean air (ambient = zero), and it contains one constant concentration source. Then the mean concentration over time in this microenvironment from this source should be equal to S_C. The mean source strength expressed in ppm/hr or µg/m³/hr is the rate of change in concentration ($\dot{C}_{source,SC}$). In equilibrium,

$$C_S = \frac{\dot{C}_{source,\,SC}}{R_{air\ exchange} + R_{removal}} \qquad (5A\text{-}12)$$

where, Cs is the mean increase in concentration over time in the microenvironment due to the source $\dot{C}_{source,SC}$. Thus, $\dot{C}_{source,SC}$ can be expressed as

$$\dot{C}_{source,\,SC} = C_S \times R_{mean} \qquad (5A\text{-}13)$$

where R_{mean} is the chemical removal rate. From equation (5A-13), R_{mean} is equal to the sum of the air exchange rate and the removal rate ($R_{air\ exchange} + R_{removal}$) under equilibrium conditions. In general, however, the microenvironment will not be in equilibrium, but in such conditions there is no clear meaning to attach to $\dot{C}_{source,SC}$ since there is no fixed emission rate that will lead to a fixed increase in concentration. The simplest solution is to use $R_{mean} = R_{air\ exchange} + R_{removal}$. However, the user is given the option of specifically specifying R_{mean} (see discussion of parameters below). This may be used to generate a truly constant source strength

$\dot{C}_{source,SC}$ by making S_C and R_{mean} both constant in time. If this is not done, then R_{mean} is simply set to the sum of ($R_{air\ exchange} + R_{removal}$). If these parameters change over time, then $\dot{C}_{source,SC}$ also changes. Physically, the reason for this is that in order to maintain a fixed elevation of concentration over the base conditions, then the source emission rate would have to rise if the air exchange rate were to rise.

Multiple emission and concentration sources within a single microenvironment are combined into the final total source term by combining equations (5A-11) and (5A-13):

$$\dot{C}_{source} = \dot{C}_{source,SE} + \dot{C}_{source,SC} = \frac{1}{V}\sum_{i=1}^{n_e} E_{S_i} + R_{mean}\sum_{i=1}^{n_c} C_{S_i} \qquad (5A\text{-}14)$$

where,

S_{Ei}	=	Emission source strength for emission source i (µg/hr, irrespective of the concentration units)
S_{Ci}	=	Emission source strength for concentration source i (µg/m^3)
n_e	=	Number of emission sources in the microenvironment
n_c	=	Number of concentration sources in the microenvironment

In equations (5A-11) and (5A-14), if the units of air quality are ppm rather than µg/m^3, $1/V$ is replaced by f/V, where f = ppm / µg/m^3 = gram molecular weight / 24.45. (24.45 is the volume (liters) of a mole of the gas at 25°C and 1 atmosphere pressure.)

Equations (5A-7), (5A-9), (5A-10), and (5A-14) can now be combined with equation (5A-6) to form the differential equation for the microenvironmental concentration C(t). Within the time period of a time step (at most 1 hour), $\dot{C}_{source}$ and $\dot{C}_{in}$ are assumed to be constant. Using $\dot{C}_{combined} = \dot{C}_{source} + \dot{C}_{in}$ leads to:

$$\begin{aligned}\frac{dC(t)}{dt} &= \dot{C}_{combined} - R_{air\ exchange}C(t) - R_{removal}C(t) \\ &= \dot{C}_{combined} - R_{mean}C(t)\end{aligned} \qquad (5A\text{-}15)$$

Solving this differential equation leads to:

$$C(t) = \frac{\dot{C}_{combined}}{R_{mean}} + \left(C(t_0) - \frac{\dot{C}_{combined}}{R_{mean}} \right) e^{-R_{mean}(t-t_0)} \qquad (5A\text{-}16)$$

where,

$C(t_0)$	=	Concentration of a pollutant in a microenvironment at the beginning of a time step (µg/m^3)

$C(t)$ = Concentration of a pollutant in a microenvironment at time t within the time step ($\mu g/m^3$).

Based on equation (5A-16), the following three concentrations in a microenvironment are calculated:

$$C_{equil} = C(t \rightarrow \infty) = \frac{\dot{C}_{combined}}{R_{mean}} = \frac{\dot{C}_{source} + \dot{C}_{in}}{R_{air\,exchange} + R_{removal}} \tag{5A-17}$$

$$C(t_0 + T) = C_{equil} + \left(C(t_0) - C_{equil}\right)e^{-R_{mean}T} \tag{5A-18}$$

$$C_{mean} = \frac{1}{T}\int_{t_0}^{t_0+T} C(t)\,dt = C_{equil} + \left(C(t_0) - C_{equil}\right)\frac{1 - e^{-R_{mean}T}}{R_{mean}T} \tag{5A-19}$$

where,

C_{equil} = Concentration in a microenvironment ($\mu g/m^3$) if $t \rightarrow \infty$ (equilibrium state).

$C(t_0)$ = Concentration in a microenvironment at the beginning of the time step ($\mu g/m^3$)

$C(t_0+T)$ = Concentration in a microenvironment at the end of the time step ($\mu g/m^3$)

C_{mean} = Mean concentration over the time step in a microenvironment ($\mu g/m^3$)

R_{mean} = $R_{air\,exchange} + R_{removal}$ (hr^{-1})

At each time step of the simulation period, APEX uses equations (5A-17), (5A-18), and (5A-19) to calculate the equilibrium, ending, and mean concentrations, respectively. The calculation continues to the next time step by using $C(t_0+T)$ for the previous hour as $C(t_0)$.

5A-7.2. Factors Model

The factors model is simpler than the mass balance model. In this method, the value of the concentration in a microenvironment is not dependent on the concentration during the previous time step. Rather, this model uses the following equation to calculate the concentration in a microenvironment from the user-provided hourly air quality data:

$$C_{mean} = C_{ambient}\, f_{proximity}\, f_{penetration} + \sum_{i=1}^{n_c} S_{Ci} \tag{5A-20}$$

where,

C_{mean} = Mean concentration over the time step in a microenvironment ($\mu g/m^3$)
$C_{ambient}$ = The concentration in the ambient (outdoor) environment ($\mu g/m^3$)
$f_{proximity}$ = Proximity factor (unitless)
$f_{penetration}$ = Penetration factor (unitless)
S_{Ci} = Mean air concentration resulting from source i ($\mu g/m^3$)
n_c = Number of concentration sources in the microenvironment

The user may specify distributions for proximity, penetration, and any concentration source terms. All of the parameters in equation (5A-20) are evaluated for each time step, although these values might remain constant for several time steps or even for the entire simulation.

The ambient air quality data are supplied as time series over the simulation period at several locations across the modeled region. The other variables in the factors and mass balance equations are randomly drawn from user-specified distributions. The user also controls the frequency and pattern of these random draws. Within a single day, the user selects the number of random draws to be made and the hours to which they apply. Over the simulation, the same set of 24 hourly values may either be reused on a regular basis (for example, each winter weekday), or a new set of values may be drawn. The usage patterns may depend on day of the week, on month, or both. It is also possible to define different distributions that apply if specific conditions are met. The air exchange rate is typically modeled with one set of distributions for buildings with air conditioning and another set of distributions for those which do not. The choice of a distribution within a set typically depends on the outdoor temperature and possibly other variables. In total there are eleven such *conditional variables* which can be used to select the appropriate distributions for the variables in the mass balance or factors equations.

For example, the hourly emissions of CO from a gas stove may be given by the product of three random variables: a binary on/off variable that indicates if the stove is used at all during that hour, a usage duration sampled from a continuous distribution, and an emission rate per minute of usage. The binary on/off variable may have a probability for *on* that varies by time of day and season of the year. The usage duration could be taken from a truncated normal or lognormal distribution that is resampled for each cooking event, while the emission rate could be sampled just once per stove.

5A-8. EXPOSURE AND DOSE TIME SERIES CALCULATIONS

The activity diaries provide the time sequence of microenvironments visited by the simulated individual and the activities performed by each individual. The pollutant concentration in the air in each microenvironment is assumed to be spatially uniform throughout the microenvironment and unchanging within each diary event and is calculated by either the factors or the mass balance method, as specified by the user. The exposure of the individual is given by the time sequence of airborne pollutant concentrations that are encountered in the microenvironments visited. Figure 5A-2 illustrates the exposures for one simulated 12-year old child over a 2-day period. On both days the child travels to and from school in an automobile, goes outside to a playground in the afternoon while at school, and spends time outside at home in the evening.

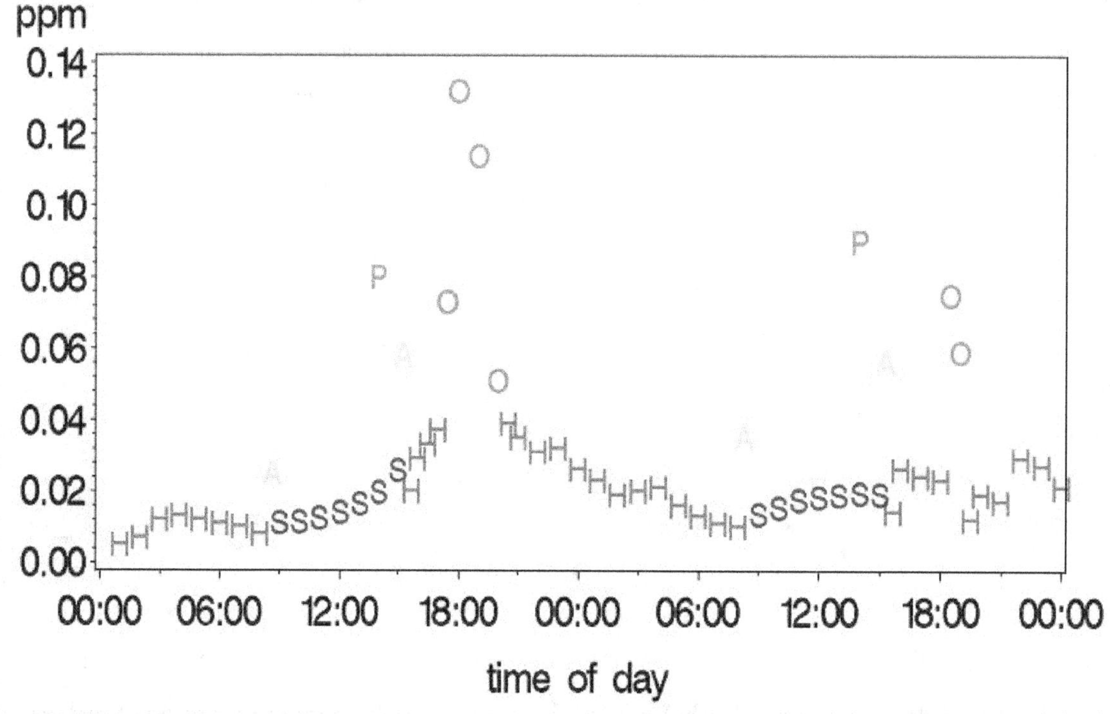

Figure 5A-2. Example of microenvironmental and exposure concentrations for a simulated individual over a 48 hours simulation. (H: home, A: automobile, S: school, P: playground, O: outdoors at home).

In addition to exposure, APEX models breathing rates based on the physiology of each individual and the exertion levels associated with the activities performed. For each activity type in CHAD, a distribution is provided for a corresponding normalized metabolic equivalent of work or METs (McCurdy, 2000). METs are derived by dividing the metabolic energy requirements for the specific activity by a persons resting, or basal, metabolic rate. The MET ratios have less interpersonal variation than do the absolute energy expenditures. Based on age and gender, the resting metabolic rate, along with other physiological variables is determined for each individual as part of their anthropometric characteristics. Because the MET ratios are sampled independently from distributions for each diary event, it would be possible to produce time-series of MET ratios that are physiologically unrealistic. APEX employs a MET adjustment algorithm based on a modeled oxygen deficit to prevent such overestimation of MET and breathing rates (Isaacs et al., 2008). The relationship between the oxygen deficit and the applied limits on MET ratios are nonlinear and are derived from published data on work capacity and oxygen consumption. The resulting combination of microenvironmental concentration and breathing ventilation rates provides a time series of inhalation intake dose for most pollutants.

5A-9. MODEL OUTPUT

APEX calculates the exposure and dose time series based on the events as listed on the activity diary with a minimum of one event per hour but usually more during waking hours. APEX can aggregate the event level exposure and dose time series to output hourly, daily, monthly, and annual averages. The types of output files are selected by the user, and can be as detailed as event-level data for each simulated individual (note, Figure 5A-2 was produced from the event output file). A set of summary tables are produced for a variety of exposure and dose measures. These include tables of person-minutes at various exposure levels, by microenvironment, a table of person-days at or above each average daily exposure level, and tables describing the distributions of exposures for different groups. An example of how APEX results can be depicted is given in Figure 5A-3, which shows the percent of children with at least one 8-hour average exposure at or above different exposure levels, concomitant with moderate or greater exertion. These are results from a simulation of O_3 exposures for the greater Washington, D.C. metropolitan area for the year 2002. From this graph ones sees, for example, that APEX estimates 30 percent of the children in this area experience exposures above 0.08 ppm-8hr while exercising, at least once during the year.

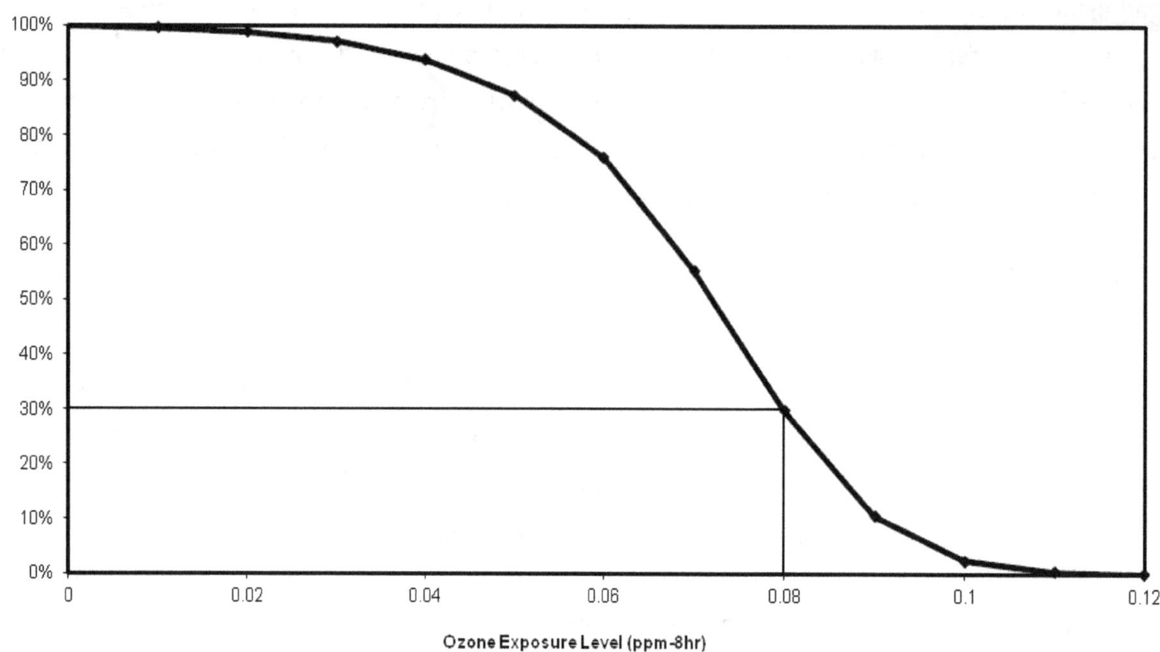

Figure 5A-3. The percent of simulated children (ages 5-18) at or above 8-hour average O_3 exposures while at moderate or greater exertion.

5A-10. REFERENCES

Burmaster, D.E. (1998). LogNormal distributions for skin area as a function of body weight. *Risk Analysis*. 18(1):27-32.

Glen, G., Smith, L., Isaacs, K., McCurdy, T., Langstaff, J. (2008). A new method of longitudinal diary assembly for human exposure modeling. *J Expos Sci Environ Epidem.* 18:299-311.

Graham, S.E., McCurdy, T. (2005). Revised ventilation rate (VE) equations for use in inhalation-oriented exposure models. Report no. EPA/600/X-05/008 is Appendix A of US EPA (2009). Metabolically Derived Human Ventilation Rates: A Revised Approach Based Upon Oxygen Consumption Rates (Final Report). Report no. EPA/600/R-06/129F. Available at: http://cfpub.epa.gov/ncea/cfm/recordisplay.cfm?deid=202543.

Isaacs, K., Glen, G., McCurdy, T., Smith, L. (2008). Modeling energy expenditure and oxygen consumption in human exposure models: accounting for fatigue and EPOC. *J Expos Sci Environ Epidemiol.* 18:289-298.

Isaacs, K., Smith, L. (2005). New Values for Physiological Parameters for the Exposure Model Input File Physiology.txt. Memorandum submitted to the U.S. Environmental Protection Agency under EPA Contract EP-D-05-065. NERL WA 10. Alion Science and Technology. Found in US EPA . (2009). Risk and Exposure Assessment to Support the Review of the SO_2 Primary National Ambient Air Quality Standard. EPA-452/R-09-007. August 2009. Available at http://www.epa.gov/ttn/naaqs/standards/so2/data/200908SO2REAFinalReport.pdf.

Johnson, T., Capel, J., McCoy, M. (1996a). Estimation of Ozone Exposures Experienced by Urban Residents Using a Probabilistic Version of NEM and 1990 Population Data. Prepared by IT Air Quality Services for the Office of Air Quality Planning and Standards, U.S. Environmental Protection Agency, Research Triangle Park, North Carolina, September.

Johnson, T., Capel, J., Mozier, J., McCoy, M. (1996b). Estimation of Ozone Exposures Experienced by Outdoor Children in Nine Urban Areas Using a Probabilistic Version of NEM. Prepared for the Air Quality Management Division under Contract No. 68-DO-30094, April.

Johnson, T., Capel, J., McCoy, M., Mozier, J. (1996c). Estimation of Ozone Exposures Experienced by Outdoor Workers in Nine Urban Areas Using a Probabilistic Version of NEM. Prepared for the Air Quality Management Division under Contract No. 68-DO-30094, April.

Johnson, T. (2002). A Guide to Selected Algorithms, Distributions, and Databases Used in Exposure Models Developed By the Office of Air Quality Planning and Standards. Revised Draft. Prepared for U.S. Environmental Protection Agency under EPA Grant No. CR827033.

McCurdy, T. (2000). Conceptual basis for multi-route intake dose modeling using an energy expenditure approach. *J Expo Anal Environ Epidemiol.* 10:1-12.

McCurdy, T., Glen, G., Smith, L., Lakkadi, Y. (2000). The National Exposure Research Laboratory's Consolidated Human Activity Database. *J Exp Anal Environ Epidemiol.* 10:566-578.

Rosenbaum, A. S. (2008). The Cluster-Markov algorithm in APEX. Memorandum prepared for Stephen Graham, John Langstaff. USEPA OAQPS by ICF International.

Schofield, W. N. (1985). Predicting basal metabolic rate, new standards, and review of previous work. *Hum Nutr Clin Nutr.* 39C(S1):5-41.

US EPA. (2002). Consolidated Human Activities Database (CHAD) Users Guide. Database and documentation available at: http://www.epa.gov/chadnet1/.

US EPA. (2012a). Total Risk Integrated Methodology (TRIM) - Air Pollutants Exposure Model Documentation (TRIM.Expo / APEX, Version 4.4) Volume I: User's Guide. Office of Air Quality Planning and Standards, U.S. Environmental Protection Agency, Research Triangle Park, NC. EPA-452/B-12-001a. Available at: http://www.epa.gov/ttn/fera/human_apex.html

US EPA. (2012b). Total Risk Integrated Methodology (TRIM) - Air Pollutants Exposure Model Documentation (TRIM.Expo / APEX, Version 4.4) Volume II: Technical Support Document. Office of Air Quality Planning and Standards, U.S. Environmental Protection Agency, Research Triangle Park, NC. EPA-452/B-12-001b. Available at: http://www.epa.gov/ttn/fera/human_apex.html

Appendix 5-B

Inputs to the APEX Exposure Model

Table of Contents

5B-1 POPULATION DEMOGRAPHICS **3**

5B-2 POPULATION COMMUTING PATTERNS **4**

5B-3 ASTHMA PREVALENCE RATES **5**

5B-4 HUMAN ACTIVITY DATA **6**

5B-4.1 CHAD UPDATES SINCE THE 2007 OZONE NAAQS REVIEW 8

5B-4.2 LONGITUDINAL ACTIVITY PATTERN METHODOLOGY 10

5B-5 PHYSIOLOGICAL AND METABOLIC EQUIVALENTS DATA **13**

5B-6 MICROENVIRONMENTS MODELED **14**

5B-6.1 AIR EXCHANGE RATES FOR INDOOR RESIDENTIAL MICROENVIRONMENTS 16

5B-6.2 AIR CONDITIONING PREVALENCE FOR INDOOR RESIDENTIAL MICROENVIRONMENTS 18

5B-6.3 AER DISTRIBUTIONS FOR OTHER INDOOR ENVIRONMENTS 20

5B-6.4 PROXIMITY AND PENETRATION FACTORS FOR IN-VEHICLE AND NEAR-ROAD MICROENVIRONMENTS 21

5B-6.5 PROXIMITY AND PENETRATION FACTORS FOR OUTDOOR MICROENVIRONMENTS 22

5B-6.6 OZONE DECAY AND DEPOSITION RATES 22

5B-7 AMBIENT OZONE CONCENTRATIONS **24**

5B-8 METEOROLOGICAL DATA **33**

5B-9 REFERENCES **38**

List of Tables

Table 5B-1. Consolidated Human Activity Database (CHAD) study information and diary-days used by APEX. 11
Table 5B-2. Microenvironments modeled and calculation method used. 15
Table 5B-3. AERs for indoor residential microenvironments (ME-1) with A/C by study area and temperature. 17
Table 5B-4. AERs for indoor residential microenvironments (ME-1) without A/C by study area and temperature. 18
Table 5B-5. American Housing Survey A/C prevalence from Current Housing Reports (Table 1-4) for selected urban areas. 19
Table 5B-6. Parameter values for distributions of penetration and proximity factors used for estimating in-vehicle microenvironmental concentrations. 21
Table 5B-7. VMT fractions of interstate, urban, and local roads in the study areas used to select in-vehicle proximity factor distributions. 23
Table 5B-8. Counties and air districts modeled in each study area. 26
Table 5B-9. Ambient monitors used to define exposure modeling domain and the population modeled in each study area. 27
Table 5B-10. Study area meteorological stations, locations, and hours of missing data. 35

List of Figures

Figure 5B-1. Illustration of APEX exposure modeling domains (2000 US Census tract centroids) for Atlanta, Boston, Baltimore and Chicago study areas. 29
Figure 5B-2. Illustration of APEX exposure modeling domains (2000 US Census tract centroids) for Cleveland, Dallas, Denver and Detroit study areas. 30
Figure 5B-3. Illustration of APEX exposure modeling domains (2000 US Census tract centroids) for Houston, Los Angeles, New York and Philadelphia study areas. 31
Figure 5B-4. Illustration of APEX exposure modeling domains (2000 US Census tract centroids) for Sacramento, St. Louis and Washington DC study areas. 32

The APEX model inputs require extensive analysis and preparation to ensure the model outputs are appropriate as intended, reasonable, and relevant. This Appendix describes the preparation and the sources of data for the APEX input files.

5B-1 POPULATION DEMOGRAPHICS

APEX accounts for important population characteristics in representing study area demographics. Population counts and employment probabilities by age and gender are used to develop representative profiles of hypothetical individuals for the simulation. For the main-body results of the REA, we estimated population-based exposures using US Census tract-level population counts stratified by age in one-year increments, from birth to 99 years, and were obtained from the 2000 Census of Population and Housing Summary File 1 (SF1).[1] The SF1 contains the 100-percent data, which is the information compiled from the questions asked of all people and about every housing unit.

Three standard APEX input files are used for the current O_3 assessment:

- *pop_geo2000_011403.txt*: census tract ID's, their latitudes and longitudes
- *pop_fall2000_043003.txt*: tract-level population counts for females by age
- *pop_fall2000_043003.txt*: tract-level population counts for males by age

Census tract employment rates were developed using the Employment Status: 2000-Supplemental Tables.[2] The file input to APEX is stratified by gender and age group, so that each gender/age group combination is given an employment probability fraction (ranging from 0 to 1) within each census tract. The age groupings in this employment file are: 16-19, 20-21, 22-24, 25-29, 30-34, 35-44, 45-54, 55-59, 60-61, 62-64, 65-69, 70-74, and >75. Children under 16 years of age are assumed to not be employed.[3]

One standard APEX input file is used for the current O_3 assessment:

- *Employment2000_043003.txt*: census tract employment probabilities by age groups

---

[1] http://www.census.gov/census2000/sumfile1.html.

[2] http://www.census.gov/population/www/cen2000/phc-t28 html.

[3] While children can be employed at ages <16, staff feel that when modeling population-based exposures for these young children regardless of whether or not they have been designated as being employed, it is likely the overall study group exposure results would not be significantly affected given the small fraction of the population that may be employed at these ages and that the principal factor influencing high O_3 exposure concentrations is afternoon time spent outdoors. In such a simulation that included employed children <16, only their home tract to work tract commuting would be affected and unless they were employed as an outdoor worker (also a further subdivision of those employed), substantial time spent outdoors is unlikely to occur at work.

5B-2 POPULATION COMMUTING PATTERNS

To more realistically simulate human behavior, APEX incorporates workplace patterns into the assessment by use of home-to-work commuting data. By design, commuting is only used for those simulated individuals who are employed (i.e., ≥ 16 years old). The commuting data were derived from the 2000 Census Transportation Planning Package (CTPP) Part 3-Journey-to-Work (JTW) files.[4] These files contain counts of individuals commuting from home to work locations at varying geographic scales. These data were processed to calculate fractions for each tract-to-tract flow to create a national commuting flow file distributed with APEX. This database contains commuting data for each of the 50 states and Washington, D.C. Important processing and application assumptions include the following:

- **Commuting within the Home Tract**: the APEX commuting database does not differentiate people that work at home from those that commute within their home tract.
- **Commuting Distance Cutoff**: all persons in home-work flows up to 120 km are daily commuters and no persons in more widely separated flows commute daily, thus the list of destinations for each home tract was restricted to only those work tracts that are within 120 km of the home tract.[5]
- **Eliminated Records**: tract-to-tract pairs that represented workers who either worked outside of the U.S. (9,631 tract pairs with 107,595 workers) or worked in an unknown location (120,830 tract pairs with 8,940,163 workers) were eliminated. An additional 515 workers in the commuting database whose data were missing from the original files, possibly due to privacy concerns or errors, were also deleted.
- **Simulation of Leavers**: we restricted the simulated population to those who do not commute to destinations outside the study area because we have not estimated ambient concentrations of O_3 in counties outside of the modeled areas.

[4] Files downloaded from http://transtats.bts.gov/.

[5] Plotting log(flows) versus log(distance) indicates a near-constant slope out to a distance of approximately 120 kilometers. Beyond that distance, the relationship also had a fairly constant, though less inclined, slope. A simple interpretation is that for distances up to 120 km, the majority of the flow was due to persons traveling between home and work tracts daily, with the numbers of such persons decreasing rapidly with increasing distance. Beyond 120 km, the majority of the flow is made up of persons who stay at the workplace for extended times, in which case the separation distance is not as crucial in determining the flow.

An additional commuting input file was recently developed as a companion to the APEX commuting flow file. Also derived from the 2000 census are tract-level population counts of one-way commute times, and given in 13 time bins (in minutes): < 5, 5 to 9, 10 to 14, 15 to 19, 20 to 24, 25 to 29, 30 to 34, 35 to 39, 40 to 44, 45 to 59, 60 to 89, 90-120, works at home (0 minutes commuting time). APEX uses these time bins to create a cumulative probability distribution of commuting times for each tract, which it then uses in conjunction with the distribution of commuting distances to assign a profile-level one-way commuting time variable to each employed person in the population. This commuting time profile variable is then used to select for CHAD diaries having appropriate commute times in their daily activity pattern (i.e., a total time spent in travel locations or activities before and after work activities) to represent the simulated individual.

Two standard APEX input files are used for the current O_3 assessment:

- *Commuting2000_010505.txt*: home/work census tract ID's, cumulative probabilities of commuting to work tract from home tract, distances of home to work tract (km)
- *CommutingTimes2000_050610.txt*: tract-level counts of all workers, commuters, and commute time bins

5B-3 ASTHMA PREVALENCE RATES

One of the important study group in the exposure assessment is asthmatic school-age children (ages 5-18). Modeling exposures for this study group with APEX requires the estimation of children's asthma prevalence rates. The estimates are based on children's asthma prevalence data from the 2006-2010 National Health Interview Survey (NHIS). Briefly, 2000 US census tract level asthma prevalence was estimated for children (by single age years) and adults (by age groups), also stratified by gender and family income/poverty ratio (i.e., whether the family income was considered below or at/above the US Census estimate of poverty level for the given year). Given the significant differences in asthma prevalence by age, gender, region, and poverty status, the variability in the spatial distribution of poverty status across census tracts (and also stratified by age), and the spatial variability in local scale ambient concentrations of many air pollutants, the goal was to better represent the variability in population-based exposures when accounting for and modeling these newly refined attributes of this study group. A detailed description of how the NHIS data were processed for input to APEX is provided in Appendix 5C.

One standard APEX input file is used for the current O_3 assessment:

- *AsthmaPrevalence053112.txt*: tract-level asthma prevalence by age (for ages <18) and age groups (for ages > 17)

5B-4 HUMAN ACTIVITY DATA

Exposure models use human activity pattern data to predict and estimate exposure to pollutants. Different human activities, such as outdoor exercise, indoor reading, or driving, would lead to varying pollutant exposures. In addition, different human activities require different energy expenditures, and thus, higher exposure media consumption rates lead to higher doses received. To accurately model individuals and their exposure to pollutants, it is critical to have a firm understanding of the locations where people spend time and the activities performed in such locations.

The Consolidated Human Activity Database (CHAD) provides time series data on human activities through a database system of collected human diaries, or daily time location activity logs (US EPA, 2002). The purpose of CHAD is to provide a basis for conducting multi-route, multi-media exposure assessments (McCurdy et al., 2000). The data contained within CHAD come from multiple surveys with somewhat variable study-specific structure (e.g., minute-by-minute versus time-block averaged sequence of diary events), though common to all studies included, individuals provided information on their locations visited and activities performed for each survey day. Personal attribute data for these surveyed individuals, such as age and gender, are included in CHAD as well. The latest version of CHAD master (071113) contains data for 54,373 person-days.

The CHAD served as the primary source of time location activity pattern data and was processed to retain appropriate diary data for use by APEX. Diaries with missing personal attribute data (i.e., age, gender), missing diary day information (i.e., either daily mean/ maximum temperature, day-of-week), or having 3-hours or more of missing location and/or activity information are not used by APEX. For the latter case, CHAD diaries were evaluated for instances where a diary may contain enough information for the purposes of this exposure assessment allowing it to be adjusted to reduce the missing information to less than 3 hours on a given day. For example, the diary structure of the ozone averting behavior (OAB) study resulted in nearly all of the diary days (n=2,776) having no diary information between the hours of 8PM and midnight. In processing the CHAD data for this subset of diaries, the location was assumed by staff to be indoors at their residence and persons were engaged in a sleep activity. This substitution was judged by staff as a reasonable approximation based on the limited likelihood of a person's highest O_3 exposures occurring at this time of day, while still retaining the relevant activity pattern data of interest (e.g., locations visited and activities performed during the daytime hours).

The following is a list of adjustments made to CHAD diary data where study specific structure was a factor in missing data or diary information was present in either CHAD location or activity codes to infer specific information where data were missing.

- OAB (a children's study) missing location and activity events from 8PM – 12AM were set to 'indoor residence' and 'sleep';
- BAL missing activity events at 8AM occurring indoors were set to 'personal care';
- ISR missing activity events occurring when attending school were set to either 'attend K-12' (ages 5-18) or 'attend day-care' (ages <5);
- NSA (an adults study) missing activity events at 8PM - 12AM occurring indoor residences were set to 'leisure, general';
- Locations missing for a number of staff judged outdoor activities[6] were set to 'outdoor, general';
- Locations missing for a number of staff judged indoor residential activities[7] were set to "indoor, residence"; and
- Locations missing for a number of staff judged general indoor activities[8] were set to "indoor, other".

Three standard APEX input files are used for the current O_3 assessment:

- *CHADQuest_013013B.txt*: personal (e.g., age, gender, employment status, county of residence, etc.) and day (e.g., daily maximum temperature, day-of-week) attribute meta data for each diary day
- *CHADEvents_013013B.txt*: time sequence of locations visited and activities performed by individuals for each diary day
- *CHADSTATSOutdoor_013013B.txt*: time spent outdoors for each diary day

Table 5B-1 summarizes the studies and number of diary days used by APEX in this modeling analysis, providing over 41,000 diary-days of activity data (nearly 18,000 diary-days for ages 4-18) collected between 1982 and 2010.

---

[6] For CHAD activity codes (US EPA, 2002) "11300", "11630", "17100", "17110", "17112", "17120", "17131" or "17170".

[7] For CHAD activity codes (US EPA, 2002) "11100", "11110", "11200", "11210", "11220", "14000", "14100", "14110", "14120", "14300", "14400", "14500", "14600", or "17223".

[8] For CHAD activity codes (US EPA, 2002) "13300", "13400", "15400", "16300", "16400", or "16500".

5B-4.1 CHAD Updates Since the 2007 Ozone NAAQS Review

Since the time of the prior O_3 NAAQS review conducted in 2007, there have been a number new data sets incorporated into CHAD and used in our current exposure assessment, most of which were from recently conducted studies. The data from these eight additional studies incorporated in CHAD and available for use by APEX have more than doubled the total activity pattern data used for O_3 exposure modeling in 2007 and has increased the number of children diaries by a factor of five. The studies from which these new data were derived are briefly described below.

- **DEA.** The diaries are from 2 seasons of the 6-season sampling period (2004-2007) used by EPA in the Detroit Exposure and Aerosol Research Study (DEARS) (Williams et al., 2008). The intent was to obtain environmental samples and time use data for 10 days—5 in each of 2 seasons per participant located in 6 areas in Wayne County, Michigan (in and around Detroit). A 15-minute block diary approach was used to collect activity data. Participants were all adults and activity data was collected from Tuesday through Saturday. Just over 300 diary-days from DEARS are used by APEX.

- **EPA.** The diaries were collected as part of an ongoing longitudinal internal EPA study by EPA scientists, and in some cases, their families. This dataset contains two long-term longitudinal diaries: one by a 60 year-old-male in 1999-2000 (McCurdy and Graham, 2003), and one by a 35 year old male in 2002. Additional longitudinal diaries were kept for a 35-year-old female and her infant daughter in 2008 (though the infant data are not used here). The remaining diaries are from a study of a group of 9 adults (Isaacs et al. 2012). In this portion of the study, all subjects were studied for approximately 17 consecutive days in each of 4 seasons in 2006 and 2007. Approximately of 1,400 diary-days are used by APEX.

- **ISR.** The diaries are from phase I (1997), phase II (2002-03), and phase III (2007-08) of the University of Michigan's Panel Study of Income Dynamics (PSID), respectively (University of Michigan, 2012). Nationally representative activity pattern data from nearly 11,000 children ages 0-13 (phase I), ages 5-19 (phase II), and ages 10-19 (phase III) were added to the APEX activity pattern data. For each child, time use data were reported by primary care-givers, school teachers, and/or the children themselves on two nonconsecutive days in a single week, in no particular season, though mostly occurring during the spring and fall (phase I), winter (phase II), and spring, fall and winter (phase III) months.

- **NSA.** The diaries were collected as part of the National-Scale Activity Survey (NSAS), an EPA-funded study of averting behavior related to air quality alerts (Knowledge Networks, 2009). Data were collected from about 1,200 adults aged 35-92 in seven metropolitan areas (Atlanta, St. Louis, Sacramento, Washington DC, Dallas, Houston, and Philadelphia). Data were collected over 1-15 (partially consecutive) days across the 2009 ozone season, providing approximately 7,000 person days of data for use by APEX.

- **OAB.** The diaries were collected in a study of children's activities on high and low ozone days during the 2002 ozone season (Mansfield et al., 2009). Children ages 2-12 from 35 U.S. metropolitan areas having the worst O_3 pollution were studied, and of whom, about half of were asthmatics. Activity data were collected on 6 nonconsecutive days from each subject, with some subjects providing fewer days, providing nearly 2,200 persons days of data to APEX.

- **SEA.** The diaries are from a particulate matter (PM) exposure study of susceptible study groups living in Seattle, WA between 1999 and 2002 (Liu et al., 2003). Two cohorts were studied: an older adult group with either chronic obstructive pulmonary disease (COPD) or coronary heart disease and a children's group (ages 6-13) with asthma. Activity data were collected on 10 consecutive days from each subject, with some subjects providing fewer days. Over 1,600 adult diaries and more than 300 children diaries were included in the APEX activity pattern file.
- **SUP.** The diaries are from the SUPERB study (Study of Use of Products and Exposure-Related Behaviors) undertaken by researchers from the University of California at Davis Bennett et al., 2012a; Hertz-Picciotto et al., 2012). The study focused on the use of household and personal care products from 47 California households, 30 with children (ages 1-18) living in 22 counties in northern California, and 17 with an older adult (>55 y) living in 3 central California counties. Two days of activity data were obtained via the internet for each participant—a weekday and a weekend day. Approximately 2,500 diary-days from SUPERB met appropriate criteria for use in APEX.
- **RTP**. The diaries were collected in a panel study of PM exposure in the Research Triangle Park (RTP), NC area (Williams et al., 2003a, b). Two older adult cohorts (ages 55-85) were studied: a cohort having implanted cardiac defibrillators living in Chapel Hill, NC and a second group of 30 people having controlled hypertension and residing in a low-to-moderate SES neighborhood in Raleigh, NC. Data were collected on approximately 8 consecutive days in 4 consecutive calendar seasons in 2000-2001. Approximately 900 diary-days were included from this study.

5B-4.2 Longitudinal Activity Pattern Methodology

An important issue in this assessment is the approach used for creating an O_3-season or year-long activity sequence for each simulated individual based on a largely cross-sectional activity database of 24-hour records. The typical subject in the time location activity studies in CHAD provided about two days of diary data. For this reason, the construction of a season-long activity sequence for each individual requires some combination of repeating the same data from one subject and using data from multiple subjects. The best approach would reasonably account for the day-to-day and week-to-week repetition of activities common to individuals (though recognizing even these diary sequences are not entirely correlated) while maintaining realistic variability among individuals comprising each study group.

The method currently used in APEX for creating longitudinal diaries was designed to capture the tendency of individuals to repeat activities, based on reproducing realistic variation in a key diary variable, which is a user selected function of diary variables. For this O_3 analysis, the key variable selected is the amount of time an individual spends outdoors each day, one of the most important determinants of exposure to high levels of O_3. The actual diary construction method targets two statistics, a population diversity statistic (*D*) and a within-person autocorrelation statistic (*A*). The *D* statistic reflects the relative importance of within- and between-person variance in the key variable. The *A* statistic quantifies the lag-one (day-to-day) key variable autocorrelation. Further details regarding the longitudinal methodology can be found in US EPA (2013a, b).

Desired *D* and *A* values for the key variable are selected by the user and set in the APEX parameters file, and the method algorithm constructs longitudinal diaries that preserve these parameters. Longitudinal diary data from a limited field study of children ages 7-12 (Geyh et al., 2000; Xue et al., 2004) estimated values of approximately 0.2 for *D* and 0.2 for *A*. In the absence of data for estimating these statistics for younger children and others outside the study age range, and since APEX appears to underestimate repeated activities, values of 0.5 for *D* and 0.2 for *A* are used for all ages.

Table 5B-1. Consolidated Human Activity Database (CHAD) study information and diary-days used by APEX.

Study Name (CHAD Abbreviation)	Geographic Coverage	Study Dates	Study Subject Ages	APEX Diary-days (ages 4-94)	APEX Diary-days (ages 4-18)	Diary Type, Time Format, Survey Design	Study Reference
Baltimore Retirement Home Study (BAL)	One building in Baltimore, MD	1/1997 to 2/1997; 7/1998 to 8/1998	72 - 93	304	0	Diary, 15 Minute Block, Panel	Williams et al. (2000)
California Youth Activity Patterns Study (CAY)	California	10/1987 to 9/1988	12 - 17	182	182	Recall, Event, Random	Robinson et al. (1989), Wiley et al. (1991a)
California Adults Activity Patterns Study (CAA)	California	10/1987 to 9/1988	18 - 94	1,555	36	Recall, Event, Random	Robinson et al. (1989), Wiley et al. (1991a)
California Children Activity Patterns Study (CAC)	California	4/1989 to 2/1990	<1 - 11	1,195	771	Recall, Event, Random	Wiley et al. (1991b)
Cincinnati Activity Patterns Study (CIN)	Cincinnati, OH metro. area	3/1985 to 4/1985; 8/1985	<1 - 86	2,449	727	Diary, Event, Random	Johnson (1989)
Detroit Exposure and Aerosol Research Study (DEA)	Detroit, MI metro. area	7/2005 to 8/2005; 7/2006 to 8/2006	18 - 74	331	5	Recall, 15 Minute Block, Panel	Williams et al. (2008)
Denver CO Personal Exposure Study (DEN)	Denver, CO metro. area	11/1982 to 2/1983	18 - 70	714	7	Diary, Event, Random	Johnson (1984), Akland et al. (1985)
EPA Longitudinal Studies (EPA)	RTP, NC	2/1999 to 2/2000; 2/2002 to 8/2002; 7/2006 to 6/2008	<1 - 60	1,417	0	Diary, Event, Panel	Isaacs et al. (2012)
Los Angeles Ozone Exposure Study: Elementary School (LAE)	Los Angeles, CA	10/1989	10 - 12	50	50	Diary, Event, Panel	Spier et al. (1992)
Los Angeles Ozone Exposure Study: High School (LAH)	Los Angeles, CA	9/1990 to 10/1990	13 - 17	42	42	Diary, Event, Panel	Spier et al. (1992)
National Human Activity Pattern Study: Air (NHA)	National	9/1992 to 10/1994	<1 - 93	4,329	693	Recall, Event, Random	Klepeis et al. (1996), Tsang and Klepeis (1996)

Study Name (CHAD Abbreviation)	Geographic Coverage	Study Dates	Study Subject Ages	APEX Diary-days (ages 4-94)	APEX Diary-days (ages 4-18)	Diary Type, Time Format, Survey Design	Study Reference
National Human Activity Pattern Study: Water (NHW)	National	9/1992 to 10/1994	<1 - 93	4,329	745	Recall, Event, Random	Klepeis et al. (1996), Tsang and Klepeis (1996)
National-Scale Activity Survey (NSA)	7 US metro. areas	6/2009 to 9/2009	35 - 92	6,825	0	Recall, 15 Minute Block, Random	Knowledge Networks (2009)
Population Study of Income Dynamics PSID I (ISR)	National	2/1997 to 12/1997	<1 - 13	4,978	3,507	Recall, 15 Minute Block, Random/Panel	University of Michigan (2012)
Population Study of Income Dynamics PSID II (ISR)	National	1/2002 to 12/2003	5 - 19	4,800	4,793	Recall, 15 Minute Block, Random/Panel	University of Michigan (2012)
Population Study of Income Dynamics PSID III (ISR)	National	10/2007 to 4/2008	10 - 19	2,650	2,614	Recall, 15 Minute Block, Random/Panel	University of Michigan (2012)
RTI Ozone Averting Behavior (OAB)	35 US metro. areas	7/2002 to 8/2003	2 - 12	2,872	2,187	Recall, 15 Minute Block, Random	Mansfield et al. (2006, 2009)
RTP Panel (RTP)	RTP, NC	6/2000 to 5/2001	55 - 85	871	0	Diary, 15 Minute Block, Panel	Williams et al. (2003a,b)
Seattle (SEA)	Seattle, WA	10/1999 to 3/2002	6 - 91	1,624	317	Diary, 15 Minute Block, Panel	Liu et al. (2003)
Study of Use of Products and Exposure Related Behavior (SUP)	Broader Sacramento & San Francisco, CA Counties	7/2006 to 3/2010	1 - 88	3,456	994	Recall, 15 Minute Block, Panel	Bennett et al. (2012a), Hertz-Picciotto et al. (2010)
Washington, D.C. (WAS)	Wash. DC metro. area	11/1982 to 2/1983	18 - 71	686	10	Diary, Event, Random	Hartwell et al. (1984), Akland et al. (1985)
Totals		**1982 - 2010**	**<1 - 94**	**41,474**	**17,680**		

5B-5 PHYSIOLOGICAL AND METABOLIC EQUIVALENTS DATA

APEX requires several physiological parameters to accurately model processes that affect pollutant intake rate for individuals. This is because differences in physiology may cause people with the same exposure and activity scenarios to have different pollutant intake levels. The physiological parameters file used by APEX contains individual data or data distributions stratified by age and gender for maximum ventilatory capacity (in terms of age- and gender-specific maximum oxygen consumption potential, NVO_2max), body mass (BM), resting metabolic rate (RMR), body surface area (BSA), maximum oxygen deficits (MOXD) and associated recovery time (RECTIME), height, and oxygen consumption-to-ventilation rate relationships (ECF), among a few others not used for estimating O_3 exposure and dose).

APEX also uses an input file containing the metabolic equivalents for work (METS) to estimate the specific energy expended for each activity listed in the diary file. These METS values are commonly in the form of distributions and were originally derived as relative to an individual's RMR. Some activities are specified as a single point value (for instance, sleep), while others, such as athletic endeavors or manual labor, are normally, lognormally, or otherwise statistically distributed. APEX samples from these distributions and calculates values to simulate the variable nature of activity levels among different people. These personal- and activity-level physiological variables are ultimately used to estimate ventilation rate (VE) and decrements in forced expiratory volume, in one second ($dFEV_1$).

Three standard APEX input files are used for the current O_3 assessment:

- *Physiology010213_threshold.txt*: NVO2max, BM, RMR, BSA, MOXD, RECTIME, height, ECF, and $dFEV_1$ distributions and equation coefficients, by sex and age groups
- *MET_Distributions_030612.txt*: statistical form and parameters for METS distributions associated with each activity performed, some by age groups
- *Ventilation_121106.txt*: distributions and equation coefficients to estimate individual activity- specific VE by sex and age groups

5B-6 MICROENVIRONMENTS MODELED

In APEX, exposure for simulated individuals occurs in microenvironments. For exposures to be accurately estimated, it is important maintain the spatial and temporal sequence of microenvironments persons inhabit and appropriately represent the time series of concentrations that occur within them. As discussed in Appendix 5A, the two methods available in APEX for calculating pollutant concentrations within microenvironments are a mass balance model and a transfer factor approach, each of which uses an appropriate ambient pollutant concentration to estimate the microenvironmental concentration. Table 5B-2 lists the 28 microenvironments selected for this analysis and the exposure calculation method for each. The variables used and their associated parameters to calculate microenvironmental concentrations are described in subsequent subsections below.

The CHAD database has 115 locations codes, many of which go beyond the scale of the microenvironmental modeling (e.g., inside at residence in a bedroom). Therefore these more specific locations are aggregated by mapping these 115 location codes to the 28 modeled microenvironments. Further, all microenvironmental concentrations in this exposure assessement are estimated using an ambient concentration (section 5B-7), though these concentrations not only vary temporally but spatially, depending on the particular microenvironment. The mapping of locations to the 28 microenvironments also includes an identifier that designates what ambient concentration is used in the calculation of the microenvironmental concentration for each event. For this assessment, we used ambient concentration for each individual based on either their home (H), work (W), near work (NW), near home (NH), last (L, either NH or NW), other (O, average of all), or unknown (U, last ME determined) tracts.

Multiple APEX ME input files are used for the current O_3 assessment, varying by study area though given in one form. Only one ME mapping file is used:

- *ME_descriptions_28MEs_O3_CSA[studyarea.]_[date].txt*: defines calculation method, variables and their parameters used to estimate all microenvironmental concentrations
- *MicroEnv_Mapping_CHAD_to_APEX_28MEs_022613.txt*: maps 115 CHAD locations to 28 APEX microenvironments and defines tract-level ambient concentrations to use for each location

Table 5B-2. Microenvironments modeled and calculation method used.

Microenvironment (ME)	AEPX ME Number	Calculation Method	Parameters[1]
Indoor – Residence	1	Mass balance	AER and DE
Indoor – Community Center or Auditorium	2	Mass balance	AER and DE
Indoor – Restaurant	3	Mass balance	AER and DE
Indoor – Hotel, Motel	4	Mass balance	AER and DE
Indoor – Office building, Bank, Post office	5	Mass balance	AER and DE
Indoor – Bar, Night club, Café	6	Mass balance	AER and DE
Indoor – School	7	Mass balance	AER and DE
Indoor – Shopping mall, Non-grocery store	8	Mass balance	AER and DE
Indoor – Grocery store, Convenience store	9	Mass balance	AER and DE
Indoor – Metro-Subway-Train station	10	Mass balance	AER and DE
Indoor – Hospital, Medical care facility	11	Mass balance	AER and DE
Indoor – Industrial, factory, warehouse	26	Mass balance	AER and DE
Indoor – Other indoor	27	Mass balance	AER and DE
Outdoor – Residential	12	Factors	None
Outdoor – Park or Golf course	14	Factors	None
Outdoor – Restaurant or Café	15	Factors	None
Outdoor – School grounds	16	Factors	None
Outdoor – Boat	25	Factors	None
Outdoor – Other outdoor non-residential	13	Factors	None
Near-road – Metro-Subway-Train stop	17	Factors	PR
Near-road – Within 10 yards of street	18	Factors	PR
Near-road – Parking garage (covered or below ground)	19	Factors	PR
Near-road – Parking lot (open), Street parking	20	Factors	PR
Near-road – Service station	21	Factors	PR
Vehicle – Cars and Light Duty Trucks	22	Factors	PE and PR
Vehicle – Heavy Duty Trucks	28	Factors	PE and PR
Vehicle – Bus	23	Factors	PE and PR
Vehicle – Train, Subway	24	Factors	PE and PR

[1] AER = air exchange rate, DE = decay-deposition rate, PR = proximity factor, PE = penetration factor.

5B-6.1 Air Exchange Rates for Indoor Residential Microenvironments

Distributions of air exchange rates (AERs) for the indoor residential microenvironments (ME-1) were developed using data from several studies. The analysis of these data and the development of most of the distributions used in the modeling were originally described in detail in US EPA (2007) Appendix A, though recently updated by Cohen et al. (2012) and provided in Appendix 5E.

The analyses indicated that the AER distributions for the residential microenvironments depend on the type of air conditioning (A/C) and on the outdoor temperature, among other variables for which we do not have sufficient data to estimate. These analyses demonstrate that the AER distributions vary greatly across cities, A/C types, and temperatures, so that the selected AER distributions for the modeled cities should also depend on these attributes. For example, the mean AER for residences with A/C ranges from 0.38 in Research Triangle Park, NC at temperatures > 25 °C upwards to 1.244 in New York, NY considering the same temperature bin.

For each combination of A/C type, city, and temperature with a minimum of 11 AER values, exponential, lognormal, normal, and Weibull distributions were fit to the AER values and compared. Generally, the lognormal distribution was the best-fitting of the four distributions, and so, for consistency, the fitted lognormal distributions are used for all the cases. Los Angeles had an adequate number of samples and identifiers to distinguish the estimated AER distributions by central A/C and room unit A/C for the homes with A/C.

There were a number of limitations in generating study-area specific AER stratified by temperature and A/C type. For example, AER data and derived distributions were available only for selected cities, and yet the summary statistics and comparisons demonstrate that the AER distributions depend upon the city as well as the temperature range and A/C type. As a result, city-specific AER distributions were used where possible; otherwise staff selected AER data from a similar city. Another important limitation of the analysis was that distributions were not able to be fitted to all of the temperature ranges due to limited number of available measurement data in these ranges. A description of how these limitations were addressed can be found in Appendix 5E. The AER distributions used for the exposure modeling are given in Table 5B-3 (Residences with A/C) and Table 5B-4 (Residences without A/C).

Table 5B-3. AERs for indoor residential microenvironments (ME-1) with A/C by study area and temperature.

Study Area	**Daily Mean Temperature** (°C)	**Lognormal Distribution** {GM, GSD, min, max}	**Original AER Study Data Used**
Atlanta, Baltimore, Washington DC	< 10	{0.962, 1.809, 0.1, 10}	Research Triangle Park, NC
	10 - 20	{0.562, 1.906, 0.1, 10}	
	20 - 25	{0.397, 1.889, 0.1, 10}	
	> 25	{0.380, 1.709, 0.1, 10}	
Boston, New York, Philadelphia	< 10	{0.711, 2.108, 0.1, 10}	New York, NY
	10 - 25	{1.139, 2.677, 0.1, 10}	
	> 25	{1.244, 2.177, 0.1, 10}	
Chicago, Cleveland, Detroit	< 10	{0.744, 1.982, 0.1, 10}	Detroit, MI and New York, NY
	10 - 20	{0.811, 2.653, 0.1, 10}	
	20 - 25	{0.785, 2.817, 0.1, 10}	
	> 25	{0.916, 2.671, 0.1, 10}	
Dallas, Houston	< 20	{0.407, 2.113, 0.1, 10}	Houston, TX
	20 - 25	{0.467, 1.938, 0.1, 10}	
	25 - 30	{0.422, 2.258, 0.1, 10}	
	> 30	{0.499, 1.717, 0.1, 10}	
Denver, St. Louis	< 10	{0.921, 1.854, 0.1, 10}	All Cities Outside of CA
	10 - 20	{0.573, 1.990, 0.1, 10}	
	20 - 25	{0.530, 2.427, 0.1, 10}	
	25 - 30	{0.527, 2.381, 0.1, 10}	
	> 30	{0.609, 2.369, 0.1, 10}	
Los Angeles (Central A/C)	< 20	{0.577, 1.897, 0.1, 10}	Los Angeles, CA
	20 - 25	{1.084, 2.336, 0.1, 10}	
	> 25	{0.861, 2.344, 0.1, 10}	
Los Angeles (Room Unit A/C)	< 20	{0.672, 1.863, 0.1, 10}	Los Angeles, CA
	20 - 25	{1.674, 2.223, 0.1, 10}	
	> 25	{0.949, 1.644, 0.1, 10}	
Sacramento	< 25	{0.503, 1.921, 0.1, 10}	Sacramento, Riverside, San Bernardino Counties
	>25	{0.830, 2.353, 0.1, 10}	

Table 5B-4. AERs for indoor residential microenvironments (ME-1) without A/C by study area and temperature.

Study Area	Daily Mean Temperature (°C)	Lognormal Distribution {GM, GSD, min, max}	Original AER Study Data Used
Atlanta, Baltimore, Denver, St. Louis, Washington DC	< 10	{0.923, 1.843, 0.1, 10}	All Cities Outside of CA
	10 - 20	{0.951, 2.708, 0.1, 10}	
	> 20	{1.575, 2.454, 0.1, 10}	
Boston, New York, Philadelphia	< 10	{1.016, 2.138, 0.1, 10}	New York, NY
	10 - 20	{0.791, 2.042, 0.1, 10}	
	> 20	{1.606, 2.119, 0.1, 10}	
Chicago, Cleveland, Detroit	< 0	{1.074, 1.772, 0.1, 10}	Detroit, MI and New York, NY
	0 - 10	{0.760, 1.747, 0.1, 10}	
	10 - 20	{1.447, 2.950, 0.1, 10}	
	20 - 25	{1.531, 2.472, 0.1, 10}	
	> 25	{1.901, 2.524, 0.1, 10}	
Dallas, Houston	< 10	{0.656, 1.679, 0.1, 10}	Houston, TX
	10 - 20	{0.625, 2.916, 0.1, 10}	
	> 20	{0.916, 2.451, 0.1, 10}	
Los Angeles	< 20	{0.744, 2.057, 0.1, 10}	Los Angeles, CA
	20 - 25	{1.448, 2.315, 0.1, 10}	
	> 25	{0.856, 2.018, 0.1, 10}	
Sacramento	< 10	{0.526, 3.192, 0.1, 10}	Sacramento, Riverside, San Bernardino Counties
	10 - 20	{0.665, 2.174, 0.1, 10}	
	20 - 25	{1.054, 1.711, 0.1, 10}	
	> 25	{0.827, 2.265, 0.1, 10}	

5B-6.2 Air Conditioning Prevalence for Indoor Residential MicroEnvironments

The selection of an AER distribution is conditioned on the presence or absence of A/C. We assigned this housing attribute to indoor residential microenvironments (ME-1) using A/C prevalence data from the American Housing Survey (AHS)[9]. A/C prevalence is noted as distinct from usage rate, the latter represented by the AER distribution and dependent on temperature. The A/C prevalence data were assigned to our study areas where the AHS data best matched our exposure simulation years (Table 5B-5). Because we were able to stratify the AER distributions by three A/C types in Los Angeles, both the individual central and room unit values were used. In all other study areas, the sum of room unit and central A/C prevalence was used.

[9] Available at: http://www.census.gov/housing/ahs/data/metro.html.

Table 5B-5. American Housing Survey A/C prevalence from Current Housing Reports (Table 1-4) for selected urban areas.

Metropolitan Area	Area[1]	Study Years	Total Occupied Housing Units (x1000)	Number of Occupied Housing Units (x1000) with:					% of Occupied Housing Units with:[2]		
				Central A/C	>1 Central A/C	1 Room Unit	2 Room Units	3+ Room Units	Central A/C	Window Units	Central & Window A/C
Atlanta	MA	2004	1595.8	1473.8	245.9	39.8	32.5	18.3	92	6	98
Baltimore	MSA	2007	1012.3	785.4	44.4	55.7	72.4	65.4	78	19	97
Boston	CMSA	2007	1057.1	291.5	20.3	259.4	198.7	156.0	28	58	86
Chicago	PMSA	2009	3010.7	2050.6	116.2	412.0	265.1	124.4	68	27	95
Cleveland	PMSA	2004	769.3	416.4	14.1	132.9	47.0	17.6	54	26	80
Dallas	PMSA	2002	1235.3	1146.3	171.6	25.5	29.8	27.8	92.8	6.7	99.5
Denver	MA	2004	855.7	425.1	16.8	123.8	20.3	3.9	50	17	67
Detroit	PMSA	2009	1672.5	1194.3	46.5	192.3	82.8	29.2	71	18	90
Houston	PMSA	2007	1872.0	1682.5	153.7	46.0	59.9	60.6	90	9	99
Los Angeles[3]	PMSA	2003	5152.4	2448.4	161.6	702.1	118.6	46.9	47.5	16.8	64.4
New York[4]	PMSA	2009	4493.3	872.4	38.2	1036.9	1184.1	812.6	19	68	87
Philadelphia	PMSA	2009	1916.2	1095.9	52.3	197.9	260.8	265.8	57	38	95
Sacramento	PMSA	2004	669.4	549.5	30.7	57.1	12.3	2.4	82	11	93
St. Louis	MA	2004	1139.6	974.4	53.7	65.8	43.5	16.6	86	11	97
Washington, DC	MA	2007	1949.1	1729.6	145.7	69.2	64.8	61.2	89	10	99

[1] MA – metropolitan area; CMSA – consolidated metropolitan statistical area; PMSA – primary metropolitan statistical area.
[2] Shaded areas indicate final values used in APEX functions files to select AER distributions used for indoor residential microenvironments (ME-1).
[3] Los Angeles includes Los Angeles-Long Beach, Riverside-San Bernardino-Ontario, and Anaheim-Santa Ana MSA's.
4 New York is represented by the NY-Nassau-Suffolk-Orange MSA.

5B-6.3 AER DISTRIBUTIONS FOR OTHER INDOOR ENVIRONMENTS

To estimate AER distributions for non-residential, indoor environments (e.g., offices, libraries), we obtained and analyzed two AER data sets: "Turk" (Turk et al., 1989); and "Persily" (Persily and Gorfain, 2004; Persily et al., 2005). The Turk data set includes 40 AER measurements from offices (25 values), schools (7 values), libraries (3 values), and multi-purpose buildings (5 values), each measured using an SF_6 tracer over two or four hours in different seasons of the year. The Persily data were derived from the US EPA Building Assessment Survey and Evaluation (BASE) study, which was conducted to assess indoor air quality, including ventilation, in a large number of randomly selected office buildings throughout the US. This data base consists of 390 AER measurements in 96 large, mechanically ventilated offices. AERs were measured both by a volumetric method and by a CO_2 ratio method, and included their uncertainty estimates. For these analyses, we used the recommended "Best Estimates" defined by the values with the lower estimated uncertainty; in the vast majority of cases the best estimate was from the volumetric method.

Due to the small sample size of the Turk data, the data were analyzed without stratification by building type and/or season. For the Persily data, the AER values for each office space were averaged, rather using the individual measurements, to account for the strong dependence of the AER measurements for the same office space over a relatively short period. The mean values are similar for the two studies, but the standard deviations are about twice as high for the Persily data. We fitted exponential, lognormal, normal, and Weibull distributions to the 96 office space average AER values from the more recent Persily data, and the best fitting of these was the lognormal. The fitted parameters for this distribution are a geometric mean of 1.109, geometric standard deviation of 3.015, and bounded by the lower and upper values of the sample data set {0.07, 13.8}. These are used for AER distributions for several indoor non-residential microenvironments (ME-2, ME-4, ME-5, ME-8, ME-9, ME-10, ME-11, ME-26) except for indoor schools (ME-7) and indoor restaurants, bars, night clubs, and cafés (ME-3 and ME-6).

The AER distribution used for indoor schools (ME-7) is a discrete distribution {0.8, 1.3, 1.8, 2.19, 2.2, 2.21, 3.0, 0.6, 0.1, 0.6, 0.2, 1.8, 1.3, 1.2, 2.9, 0.9, 0.9, 0.9, 0.9, 0.4, 0.4, 0.4, 0.4, 0.9, 0.9, 0.9, 0.9, 0.3, 0.3, 0.3, 0.3} developed using data from Turk et al. (1989) and Shendell et al. (2004).

The AER distribution used for indoor restaurants, bars, night clubs, and cafés (ME-3, ME-6) is a fitted lognormal distribution, having a geometric mean = 3.712, geometric standard deviation = 1.855 and bounded by the lower and upper values of the sample data set {1.46, 9.07}. This distribution was developed using data from Bennett et al. (2012b), who measured these six values in restaurants (details on derivation provided in Appendix 5E).

5B-6.4 Proximity and Penetration Factors for In-vehicle and Near-Road Microenvironments

For the in-vehicle proximity and penetration factors (ME-22, ME-23, ME-24, ME-28), we use distributions developed from the Cincinnati Ozone Study (American Petroleum Institute, 1997, Appendix B; Johnson et al., 1995). This field study was conducted in the greater Cincinnati metropolitan area in August and September, 1994. Vehicle tests were conducted according to an experimental design specifying the vehicle type, road type, vehicle speed, and ventilation mode. Vehicle types were defined by the three study vehicles: a minivan, a full-size car, and a compact car. Road types were interstate highways (interstate), principal urban arterial roads (urban), and local roads (local). Nominal vehicle speeds (typically met over one minute intervals within 5 mph) were at 35 mph, 45 mph, or 55 mph. Ozone concentrations were measured inside the vehicle, outside the vehicle, and at six fixed-site monitors in the Cincinnati area. Table 5B-6 lists the parameters of the normal distributions developed for penetration and proximity factors for in-vehicle microenvironments used in this modeling analysis.

Table 5B-6. Parameter values for distributions of penetration and proximity factors used for estimating in-vehicle microenvironmental concentrations.

Microenvironmental Factor	Road type	Arithmetic Mean	Standard Deviation	Lower Bound[1]	Upper Bound
Penetration	All	0.300	0.232	0.100	1.0
Proximity	Local	0.755	0.203	0.422	1.0
	Urban	0.754	0.243	0.355	1.0
	Interstate	0.364	0.165	0.093	1.0

[1] A 5th percentile value estimated using a normal approximation as Mean – 1.64 × standard deviation.

The Vehicle Miles of Travel (VMT) fractions[10] provided by the U.S. Department of Transportation (DOT) are used to generate daily conditional variables that determine the selection of which proximity factor distributions are used to estimate in-vehicle microenvironmental concentrations (Table 5B-7). For local and interstate road types, the VMT for the same DOT categories were used. For urban roads, the VMT for all other DOT road types were summed (i.e., other freeways/expressways, other principal arterial, minor arterial, and collector). At the time of this writing, data were only available for four of our modeled years, 2006-2008 and 2010. Staff assumed that values for 2009 would be best represented by averaging 2008 and 2010.

For all outdoors-near-road microenvironments (ME-17, ME-18, ME-19, ME-20, ME-21) we employed the distribution for local roads (i.e., a normal distribution {0.755, 0.203}, bounded by 0.422 and 1.0), based on the assumption that most of the outdoors-near-road ozone exposures will occur proximal to local roads.

5B-6.5 Proximity and Penetration Factors for Outdoor Microenvironments

All outdoor microenvironments (ME-12, ME-13, ME-14, ME-15, ME-16, ME-25) are assumed well represented by the census tract level O_3 concentrations. Therefore, both the penetration factor and proximity factor for this microenvironment were set to equal 1.

5B-6.6 Ozone Decay and Deposition Rates

A distribution for combined O_3 decay and deposition rates was obtained from the analysis of measurements from a study by Lee et al. (1999). This study measured decay rates in the living rooms of 43 residences in Southern California. Measurements of decay rates in a second room were made in 24 of these residences. The 67 decay rates range from 0.95 to 8.05 hour^{-1}. A lognormal distribution was fit to the measurements from this study, yielding a geometric mean of 2.51 and a geometric standard deviation of 1.53. These values are constrained to lie between 0.95 and 8.05 hour^{-1}. This distribution was used for all indoor microenvironments.

---

[10] U.S. Department of Transportation, Federal Highway Administration. Annual *Highway Statistics*, Table HM-71: Urbanized Areas - Miles And Daily Vehicle Miles Of Travel. For example, 2010 data available at: www.fhwa.dot.gov/policyinformation/statistics/2010/xls/hm71.xls

Table 5B-7. VMT fractions of interstate, urban, and local roads in the study areas used to select in-vehicle proximity factor distributions.

	2006			2007			2008			2009			2010		
Study Area	Inter-state	Urban	Local	Inter-state	Urban	Local	Inter-state	Urban	Local	Inter-state	Urban	Local	Inter-state	Urban	Local
Atlanta	0.34	0.46	0.20	0.34	0.47	0.19	0.32	0.45	0.23	0.31	0.44	0.25	0.30	0.43	0.27
Baltimore	0.34	0.59	0.07	0.34	0.59	0.07	0.34	0.59	0.07	0.34	0.59	0.07	0.34	0.59	0.07
Boston	0.32	0.55	0.13	0.32	0.55	0.13	0.32	0.54	0.14	0.32	0.54	0.14	0.32	0.54	0.14
Chicago	0.30	0.58	0.12	0.30	0.58	0.12	0.31	0.57	0.12	0.30	0.57	0.13	0.31	0.56	0.13
Cleveland	0.40	0.44	0.16	0.40	0.44	0.16	0.39	0.45	0.16	0.39	0.45	0.16	0.38	0.46	0.16
Dallas	0.30	0.66	0.04	0.30	0.66	0.04	0.30	0.65	0.05	0.30	0.66	0.04	0.29	0.67	0.04
Denver	0.23	0.67	0.10	0.24	0.66	0.10	0.25	0.65	0.10	0.25	0.65	0.10	0.25	0.65	0.10
Detroit	0.26	0.64	0.10	0.25	0.65	0.10	0.24	0.66	0.10	0.25	0.65	0.10	0.26	0.63	0.11
Houston	0.24	0.72	0.04	0.24	0.72	0.04	0.24	0.72	0.04	0.24	0.72	0.04	0.24	0.72	0.04
Los Angeles	0.29	0.66	0.05	0.29	0.67	0.04	0.28	0.67	0.05	0.29	0.66	0.05	0.29	0.66	0.05
New York	0.19	0.66	0.15	0.19	0.65	0.16	0.19	0.66	0.15	0.19	0.66	0.15	0.19	0.65	0.16
Philadelphia	0.23	0.65	0.12	0.24	0.65	0.11	0.24	0.65	0.11	0.20	0.59	0.21	0.18	0.52	0.30
Sacramento	0.25	0.72	0.03	0.25	0.70	0.05	0.23	0.69	0.08	0.23	0.69	0.08	0.23	0.69	0.08
St. Louis	0.36	0.45	0.19	0.37	0.45	0.18	0.38	0.45	0.17	0.37	0.45	0.18	0.36	0.44	0.20
Wash., DC	0.31	0.61	0.08	0.31	0.61	0.08	0.30	0.61	0.09	0.30	0.61	0.09	0.30	0.61	0.09

A few individual fractions have been adjusted to yield an annual sum of 1.00.

5B-7 AMBIENT OZONE CONCENTRATIONS

To estimate exposure in this assessment, APEX requires hourly ambient O_3 concentrations at a set of locations (or air districts) within study area. We used hourly ambient monitoring data along with a statistical approach (VNA) to better approximate spatial heterogeneity (where such heterogeneity might be present) across each study area (O_3 REA, Chapter 4). General processing steps performed to generate the final APEX ambient concentration input files that were used were as follows.

After identifying the 15 study areas to be modeled in this assessment, staff defined a broad air quality modeling domain for each study area, specifically bounding where exposures were to be estimated. We evaluated 1) counties modeled in the previous 2007 O_3 NAAQS review common to current study areas, 2) political/statistical county aggregations (MSA, PMSA), and 3) if the study area was designated as a non-attainment area (NAA), the counties that were part of the NAA list. A final list of counties was generated using this information (Table 5B-8), then hourly O_3 concentrations were estimated at every census tract within the counties that comprised each study area (O_3 REA, Chapter 4). These data served as the air quality input to APEX with some exception (see below), though note also, not all of the estimated hourly concentrations would be used in the exposure simulation even if supplied to APEX.

As was done in the first draft REA, a 30 km radius of influence was used for each monitoring site within the above county-level defined study. All census tracts that fell within the 30 km radius of each ambient monitor used to estimate the air quality concentration fields were selected, then any tracts/monitor radii that were largely outside of the urban core were removed, thus defining a final exposure modeling domain in each study area (Table 5B-8).

Because APEX uses 2000 census population data and the air concentrations were modeled to 2010 census tracts, some of the air district locations differed slightly from that of the exposure tracts, resulting in different numbers of air districts when compared with the number of census tracts used in simulating exposures. This difference is expected to have a negligible effect on exposure and risk results because APEX always uses the air district nearest to the tract to be modeled, the distances between any two air district centroids within these urban study areas (census tract level) is expected to be small, and the concentration gradient across that said distance is also expected to not be significant.

Further, staff had computational difficulty in simulating the large number of tracts and air districts for the Los Angeles, New York, and Chicago study areas (number and size of arrays needed in APEX calculations was beyond the standard PC capabilities); based on simulations that ran to completion, the maximum number of air districts possible using a standard 32-bit PC was estimated as 1,900 – 2,000. Thus, to make the analysis more tractable for these study areas, first staff reduced the number of air districts originally modeled (i.e., all year 2010 US tracts in the broad county domain) to the number needed for the actual year 2000 census tracts in the exposure model domain (i.e., all tracts within 30 km of ambient monitors in the broad county domain). Using this approach, the number of air districts was reduced to the following: Chicago (1,882), Los Angeles (3,268), and New York (4,646). For Los Angeles and New York, the number of air districts was reduced to 2,000 and 1,900 using simple random sampling of these tracts using SAS's SURVEYSELECT procedure; the number of air districts for Chicago remained at 1,882. While we estimated this number of districts would run on a standard PC, these three study areas would only run on a 64-bit PC.

The final list of year 2000 census tract IDs where exposure was modeled is within the APEX control files. The final list of 2010 census tract IDs where ambient concentrations were estimated is within the APEX air districts files. Table 5B-9 contains the final list of counties, the number of US census tracts where exposures were estimated, the number air districts ultimately used from the air quality input files, and the population counts represented in each study area. The final list of year 2000 census tract IDs where exposure was modeled is within the APEX control files. The final list of 2010 census tract IDs where ambient concentrations were estimated is within the APEX air districts files. Figure 5B-1 through Figure 5B-4 illustrate the general exposure modeling domains (i.e., the selected census tract centroids falling within 30 km of a ambient monitor) for each of the 15 study areas.

Multiple unique APEX input files are used for the current O_3 assessment, varying by the air quality scenario, year, and study area, though generally in two forms:

- *concsCSA[studyarea]S[scenario]P[std. avg. period]Y[year].txt*: hourly concentrations for each tract, by study area, air quality scenario, standard averaging period, year
- *districtsCSA[studyarea]Y[year].txt:* tract ID's, latitudes and longitudes, start and stop dates of concentrations

Table 5B-8. Identification of U.S. counties and the number of APEX air districts included each study area.

Study Area (State Abbreviation: List of Counties[1])	**APEX Air Districts (VNA Total)**
Atlanta (GA: Barrow, Bartow, Butts, Carroll, Cherokee, Clayton, Cobb, Coweta, Dawson, De Kalb, Douglas, Fayette, Forsyth, Fulton, Gwinnett, Hall, Haralson, Heard, Henry, Jasper, Lamar, Meriwether, Newton, Paulding, Pickens, Pike, Polk, Rockdale, Spalding, Troup, Upson, Walton; *AL: Chambers*)	664 (1,019)
Baltimore (MD: Anne Arundel, Baltimore, Carroll, Harford, Howard, Queen Anne's, Baltimore (City))	603 (679)
Boston (MA: *Barnstable*, Bristol, *Dukes*, Essex, Middlesex, *Nantucket*, Norfolk, Plymouth, Suffolk, Worcester)	1,005 (1,276)
Chicago (IL: Cook, DeKalb, DuPage, Grundy, Kane, Kankakee, Kendall, Lake, McHenry, Will; IN: Jasper, Lake, LaPorte, Newton, Porter, Kenosha)	1,882 (2,267)
Cleveland (OH: Ashtabula, Cuyahoga, Geauga, Lake, Lorain, Medina, Portage, Summit)	802 (830)
Dallas (TX: Collin, Dallas, Denton, Ellis, Hunt, Johnson, Kaufman, Parker, Rockwall, Tarrant, Wise)	1012 (1,312)
Denver (CO: Adams, Arapahoe, Boulder, *Broomfield*[2], Clear Creek, Denver, Douglas, Elbert, Gilpin, Jefferson, Larimer, Park, Weld)	655 (839)
Detroit (MI: Genesee, Lapeer, Livingston, Macomb, Monroe, Oakland, St. Clair, Washtenaw, Wayne)	1,419 (1,568)
Houston (TX: *Austin*, Brazoria, Chambers, Fort Bend, Galveston, Harris, Liberty, Montgomery, San Jacinto, Waller)	779 (1,074)
Los Angeles, (CA: Los Angeles, Orange, Riverside, San Bernardino, Ventura)	2,000 (3,920)
New York (CT: Fairfield, Middlesex, New Haven; NJ: Bergen, Essex, Hudson, Hunterdon, Mercer, Middlesex, Monmouth, Morris, Passaic, Somerset, Sussex, Union, Warren; NY: Bronx, Kings, Nassau, New York, Orange, Putnam, Queens, Richmond, Rockland, Suffolk, Westchester)	1,900 (5,003)
Philadelphia (DE: New Castle, MD: Cecil; NJ: Atlantic, Burlington, Camden, Cape May, Cumberland, Gloucester, Ocean, Salem; PA: Bucks, Chester, Delaware, Montgomery, Philadelphia)	1,452 (1,735)
Sacramento (CA: El Dorado, Nevada, Placer, Sacramento, Solano, Sutter, Yolo)	447 (623)
St. Louis (IL: Bond, Calhoun, Clinton, Jersey, Macoupin, Madison, Monroe, Saint Clair, MO: *Crawford*, Franklin, Jefferson, Lincoln, Saint Charles, Saint Louis, Warren, *Washington*, St. Louis City)	494 (626)
Washington, DC (District of Columbia; MD: Calvert, Charles, Frederick, Montgomery, Prince George's, St. Mary's, VA: Arlington, Clarke, Culpeper, Fairfax, Fauquier, Frederick, Loudoun, Prince William, Spotsylvania, Stafford, Warren, Alexandria City, Fairfax City, Falls Church City, Fredericksburg City, Manassas City, Manassas Park City, Winchester City, WV: Jefferson)	1,013 (1,391)
All AREAS (Counties: 207 Exposure of 215 Air Quality)	**16,127 (24,162)**

[1] *italicized*: in air quality domain but not in exposure modeling domain; considered outside of urban core or no monitors.
[2] this county is newly defined in the 2010 census.

Table 5B-9. Ambient monitors used to define exposure modeling domain and the population modeled in each study area.

Study Area (State Abbreviation: List of Monitors[1])	Census Tracts for Exposure	Population Represented
Atlanta (GA: 130590002, 130670003, 130770002, 130850001, 130890002, 130893001, 130970004, 131130001, 131210055, 131350002, 131510002, 132230003, 132319991, 132470001)	678	3,850,951
Baltimore (MD: 240030014, 240051007, 240053001, 240130001, 240251001, 240259001, 240290002, 240313001, 240330030, 240338003, 240339991, 245100054)	618	2,209,226
Boston (MA: 250092006, 250094004, 250094005, 250095005, 250170009, 250171102, 250213003, 250250041, 250250042, 250270015, 250270024; NH: 330111011; RI: 440071010)	1,028	4,449,291
Chicago (IL: 170310001, 170310032, 170310042, 170310064, 170310072, 170310076, 170311003, 170311601, 170314002, 170314007, 170314201, 170317002, 170436001, 170890005, 170971002, 170971007, 171110001, 171971011; IN: 180890022, 180890030, 180892008, 180910005, 180910010, 181270024, 181270026, WI: 550590019, 551010017, 551270005)	2,055	8,345,373
Cleveland (OH: 390071001, 390350034, 390350060, 390350064, 390355002, 390550004, 390850003, 390850007, 390853002, 390930018, 391030003, 391030004, 391331001, 391510016, 391514005, 391530020)	879	2,692,846
Dallas (TX: 480850005, 481130069, 481130075, 481130087, 481133003, 481210034, 481211032, 481390015, 481390016, 481391044, 482210001, 482311006, 482510003, 482570005, 483670081, 483970001, 484390075, 484391002, 484392003, 484393009, 484393011)	1,036	4,698,392
Denver (CO: 080013001, 080050002, 080050006, 080130007, 080130011, 080137001, 080137002, 080190004, 080190005, 080310002, 080310014, 080310025, 080350004, 080590002, 080590005, 080590006, 080590011, 080590013, 080690007, 080690011, 080691004, 080699991, 080930001, 081190003, 081230009)	675	2,626,239
Detroit (MI: 260490021, 260492001, 260910007, 260990009, 260991003, 261250001, 261470005, 261610008, 261619991, 261630001, 261630015, 261630016, 261630019)	1,454	4,572,479
Houston (TX: 480391004, 482010024, 482010026, 482010029, 482010046, 482010047, 482010051, 482010055, 482010062, 482010066, 482010070, 482010075, 482010416, 482011015, 482011034, 482011035, 482011039, 482011050, 483390078)	802	3,925,054
Los Angeles (CA: 060370002, 060370016, 060370113, 060371002, 060371103, 060371201, 060371301, 060371302, 060371602, 060371701, 060372005, 060374002, 060374006, 060375005, 060376012, 060379033, 060590007, 060591003, 060592022, 060595001, 060650004, 060650009, 060650012, 060651010, 060651016, 060651999, 060652002, 060655001, 060656001, 060658001, 060658005, 060659001, 060659003, 060710001, 060710005, 060710012, 060710306, 060711004, 060711234, 060712002, 060714001, 060714003, 060719002, 060719004, 061110007, 061110009, 061111004, 061112002, 061112003, 061113001)	3,352	14,950,340

Study Area (State Abbreviation: List of Monitors[1])	Census Tracts for Exposure	Population Represented
New York (CT: 090010017, 090011123, 090013007, 090019003, 090070007, 090090027, 090093002; NJ: 340030005, 340030006, 340130003, 340170006, 340190001, 340210005, 340219991, 340230011, 340250005, 340273001, 340290006, 340315001; NY: 360050083, 360050110, 360050133, 360610135, 360790005, 360810098, 360810124, 360850067, 360870005, 361030002, 361030004, 361030009, 361192004)	4,889	18,520,868
Philadelphia (DE: 100031007, 100031010, 100031013, MD: 240150003; NJ: 340070003, 340071001, 340110007, 340150002, 340210005, 340219991, 340290006; PA: 420170012, 420290100, 420450002, 420910013, 421010004, 421010014, 421010024, 421010136)	1,555	5,506,954
Sacramento (CA: 060170010, 060170020, 060570005, 060610002, 060610004, 060610006, 060670002, 060670006, 060670010, 060670011, 060670012, 060670013, 060670014, 060675003, 060953003, 061010003, 061010004, 061131003)	461	1,926,598
St. Louis (IL: 170831001, 171190008, 171191009, 171193007, 171199991, 171630010; MO: 290990012, 290990019, 291130003, 291831002, 291831004, 291890004, 291890005, 291890014, 295100085, 295100086)	518	2,340,325
Washington, DC (110010025, 110010041, 110010043; MD: 240030014, 240090011, 240130001, 240170010, 240210037, 240313001, 240330030, 240338003, 240339991, 240430009; VA: 510130020, 510330001, 510590005, 510590018, 510590030, 510591005, 510595001, 510610002, 510690010, 511071005, 511390004, 511530009, 511790001, 515100009; WV: 540030003)	1,037	4,498,374
All AREAS (324 ambient monitors)	**21,037**	**85,113,310**

[1] A 30 km radius for monitors operating anytime during 2006-2010 was used to select census tracts in defining the exposure modeling domain.

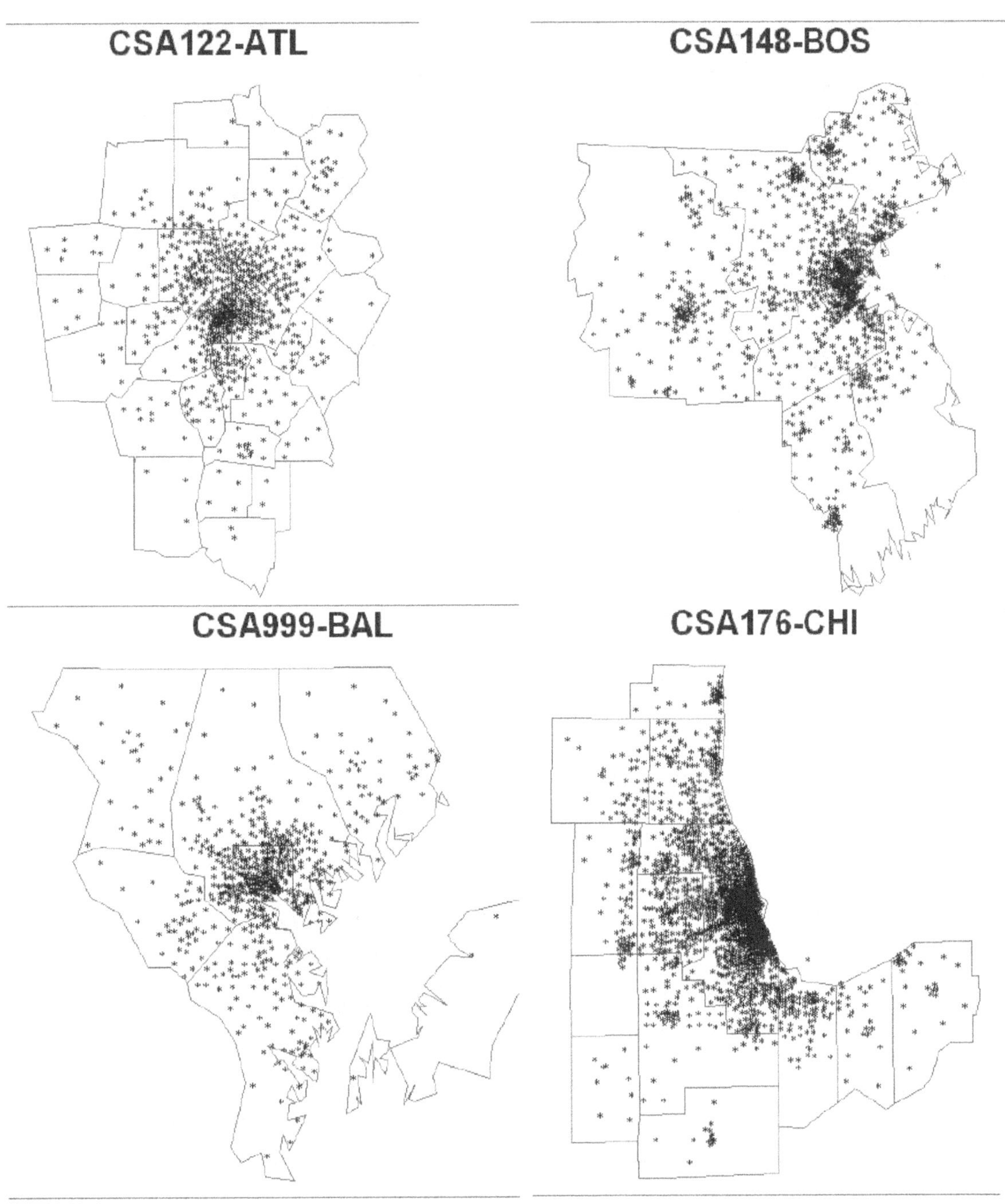

Figure 5B-1. Illustration of APEX exposure modeling domains (2000 US Census tract centroids) for Atlanta, Boston, Baltimore and Chicago study areas.

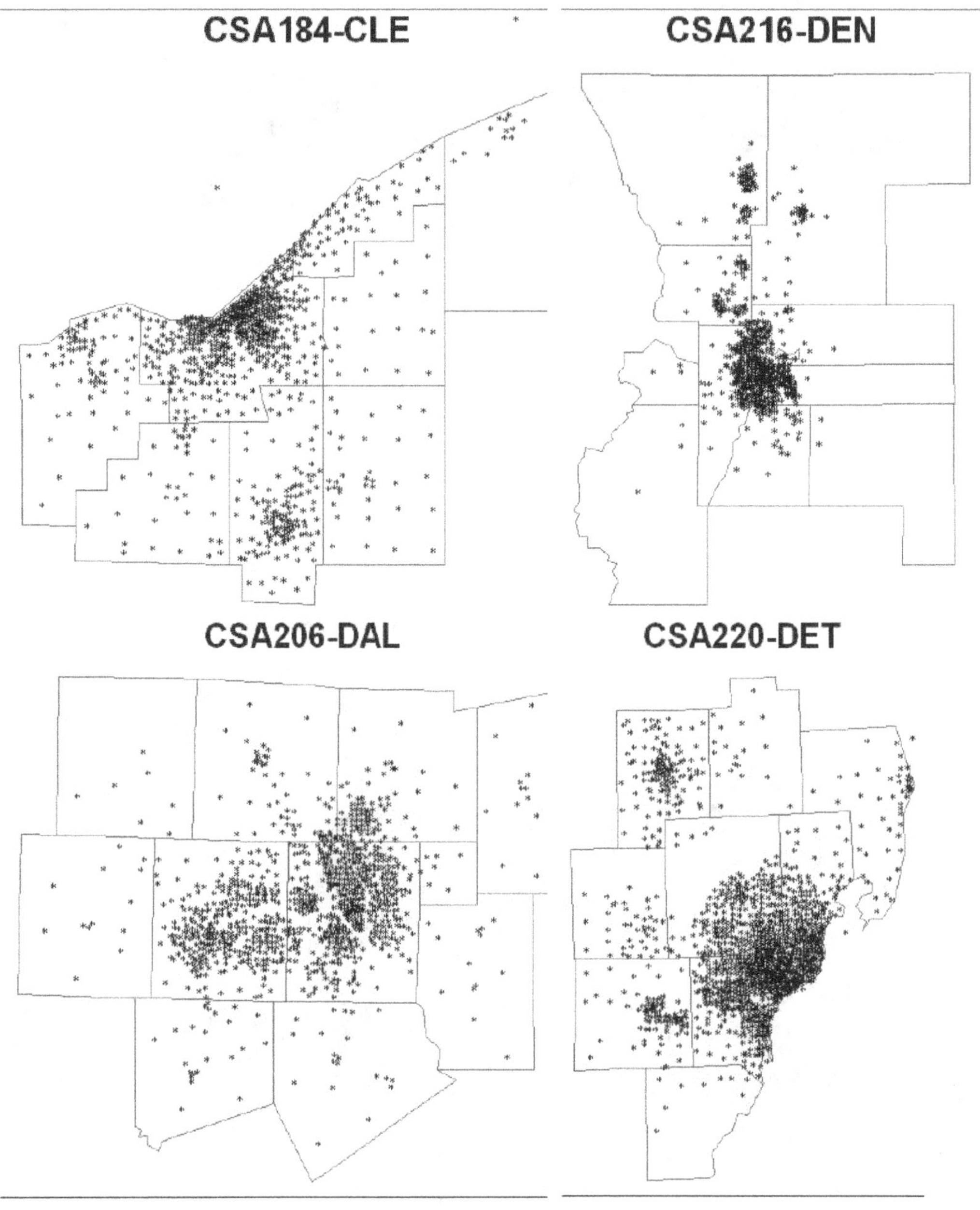

Figure 5B-2. Illustration of APEX exposure modeling domains (2000 US Census tract centroids) for Cleveland, Dallas, Denver and Detroit study areas.

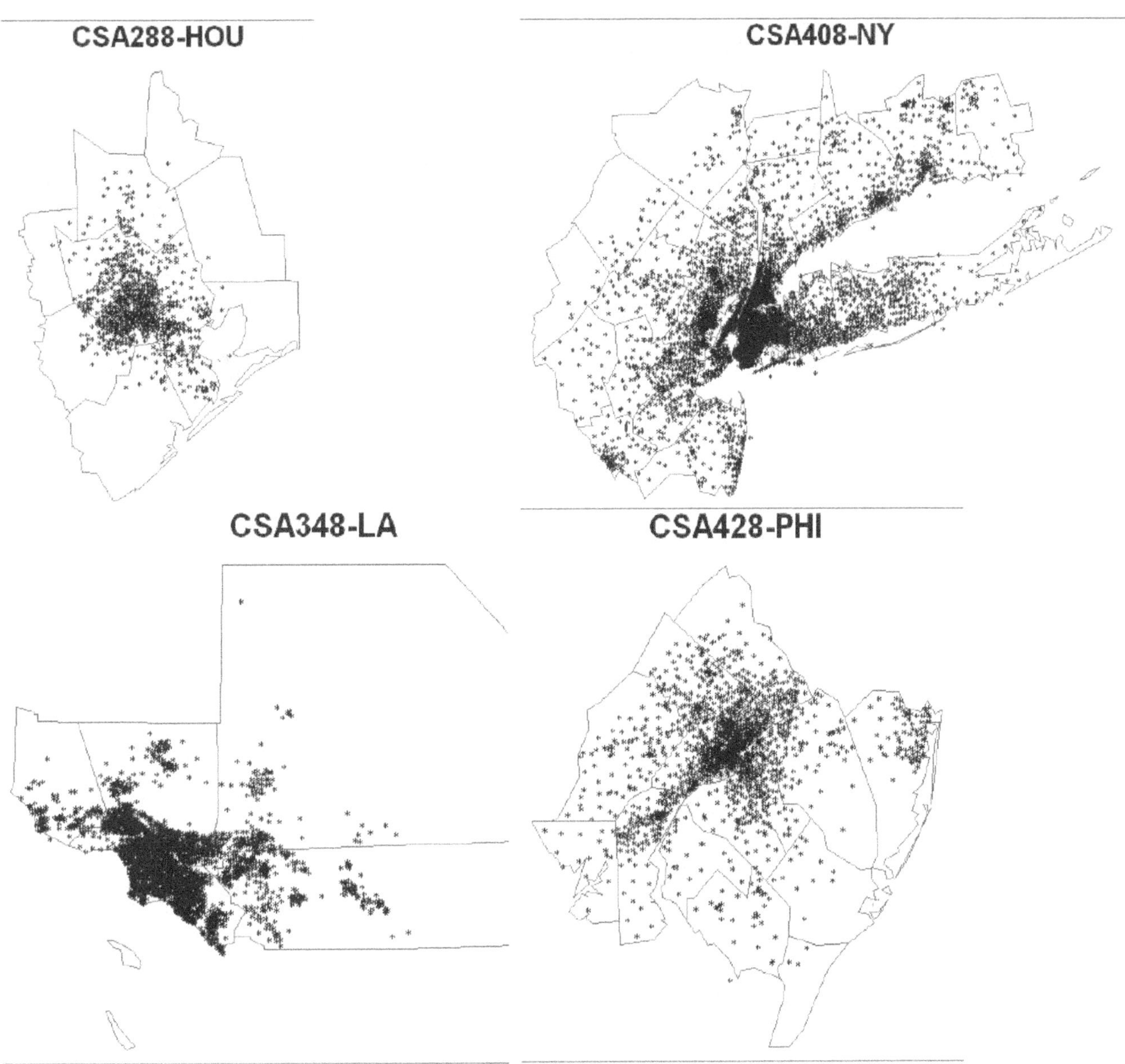

Figure 5B-3. Illustration of APEX exposure modeling domains (2000 US Census tract centroids) for Houston, Los Angeles, New York and Philadelphia study areas.

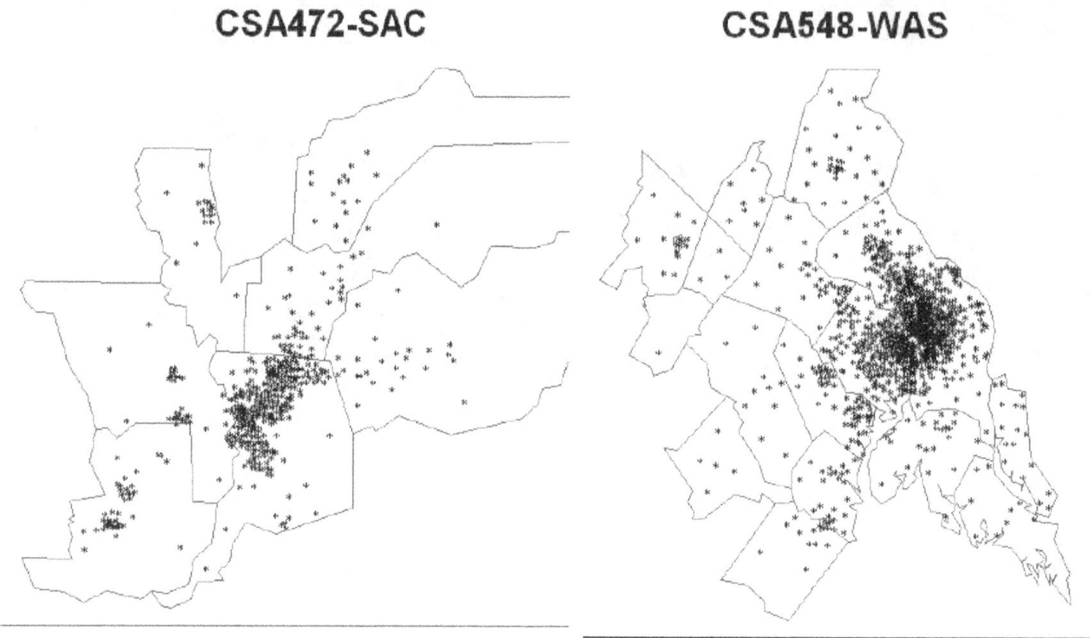

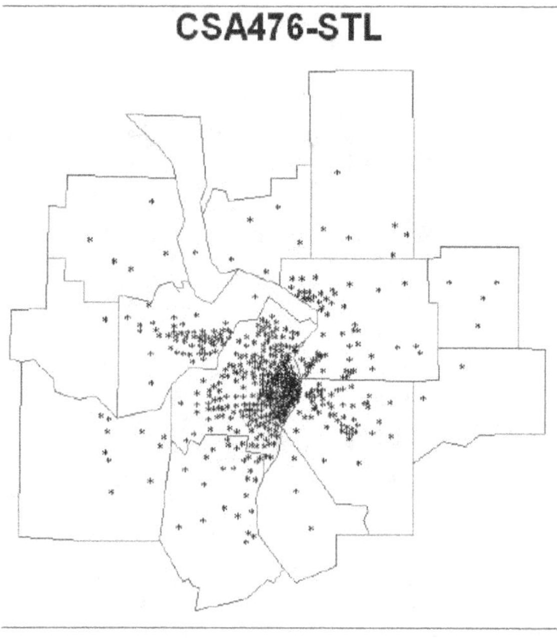

Figure 5B-4. Illustration of APEX exposure modeling domains (2000 US Census tract centroids) for Sacramento, St. Louis and Washington DC study areas.

5B-8 METEOROLOGICAL DATA

Temperature data are used by APEX in selecting human activity data and in estimating AERs for indoor residential microenvironments. Hourly surface temperature measurements were obtained from the National Weather Service ISH data files.[11] The weather stations used for each city are given in Table 5B-10. When developing profiles and selecting for am AER, APEX uses temperature data from the closest weather station to each Census tract.

Missing temperature data were estimated by the following procedure. Where there were consecutive strings of missing values (data gaps) of 4 or fewer hours, missing values were estimated by linear interpolation between the valid values at the ends of the gap. Remaining missing values at a station were estimated by fitting linear regression models for each hour of the day, with each of the other monitors, and choosing the model which maximizes R^2, for each hour of the day, subject to the constraints that R^2 be greater than 0.50 and the number of regression data values (days) is at least 60. If there were any remaining missing values at this point, for gaps of 6 or fewer hours, missing values were estimated by linear interpolation between the valid values at the ends of the gap. Any remaining missing values were replaced with the value at the closest station for that hour.

There were negligible differences between the statistically filled and the original temperature data with missing values. On average, daily mean temperatures were approximately 0.02 °C greater in the final data set used by APEX, compared with the data set having missing temperatures. The greatest positive difference occurred at station '2227013864', where the filled data had a daily average mean of about 0.72 °C greater than that of the data set with missing values. The greatest negative difference was associated with station '2403603710', where the filled data had a daily average mean of about -0.27 °C less than that of the data set with missing values. Given these small differences, the number of stations used to represent meteorological conditions in each study area and the range of values used by APEX in creating diary pools (e.g., 50 – 68 °F) or AER distributions (e.g., 55 – 84 °F) , the impact of the filled values to estimated exposures is assumed negligible.

Multiple unique APEX input files are used for the current O_3 assessment, varying by the year and study area, though generally in two forms:

- *METdataCSA[studyarea]Y[year].txt*: hourly temperature for each MET station, by study area and year
- *METlocsCSA[studyarea]Y[year].txt:* MET station ID's, latitudes and longitudes, start and stop dates of temperature data

[11] http://www.ncdc.noaa.gov/oa/climate/surfaceinventories.html

5B-9 CONDITIONAL VARIABLES

APEX has added flexiblity in using conditional variables in association with selection of the distributions used to represent input variables, across several modules (i.e., CHAD diary selection, microenvironmental concentration calculations). In this O_3 assessment, a number of temperature ranges are used in selecting the particular AER distribution (section 5B-6.1), maximum daily temperature is also used in diary selection to best match the study area MET data for the simulated individual (<55, 55-83, and ≥84; based on Graham and McCurdy, 2004), air conditionining prevalence data (section 5B-6.2), and designation of roadway type travelled based on VMT miles (section 5B-6.4).

A single unique APEX input files is used for the current O_3 assessment, varying by the year and study area:

- *Functions_O3_CSA[studyarea]Y[year]_[date].txt:* conditional variables and values used

Table 5B-10. Study area meteorological stations, locations, and hours of missing data.

Study Area	ISH ID[1]	Latitude	Longitude	2006	2007	2008	2009	2010
Atlanta	2217003813	32.683	-83.65	15	2	140	113	128
	2219013874	33.633	-84.433	0	0	101	41	18
	2219503888	33.767	-84.517	14	15	113	103	29
	2225593842	32.517	-84.95	16	4	170	52	48
	2227013864	33.917	-84.517	2502	1611	266	93	73
	2228403892	32.616	-85.433	469	114	209	77	540
	2228713871	33.583	-85.85	24	4	168	210	55
	2311013873	33.95	-83.333	4	4	271	39	45
	2320093801	34.35	-85.167	14	30	187	59	68
Baltimore	2406093721	39.167	-76.683	0	0	101	54	27
	4594013705	38.817	-76.867	34	86	173	184	42
Boston	2505464710	41.917	-71.5	275	71	362	741	315
	2506014704	41.65	-70.517	46	84	126	54	95
	2506454769	41.917	-70.733	5	10	259	143	285
	2507014765	41.717	-71.433	0	1	94	51	45
	2509014739	42.367	-71.017	0	2	97	41	34
	2509594746	42.267	-71.883	34	15	128	61	53
	4394514710	42.933	-71.433	3	10	103	49	55
Chicago	2530094846	41.983	-87.917	2	1	126	44	21
	2530594892	41.917	-88.25	71	12	170	97	64
	2534014819	41.783	-87.75	0	1	127	44	23
	2535014848	41.7	-86.333	3	0	91	46	18
	2543094822	42.2	-89.1	1	0	96	43	18
Cleveland	2521014895	40.917	-81.433	7	1	128	56	44
	2524014820	41.4	-81.85	0	1	82	38	19
	2524504853	41.517	-81.683	12	116	144	79	68
	2525014852	41.25	-80.667	0	0	119	45	28
Dallas	2258313960	32.85	-96.85	1	2	93	51	40
	2259003927	32.9	-97.017	0	0	88	46	30
	2259613961	32.817	-97.367	6	22	157	84	138
Denver	2466093037	38.817	-104.717	2	2	108	110	71
	2466693067	39.567	-104.85	2	1	104	53	45
	2469523036	39.717	-104.75	32	41	103	52	32
	2476824051	40.433	-104.633	72	393	876	134	140
	2476994062	40.45	-105.017	19	532	314	381	164
	2564024018	41.15	-104.8	1	7	129	46	66
	2565003017	39.833	-104.65	0	2	91	44	40
Detroit	2537094847	42.217	-83.35	1	1	106	40	27
	2537514822	42.4	-83	17	110	226	104	122
	2537614853	42.233	-83.533	1	7	148	107	78
	2537714804	42.617	-82.833	11	27	236	59	94
	2538404888	42.917	-82.533	40	140	94	32	55
	2539014836	42.783	-84.583	1	3	148	70	37
	2539514833	42.267	-84.467	4	35	329	49	144
	2637014826	42.967	-83.75	0	0	116	52	74
Houston	2241012917	29.95	-94.017	5	24	119	67	27

Study Area	ISH ID[1]	Latitude	Longitude	2006	2007	2008	2009	2010
	2242012923	29.3	-94.8	126	19	2468	444	186
	2243012960	30	-95.367	1	0	173	57	30
	2243512918	29.65	-95.283	0	0	160	99	33
	2244503904	30.583	-96.367	5	3	108	107	40
Los Angeles	2286023119	33.9	-117.25	13	25	103	48	29
	2288023152	34.2	-118.35	2	12	152	86	37
	2295023174	33.933	-118.4	0	0	113	44	19
	2297023129	33.833	-118.167	2	4	99	173	269
	2381523161	34.85	-116.8	69	4	196	741	241
	2381603159	34.733	-118.217	126	11	438	176	411
	2383023187	34.75	-118.717	31	3	375	276	554
	2391093111	34.117	-119.117	2046	16	265	77	792
	2392623136	34.217	-119.083	21	47	311	218	139
	4718703104	33.633	-116.167	99	18	134	161	60
	4718823158	33.617	-114.717	2	15	400	54	159
New York	2408454760	40.183	-74.067	251	475	566	183	794
	2409614706	40.017	-74.6	66	62	131	83	121
	2502014734	40.717	-74.183	1	0	105	47	22
	2502594741	40.85	-74.067	0	2	129	70	31
	2502964707	41.483	-73.133	554	292	569	842	703
	2503014732	40.783	-73.883	0	0	73	38	17
	2503504781	40.783	-73.1	2	0	174	138	18
	2503614757	41.633	-73.883	14	5	406	388	139
	2503794745	41.067	-73.717	8	0	119	93	34
	2503814714	41.5	-74.1	565	851	1144	1159	1358
	2504094702	41.183	-73.15	61	68	146	114	173
	2505894793	41.167	-71.583	687	349	586	2970	127
	2514554746	41.7	-74.8	1387	356	159	375	582
	2517014737	40.65	-75.45	5	4	148	74	51
	4486094789	40.65	-73.8	4	0	101	56	21
Philadelphia	2407093730	39.45	-74.567	4	3	142	112	161
	2407513735	39.367	-75.083	20	84	268	73	74
	2408013739	39.867	-75.233	1	0	122	57	21
	2408454760	40.183	-74.067	251	475	566	183	794
	2408594732	40.083	-75.017	0	10	143	60	38
	2408913781	39.667	-75.6	22	1	156	244	89
	2409614706	40.017	-74.6	66	62	131	83	121
	2517014737	40.65	-75.45	5	4	148	74	51
	4596603726	39	-74.917	945	500	1299	1562	1855
Sacramento	2483023232	38.5	-121.5	3	5	115	51	77
	2483993225	38.7	-121.583	0	0	116	52	42
	2584523225	39.3	-120.717	1	1	217	124	350
	4516023202	38.267	-121.933	34	110	152	78	46
St. Louis	2433813802	38.55	-89.85	40	55	129	68	41
	2434013994	38.75	-90.367	1	0	98	46	27
	2434503966	38.65	-90.65	5	33	189	62	41
	2445493996	37.767	-90.4	186	24	74	134	402

Study Area	ISH ID[1]	Latitude	Longitude	2006	2007	2008	2009	2010
Washington DC	2403093738	38.933	-77.45	3	2	102	52	28
	2403303706	38.267	-77.45	41	48	4587	145	36
	2403513773	38.5	-77.3	177	113	1937	788	610
	2403603710	38.717	-77.517	725	698	943	1631	97
	2404013721	38.3	-76.417	1414	123	322	83	86
	2405013743	38.867	-77.033	0	1	96	46	21
	2405303717	39.15	-78.15	38	7	78	32	35
	2405503714	39.083	-77.567	28	28	64	38	181
	2406093721	39.167	-76.683	0	0	101	54	27
	2417713734	39.4	-77.983	36	42	217	137	118
	4594013705	38.817	-76.867	34	86	173	184	42

[1] From the Federal Climate Complex Integrated Surface Hourly (ISH) global database.

5B-10 REFERENCES

AHS. (2003). U.S. Bureau of the Census and U.S. Department of Housing and Urban Development. 2003 American Housing Survey (AHS): National Survey Data. Available at: http://www.census.gov/hhes/www/housing/ahs/ahs.html, and http://www.huduser.org/datasets/ahs.html

Akland, G. G., Hartwell, T. D., Johnson, T. R., Whitmore, R. W. (1985). Measuring human exposure to carbon monoxide in Washington, D. C. and Denver, Colorado during the winter of 1982-83. *Environ Sci Technol.* 19: 911-918.

American Petroleum Institute. (1997). Sensitivity Testing of pNEM/O_3 Exposure to Changes in the Model Algorithms. Health and Environmental Sciences Department.

Bennett, D. H., Wu, X., Teague, C.H., Lee, K., Cassady, D. L., Ritz, B., Hertz-Picciotto, I. (2012a). Passive sampling methods to determine household and personal care product use. *J Expos Sci Environ Epidem.* 22: 148-160.

Bennett, D. H, Fisk, W., Apte, M. G., Wu, X., Trout, A., Faulkner, D., Sulivan D. (2012b). Ventilation, Temperature, and HVAC Characteristics in Small and Medium Commercial Buildings (SMCBs) in California. *Indoor Air.* 22(4): 309-320.

Geyh, A. S., Xue, J., Ozkaynak, H., Spengler, J. D. (2000). The Harvard Southern California chronic ozone exposure study: assessing ozone exposure of grade-school-age children in two southern California communities. *Environ Health Perspect.* 108: 265-270.

Hartwell, T. D., Clayton, C. A., Ritchie, R. M., Whitmore, R. W., Zelon, H. S., Jones, S. M., Whitehurst, D. A. (1984). Study of Carbon Monoxide Exposure of Residents of Washington, DC and Denver, Colorado. Research Triangle Park, NC: U.S. Environmental Protection Agency, Office of Research and Development, Environmental Monitoring Systems Laboratory. EPA-600/4-84-031.

Hertz-Picciotto, I., Cassady, D., Lee, K., Bennett, D. H., Ritz, B., Vogt, R. (2010). Study of Use of Products and Exposure-Related Behaviors (SUPERB): study design, methods, and demographic characteristics of cohorts. *Environ Health.* 9:54.

Isaacs, K. K., McCurdy, T., Glen, G., Nysewander, M., Errickson, A., Forbes, S., Graham, S., McCurdy, L., Smith, L., Tulve, N., and Vallero, D. (2012). Statistical properties of longitudinal time-activity data for use in human exposure modeling. *J Expos Sci Environ Epidemiol.* 23(3): 328-336.

Johnson, T. (1984). A Study of Personal Exposure to Carbon Monoxide in Denver, Colorado. Research Triangle Park, NC: U.S. Environmental Protection Agency, Environmental Monitoring Systems Laboratory. EPA-600/4-84-014.

Johnson, T. (1989). Human Activity Patterns in Cincinnati, Ohio. Palo Alto, CA: Electric Power Research Institute. EPRI EN-6204.

Johnson, T., Pakrasi, A., Wisbeth, A., Meiners, G., Ollison, W. (1995). Ozone exposures Within Motor Vehicles – Results of a Field Study in Cincinnati, Ohio. Proceedings 88th annual meeting and exposition of the Air & Waste Management Association, San Antonio, TX. June 18-23, 1995. Preprint paper 95-WA84A.02.

Klepeis, N. E., Tsang, A. M., Behar, J. V. (1996). Analysis of the National Human Activity Pattern Survey (NHAPS) Respondents from a Standpoint of Exposure Assessment. Washington, DC: U.S. Environmental Protection Agency, Office of Research and Development. EPA/600/R-96/074.

Knowledge Networks. (2009). Field Report: National-Scale Activity Survey (NSAS). Conducted for Research Triangle Institute. Submitted to Carol Mansfield November 13, 2009.

Lee, K., Vallarino, J., Dumyahn, T., Ozkaynak, H., and Spengler, J. D. (1999). Ozone decay rates in residences. *J Air Waste Manag Assoc*. 49: 1238-1244.

Liu, L.-J. S., Box, M., Kalman, D., Kaufman, J., Koenig J., Larson, T., Lumley, T., Sheppard, L., Wallace, L. (2003). Exposure assessment of particulate matter for susceptible populations in Seattle. *Environ Health Persp*. 111: 909-918.

Mansfield, C., Houtven, G. V., Johnson, F. R., Yang, J.-C. (2009). Environmental Risks and Behavior: Do children spend less time outdoors when ozone pollution is high? ASSA annual meeting, January 5, 2009. Update of Houtven et al. (2003), using the OAB CHAD data set, and related to Mansfield et al. (2006).

Mansfield, C., Johnson, F. R., Van Houtven, G. (2006). The missing piece: averting behavior for children's ozone exposures. *Resource Energy Econ*. 28:215-228.

McCurdy, T., Glen, G., Smith, L., Lakkadi, Y. (2000). The National Exposure Research Laboratory's Consolidated Human Activity Database. *J Expo Anal Environ Epidemiol*. 10: 566-578.

McCurdy, T., Graham, S. E. (2003). Using human activity data in exposure models: analysis of discriminating factors. *J Expos Anal Environ Epidemiol*. 13(4): 294-317.

Persily, A., Gorfain, J. (2004). Analysis of Ventilation Data from the U.S. Environmental Protection Agency Building Assessment Survey and Evaluation (BASE) Study. National Institute of Standards and Technology, NISTIR 7145, December 2004.

Persily, A., Gorfain, J., Brunner, G. (2005). Ventilation Design and Performance in U.S. Office Buildings. *ASHRAE Journal*. April 2005, 30-35.

Robinson, J. P., Wiley, J. A., Piazza, T., Garrett, K., and Cirksena, K. (1989). Activity Patterns of California Residents and their Implications for Potential Exposure to Pollution. California Air Resources Board, Sacramento, CA. CARB-A6-177-33.

Cohen, J., Mallya, H., Rosenbaum, A. (2012). Updated Analysis of Air Exchange Rate, ICF International Memo to John Langstaff. Available in Appendix 5E of O_3 REA.

Shendell, D. G., Winer, A. M., Weker, R., Colome, S. D. (2004). Evidence of inadequate ventilation in portable classrooms: results of a pilot study in Los Angeles County. *Indoor Air*. 14:154–158.

Spier, C. E., Little, D. E., Trim, S. C., Johnson, T. R., Linn, W. S., Hackney, J. D. (1992). Activity patterns in elementary and high school students exposed to oxidant pollution. *J Expo Anal Environ Epidemiol.* 2: 277-293.

Tsang, A. M., Klepeis, N. E. (1996). Descriptive Statistics Tables from a Detailed Analysis of the National Human Activity Pattern Survey (NHAPS) Data. U.S. Environmental Protection Agency. EPA/600/R-96/148.

Turk, B. H., Grimsrud, D. T., Brown, J. T., Geisling-Sobotka, K. L., Harrison, J., Prill, R. J. (1989). Commercial Building Ventilation Rates and Particle Concentrations. ASHRAE No. 3248.

US EPA. (2002). Consolidated Human Activities Database (CHAD) Users Guide. Database and documentation available at: http://www.epa.gov/chadnet1/

US EPA. (2007). Ozone Population Exposure Analysis for Selected Urban Areas. Office of Air Quality Planning and Standards, U.S. Environmental Protection Agency, Research Triangle Park, NC. Available at: http://www.epa.gov/ttn/naaqs/standards/ozone/s_o3_cr_td.html

US. EPA. (2012a). Total Risk Integrated Methodology (TRIM) Air Pollutants Exposure Model Documentation (TRIM.Expo / APEX, Version 4). Volume I: User's Guide. EPA-452/B-12-001a. Available at: http://www.epa.gov/ttn/fera/data/apex/APEX45_UsersGuide_Vol1_Aug2012.pdf.

US EPA. (2012b). Total Risk Integrated Methodology (TRIM) Air Pollutants Exposure Model Documentation (TRIM.Expo / APEX, Version 4) Volume II: Technical Support Document. EPA-452/B-12-001b. Available at: http://www.epa.gov/ttn/fera/data/apex/APEX45_UsersGuide_Vol2_Aug2012.pdf.

University of Michigan. (2012). The Panel Study of Income Dynamics (PSID). Data and documentation available at http://psidonline.isr.umich.edu/

Wiley, J. A., Robinson, J. P., Piazza, T., Garrett, K., Cirksena, K., Cheng, Y.-T., Martin, G. (1991a). Activity Patterns of California Residents: Final Report. California Air Resources Board, Sacramento, CA. ARB/R93/487. Available from: NTIS, Springfield, VA., PB94-108719.

Wiley, J. A., Robinson, J. P., Cheng, Y.-T., Piazza, T., Stork, L., Pladsen, K. (1991b). Study of Children's Activity Patterns: Final Report. California Air Resources Board, Sacramento, CA. ARB-R-93/489.

Williams, R., Suggs, J., Creason, J., Rodes, C., Lawless, P., Kwok, R., Zweidinger, R., Sheldon, L. (2000). The 1998 Baltimore particulate matter epidemiology-exposure study: Part 2. Personal exposure associated with an elderly population. *J Expo Anal Environ Epidemiol.* 10(6): 533-543.

Williams, R., Suggs J., Rea, A., Leovic, K., Vette, A., Croghan, C., Sheldon, L., Rodes, C., Thornburg, J., Ejire, A., Herbst, M., Sanders Jr., W. (2003a). The Research Triangle particulate panel study: PM mass concentrations relationships. *Atmos Environ.* 37:5349-5363.

Williams, R., Suggs, J., Rea, A., Sheldon, L., Rodes, C., Thornburg, J. (2003b). The Research Triangle particulate panel study: modeling ambient source contributions to personal and residential PM mass concentrations. *Atmos Environ.* 37:5365-5378.

Williams, R., Rea, A., Vette, A., Croghan, C., Whitaker, D., Stevens, C., McDow, S., Fortmann, R., Sheldon, L., Wilson, H., Thornburg, J., Phillips, M., Lawless, P., Rodes, C., Daughtrey, H. (2008). The design and field implementation of the Detroit Exposure and Aerosol Research Study. *J Expos Sci Environ Epidem.* 19: 643-659.

Xue, J., McCurdy, T., Spengler, J., Özkaynak, H. (2004). Understanding variability in time spent in selected locations for 7-12-year old children. *J Expo Anal Environ Epidemiol.* 14(3): 222-33.

Appendix 5-C

Generation of Adult and Child Census-tract Level Asthma Prevalence using NHIS (2006-2010) and US Census (2000) Data

Table of Contents

5C-1. OVERVIEW 2
5C-2. RAW ASTHMA PREVALENCE DATA SET DESCRIPTION 3
5C-3. LOGISTIC MODELING APPROACH USED TO ESTIMATE ASTHMA PREVALENCE 5
5C-4. APPLICATION OF LOESS SMOOTHER TO ASTHMA PREVALENCE ESTIMATION 9
5C-5. CENSUS TRACT LEVEL POVERTY RATIO DATA SET DESCRIPTION AND PROCESSING 13
5C-6. COMBINED CENSUS TRACT LEVEL POVERTY RATIO AND ASTHMA PREVALENCE DATA 14
5C-7. REFERENCES 15

List of Tables

Table 5C-1. Number of total surveyed persons from NHIS (2006-2010) sample adult and child files and the number of those responding to asthma survey questions. 5
Table 5C-2. Example of alternative logistic models evaluated to estimate child asthma prevalence using the "EVER" asthma response variable and goodness of fit test results. 8
Table 5C-3. Top 20 model smoothing fits where residual standard error at or a value of 1.0. .. 11

List of Figures

Figure 5C-1. Normal probability plot of studentized residuals generated using logistic model, smoothing set to 0.7, and the children 'EVER' asthmatic data set. 12
Figure 5C-2. Studentized residuals versus model predicted betas generated using a logistic model and using the children 'EVER' asthmatic data set, with smoothing set to 0.6. 13

5C-1 OVERVIEW

This appendix describes the generation of our census tract level children and adult asthma prevalence data developed from the 2006-2010 National Health Interview Survey (NHIS) and census tract level poverty information from the 2000 US Census. The approach is, for the most part, a reapplication of work performed by Cohen and Rosenbaum (2005), though here we incorporated a few modifications as described below. Details regarding the earlier asthma prevalence work are documented in Appendix G of US EPA (2007).

Briefly in the earlier asthma prevalence development work, Cohen and Rosenbaum (2005) calculated asthma prevalence for children aged 0 to 17 years for each age, gender, and four US regions using 2003 NHIS survey data. The four regions defined by NHIS were 'Midwest', 'Northeast', 'South', and 'West'. The asthma prevalence was defined as the probability of a 'Yes' response to the question "EVER been told that [the child] had asthma?"[1] among those persons that responded either 'Yes' or 'No' to this question.[2] The responses were weighted to take into account the complex survey design of the NHIS.[3] Standard errors and confidence intervals for the prevalence were calculated using a logistic model (PROC SURVEY LOGISTIC; SAS, 2012). A scatter-plot technique (LOESS SMOOTHER; SAS, 2012) was applied to smooth the prevalence curves and compute the standard errors and confidence intervals for the smoothed prevalence estimates. Logistic analysis of the raw and smoothed prevalence curves showed statistically significant differences in prevalence by gender and region, supporting their use as stratification variables in the final data set. These smoothed prevalence estimates were used as an input to EPA's Air Pollution Exposure Model (APEX) to estimate air pollutant exposure in asthmatic children (US EPA, 2007; 2008; 2009).

For the current asthma prevalence data set development, several years of recent NHIS survey data (2006-2010) were combined and used to calculate asthma prevalence. The current approach estimates asthma prevalence for children (by age in years) as was done previously by Cohen and Rosenbaum (2005) but now includes an estimate of adult asthma prevalence (by age groups). In addition, two sets of asthma prevalence for each adults and children were estimated here. The first data set, as was done previously, was based on responses to the question "EVER been told that [the child] had asthma". The second data set was developed using the probability of a 'Yes' response to a question that followed those that answered 'Yes' to the first question

---

[1] The response was recorded as variable "CASHMEV" in the downloaded dataset. Data and documentation are available at http://www.cdc.gov/nchs/nhis/quest_data_related_1997_forward.htm.

[2] If there were another response to this variable other than "yes" or "no" (i.e., refused, not ascertained, don't know, and missing), the surveyed individual was excluded from the analysis data set.

[3] In the SURVEY LOGISTIC procedure, the variable "WTF_SC" was used for weighting, "PSU" was used for clustering, and "STRATUM" was used to define the stratum.

regarding ever having asthma, specifically, do those persons "STILL have asthma?"[4] And finally, in addition to the nominal variables region and gender (and age and age groups), the asthma prevalence in this new analysis were further stratified by a family income/poverty ratio (i.e., whether the family income was considered below or at/above the US Census estimate of poverty level for the given year).

These new asthma prevalence data sets were linked to the US census tract level poverty ratios probabilities (US Census, 2007), also stratified by age and age groups. Given 1) the significant differences in asthma prevalence by age, gender, region, and poverty status, 2) the variability in the spatial distribution of poverty status across census tracts, stratified by age, and 3) the spatial variability in local scale ambient concentrations of many air pollutants, it is hoped that the variability in population exposures is now better represented when accounting for and modeling these newly refined attributes of this susceptible population.

5C-2 RAW ASTHMA PREVALENCE DATA SET DESCRIPTION

In this section we describe the asthma prevalence data sets used and identify the variables retained for our final data set. First, raw data and associated documentation were downloaded from the Center for Disease Control (CDC) and Prevention's National Health Interview Survey (NHIS) website.[5] The 'Sample Child' and 'Sample Adult' files were selected because of the availability of person-level attributes of interest within these files, i.e., age in years ('age_p'), gender ('sex'), US geographic region ('region'), coupled with the response to questions of whether or not the surveyed individual ever had and still has asthma. In total, five years of recent survey data were obtained, comprising over 50,000 children and 120,000 children for years 2006-2010 (**Table 5C-1**).

Information regarding personal and family income and poverty ranking are also provided by the NHIS in separate files. Five files ('INCIMPx.dat') are available for each survey year, each containing either the actual responses (where recorded or provided by survey participant) or imputed values for the desired financial variable.[6] For this current analysis, the ratio of income to poverty was used to develop a nominal variable: either the survey participant was below or at/above a selected poverty threshold. This was done in this manner to be consistent with data generated as part of a companion data set, i.e., census tract level poverty ratio probabilities stratified by age (see section 5C-5 below).

---

[4] While we estimated two separate sets of prevalence using the "STILL" and "EVER" variables, only the "STILL" data were used as input to our exposure model.

[5] See http://www.cdc.gov/nchs/nhis.htm (accessed October 4, 2011).

[6] Financial information was not collected from all persons; therefore the NHIS provides imputed data. Details into the available variables and imputation method are provided with each year's data set. For example see "Multiple Imputation of Family Income and Personal Earnings in the National Health Interview Survey: Methods and Examples" at http://www.cdc.gov/nchs/data/nhis/tecdoc_2010.pdf.

Given the changes in how income data were collected over the five year period of interest and the presence of imputed data, a data processing methodology was needed to conform each of the year's data sets to a compatible nominal variable. Briefly, for survey years 2006-2008, poverty ratios ('RAT_CATI') are provided for each person as a categorical variable, ranging from <0.5 to 5.0 by increments of either 0.25 (for poverty ratios categories between <0.5 – 2.0) and 0.50 (for poverty ratios >5.0). For 2009 and 2010 data, the poverty ratio was provided as a continuous variable ('POVRATI3') rather than a categorical variable.[7]

When considering the number of stratification variables, the level of asthma prevalence, and poverty distribution among the survey population, sample size was an important issue. For the adult data, there were insufficient numbers of persons available to stratify the data by single ages (for some years of age there were no survey persons). Therefore, the adult survey data were grouped as follows: ages 18-24, 25-34, 35-44, 45-54, 55-64, 65-74, and, ≥75.[8] To increase the number of persons within the age, gender, and four region groupings of our characterization of 'below poverty' asthmatics persons, the poverty ratio threshold was selected as <1.5, therefore including persons that were within 50% above the poverty threshold. As there were five data sets containing variable imputed poverty ratios (as well as a non varying values for where income information was reported) for each year, the method for determining whether a person was below or above the poverty threshold was as follows. If three or more of the five imputed/recorded values were <1.5, the person's family income was categorized 'below' the poverty threshold, if three or more of the 5 values were ≥1.5, the person's family income was categorized 'above' the poverty threshold. The person-level income files were then merged with the sample adult and child files using the 'HHX' (a household identifier), 'FMX' (a family identifier), and 'FPX' (an individual identifier) variables. Note, all persons within the sample adult and child files had corresponding financial survey data.

Two asthma survey response variables were of interest in this analysis and were used to develop the two separate prevalence data sets for each children and adults. The response to the first question "Have you EVER been told by a doctor or other health professional that you [or your child] had asthma?" was recorded as variable name 'CASHMEV' for children and 'AASMEV' for adults. Only persons having responses of either 'Yes' or 'No' to this question

---

[7] Actually, the 2009 data had continuous values for the poverty ratios ('POVRATI2') but the quality was determined by us to be questionable: the value varied among family members by orders of magnitude – however, it should be a constant. The income data ('FAMINCI2') provided were constant among family members, therefore we combined these data with poverty thresholds obtained from the US Census (available at: http://www.census.gov/hhes/www/poverty/data/threshld/thresh08 html) for year 2008 by family size (note, income is the annual salary from the prior year) and calculated an appropriate poverty ratio for each family member.

[8] These same age groupings were used to create the companion file containing the census tract level poverty ratio probabilities (section 5C-5).

were retained to estimate the asthma prevalence. This assumes that the exclusion of those responding otherwise, i.e., those that 'refused' to answer, instances where it was "not ascertained', or the person 'does not know', does not affect the estimated prevalence rate if either 'Yes' or 'No' answers could actually be given by these persons. There were very few persons (<0.3%) that did provide an unusable response (Table 5C-1), thus the above assumption is reasonable. A second question was asked as a follow to persons responding "Yes" to the first question, specifically, "Do you STILL have asthma?" and noted as variables 'CASSTILL' and 'AASSTILL' for children and adults, respectively. Again, while only persons responding 'Yes' and 'No' were retained for further analysis, the representativeness of the screened data set is assumed unchanged from the raw survey data given the few persons having unusable data (<0.5%).

Table 5C-1. Number of total surveyed persons from NHIS (2006-2010) sample adult and child files and the number of those responding to asthma survey questions.

Study Group/Respondents	Number of Surveyed Persons					
Children	**2010**	**2009**	**2008**	**2007**	**2006**	**TOTAL**
All Persons	11,277	11,156	8,815	9,417	9,837	50,502
Yes/No Asthma	11,256	11,142	8,800	9,404	9,815	50,417
Yes/No to Still Have + No Asthma	11,253	11,129	8,793	9,394	9,797	50,366
Adults	**2010**	**2009**	**2008**	**2007**	**2006**	**TOTAL**
All Persons	27,157	27,731	21,781	23,393	24,275	124,337
Yes/No Asthma	27,157	27,715	21,766	23,372	24,242	124,252
Yes/No to Still Have + No Asthma	27,113	27,686	21,726	23,349	24,208	124,082

5C-3 LOGISTIC MODELING APPROACH USED TO ESTIMATE ASTHMA PREVALENCE

As described in the previous section, four person-level analytical data sets were created from the raw NHIS data files, generally containing similar variables: a 'Yes' or 'No' asthma response variable (either 'EVER' or 'STILL'), an age (or age group for adults), their gender ('male' or 'female'), US geographic region ('Midwest', 'Northeast', 'South', and 'West'), and poverty status ('below' or above'). One approach to calculate prevalence rates and their uncertainties for a given gender, region, poverty status, and age is to calculate the proportion of 'Yes' responses among the 'Yes' and 'No' responses for that demographic group, appropriately weighting each response by the survey weight. This simplified approach was initially used to develop 'raw' asthma prevalence rates however this approach may not be completely appropriate. The two main issues with such a simplified approach are that the distributions of the estimated prevalence rates would not be well approximated by normal distributions and that the estimated confidence intervals based on a normal approximation would often extend outside the [0, 1] interval. A better approach for such survey data is to use a logistic transformation and fit the model:

Prob(asthma) = exp(beta) / (1 + exp(beta)),

where, *beta* may depend on the explanatory variables for age, gender, poverty status, or region. This is equivalent to the model:

Beta = logit {prob(asthma) } = log { prob(asthma) / [1 – prob(asthma)] }

The distribution of the estimated values of *beta* is more closely approximated by a normal distribution than the distribution of the corresponding estimates of prob(asthma). By applying a logit transformation to the confidence intervals for *beta*, the corresponding confidence intervals for prob(asthma) will always be inside [0, 1]. Another advantage of the logistic modeling is that it can be used to compare alternative statistical models, such as models where the prevalence probability depends upon age, region, poverty status, and gender, or on age, region, poverty status but not gender.

A variety of logistic models were fit and compared to use in estimating asthma prevalence, where the transformed probability variable beta is a given function of age, gender, poverty status, and region. I used the SAS procedure SURVEYLOGISTIC to fit the various logistic models, taking into account the NHIS survey weights and survey design (using both stratification and clustering options), as well as considering various combinations of the selected explanatory variables.

As an example, **Table 5C-2** lists the models fit and their log-likelihood goodness-of-fit measures using the sample child data and for the "EVER" asthma response variable. A total of 32 models were fit, depending on the inclusion of selected explanatory variables and how age was considered in the model. The 'Strata' column lists the eight possible stratifications: no stratification, stratified by gender, by region, by poverty status, by region and gender, by region and poverty status, by gender and poverty status, and by region, gender and poverty status. For example, "5. region, gender" indicates that separate prevalence estimates were made for each combination of region and gender. As another example, "2. gender" means that separate prevalence estimates were made for each gender, so that for each gender, the prevalence is assumed to be the same for each region. Note the prevalence estimates are independently calculated for each stratum. The 'Description' column of **Table 5C-2** indicates how beta depends upon the age:

Linear in age	Beta = $\alpha + \beta \times$ age, where α and β vary with strata.
Quadratic in age	Beta = $\alpha + \beta \times$ age + $\gamma \times$ age^2, where α β and γ vary with strata.
Cubic in age	Beta = $\alpha + \beta \times$ age + $\gamma \times$ age^2 + $\delta \times$ age^3, where α, β, γ, and δ vary with the strata.
f(age)	Beta = arbitrary function of age, with different functions for different strata

The category *f(age)* is equivalent to making age one of the stratification variables, and is also equivalent to making beta a polynomial of degree 16 in age (since the maximum age for children is 17), with coefficients that may vary with the strata.

The fitted models are listed in order of complexity, where the simplest model (i.e., model 1) is an unstratified linear model in age and the most complex model (model 32) has a prevalence that is an arbitrary function of age, gender, poverty status, and region. Model 32 is equivalent to calculating independent prevalence estimates for each of the 288 combinations of age, gender, poverty status, and region.

Table 5C-2 also includes the -2 Log Likelihood statistic, a goodness-of-fit measure, and the associated degrees of freedom (DF), which is the total number of estimated parameters. Any two models can be compared using their -2 Log Likelihood values: models having lower values are preferred. If the first model is a special case of the second model, then the approximate statistical significance of the first model is estimated by comparing the difference in the -2 Log Likelihood values with a chi-squared random variable having *r* degrees of freedom, where *r* is the difference in the DF (hence a likelihood ratio test). For all pairs of models from **Table 5C-2**, all the differences in the -2 Log Likelihood statistic are at least 600,000 and thus significant at p-values well below 1 percent. Based on its having the lowest -2 Log Likelihood value, the last model fit (model 32: retaining all explanatory variables and using *f(age)*) was preferred and used to estimate the asthma prevalence.[9]

[9] Similar results were obtained when estimating prevalence using the 'STILL' have asthma variable as well as when investigating model fit using the adult data sets. Note that because age was a categorical variable in the adult data sets it could only be evaluated using *f(age_group)*. See Attachment B, Tables 5CB-1 to 5CB-4 for all model fit results.

Table 5C-2. Example of alternative logistic models evaluated to estimate child asthma prevalence using the "EVER" asthma response variable and goodness of fit test results.

Model No.	Description	Strata	- 2 Log Likelihood	DF[1]
1	1. logit(prob) = linear in age	1. none	288740115.1	2
2	1. logit(prob) = linear in age	2. gender	287062346.4	4
3	1. logit(prob) = linear in age	3. region	288120804.1	8
4	1. logit(prob) = linear in age	4. poverty	287385013.1	4
5	1. logit(prob) = linear in age	5. region, gender	286367652.6	16
6	1. logit(prob) = linear in age	6. region, poverty	286283543.6	16
7	1. logit(prob) = linear in age	7. gender, poverty	285696164.7	8
8	1. logit(prob) = linear in age	8. region, gender, poverty	284477928.1	32
9	2. logit(prob) = quadratic in age	1. none	286862135.1	3
10	2. logit(prob) = quadratic in age	2. gender	285098650.6	6
11	2. logit(prob) = quadratic in age	3. region	286207721.5	12
12	2. logit(prob) = quadratic in age	4. poverty	285352164	6
13	2. logit(prob) = quadratic in age	5. region, gender	284330346.1	24
14	2. logit(prob) = quadratic in age	6. region, poverty	284182547.5	24
15	2. logit(prob) = quadratic in age	7. gender, poverty	283587631.7	12
16	2. logit(prob) = quadratic in age	8. region, gender, poverty	282241318.6	48
17	3. logit(prob) = cubic in age	1. none	286227019.6	4
18	3. logit(prob) = cubic in age	2. gender	284470413	8
19	3. logit(prob) = cubic in age	3. region	285546716.1	16
20	3. logit(prob) = cubic in age	4. poverty	284688169.9	8
21	3. logit(prob) = cubic in age	5. region, gender	283662673.5	32
22	3. logit(prob) = cubic in age	6. region, poverty	283404487.5	32
23	3. logit(prob) = cubic in age	7. gender, poverty	282890785.3	16
24	3. logit(prob) = cubic in age	8. region, gender, poverty	281407414.3	64
25	4. logit(prob) = f(age)	1. none	285821686.2	18
26	4. logit(prob) = f(age)	2. gender	283843266.2	36
27	4. logit(prob) = f(age)	3. region	284761522.8	72
28	4. logit(prob) = f(age)	4. poverty	284045849.2	36
29	4. logit(prob) = f(age)	5. region, gender	282099156.1	144
30	4. logit(prob) = f(age)	6. region, poverty	281929968.5	144
31	4. logit(prob) = f(age)	7. gender, poverty	281963915.7	72
32	4. logit(prob) = f(age)	8. region, gender, poverty	278655423.1	288

[1] model degrees of freedom.

The SURVEYLOGISTIC procedure produces estimates of the beta values and their 95% confidence intervals for each combination of age, region, poverty status, and gender. By applying the inverse logit transformation,

$$Prob(asthma) = exp(\, beta) \,/\, (1 + exp(beta)\,),$$

one can convert the beta values and associated 95% confidence intervals into predictions and 95% confidence intervals for the prevalence. The standard error for the prevalence was estimated as

$$Std\ Error\ \{Prob(asthma)\} = Std\ Error\ (beta) \times exp(-\, beta) \,/\, (1 + exp(beta)\,)^2$$

which follows from the delta method (i.e., a first order Taylor series approximation). Estimated asthma prevalence using this approach and termed here as 'unsmoothed' are provided in Attachment A. Asthma prevalence for children is provided in Attachment A, Tables 5CA-1 ('EVER' had Asthma) and 5CA-2 ('STILL' have asthma) while adult asthma prevalence is provided in Attachment A, Tables 5CA-3 ('EVER' had Asthma) and 5CA-4 ('STILL' have asthma). Graphical representation of each study group is also provided in a series of plots within Attachment A, Figures 5CA-1 to 5CA-4. The variables provided in the tabular presentation are:

- Region
- Gender
- Age (in years) or Age_group (age categories)
- Poverty Status
- Prevalence = predicted prevalence
- SE = standard error of predicted prevalence
- LowerCI = lower bound of 95 % confidence interval for predicted prevalence
- UpperCI = upper bound of 95 % confidence interval for predicted prevalence

5C-4 APPLICATION OF LOESS SMOOTHER TO ASTHMA PREVALENCE ESTIMATION

The estimated prevalence curves shows that the prevalence is not necessarily a smooth function of age. The linear, quadratic, and cubic functions of age modeled by SURVEYLOGISTIC were identified as a potential method for smoothing the curves, but they did not provide the best fit to the data. One reason for this might be due to the attempt to fit a global regression curve to all the age groups, which means that the predictions for age *A* are affected by data for very different ages. A local regression approach that separately fits a regression curve to each age A and its neighboring ages was used, giving a regression weight of 1 to the age *A*, and lower weights to the neighboring ages using a tri-weight function:

$$Weight = \{1 - [\ |age - A| / q\ ]^3\},\ \ where\ |\ age - A| <= q$$

The parameter *q* defines the number of points in the neighborhood of the age *A*. Instead of calling *q* the smoothing parameter, SAS defines the smoothing parameter as the proportion of points in each neighborhood. A quadratic function of age to each age neighborhood was fit separately for each gender and region combination. These local regression curves were fit to the beta values, the logits of the asthma prevalence estimates, and then converted them back to estimated prevalence rates by applying the inverse logit function exp(beta) / (1 + exp(beta)). In addition to the tri-weight variable, each beta value was assigned a weight of 1 / [std error (beta)]^2, to account for their uncertainties.

In this application of LOESS, weights of 1 / [std error (beta)]^2 were used such that $\sigma^2 = 1$. The LOESS procedure estimates σ^2 from the weighted sum of squares. Because it is assumed $\sigma^2 = 1$, the estimated standard errors are multiplied by 1 / estimated σ and adjusted the widths of the confidence intervals by the same factor.

One data issue was an overly influential point that needed to be adjusted to avoid imposing wild variation in the "smoothed" curves: for the West region, males, age 0, above poverty threshold, there were 249 children surveyed that all gave 'No' answers to the asthma question, leading to an estimated value of -14.203 for beta with a standard error of 0.09. In this case the raw probability of asthma equals zero, so the corresponding estimated beta would be negative infinity, but SAS's software gives -14.203 instead. To reduce the excessive impact of this single data point, we replaced the estimated standard error by 4, which is approximately four times the maximum standard error for all other region, gender, poverty status, and age combinations.

There are several potential values that can be selected for the smoothing parameter; the optimum value was determined by evaluating three regression diagnostics: the residual standard error, normal probability plots, and studentized residuals. To generate these statistics, the LOESS procedure was applied to estimated smoothed curves for beta, the logit of the prevalence, as a function of age, separately for each region, gender, and poverty classification. For the children data sets, curves were fit using the choices of 0.4, 0.5, 0.6, 0.7, 0.8, 0.9, and 1.0 for the smoothing parameter. This selected range of values was bounded using the following observations. With only 18 points (i.e., the number of ages), a smoothing parameter of 0.2 cannot be used because the weight function assigns zero weights to all ages except age *A*, and a quadratic model cannot be uniquely fit to a single value. A smoothing parameter of 0.3 also cannot be used because that choice assigns a neighborhood of 5 points only ($0.3 \times 18 = 5$, rounded down), of which the two outside ages have assigned weight zero, making the local quadratic model fit exactly at every point except for the end points (ages 0, 1, 16 and 17). Usually one uses a smoothing parameter below 1 so that not all the data are used for the local regression at a given x value. Note also that a smoothing parameter of 0 can be used to generate the unsmoothed prevalence. The selection of the smoothing parameter used for the adult curves would follow a similar logic, although the lower bound could effectively be extended only to 0.9 given the number of age groups. This limits the selection of smoothing parameter applied to the two adult data sets to a value of 0.9, though values of 0.8 to 1.0 were nevertheless compared for good measure.

The first regression diagnostic used was the residual standard error, which is the LOESS estimate of σ. As discussed above, the true value of σ equals 1, so the best choice of smoothing parameter should have residual standard errors as close to 1 as possible. Attachment B, Tables 5CB-5 to 5CB-8 contain the residual standard errors output from the LOESS procedure, considering region, gender, poverty status and each data set examined. For children 'EVER' having asthma and when considering the best 20 models (of the 112 possible) using this criterion (note also within 0.06 RSE units of 1), the best choice varies with gender, region, and poverty status between smoothing parameters of 0.6, 0.7, and 0.8 (**Table 5C-3**). Similar results were observed for the 'STILL' data set, though a value of 0.6 would be slightly preferred. Either adult data set could be smoothed using a value of 0.8 or 0.9 given the limited selection of smoothing values, though 0.9 appears a better value for the 'STILL' data set.

Table 5C-3. Top 20 model smoothing fits where residual standard error at or a value of 1.0.

Data Set	Asthma	LOESS Smoothing Parameter						
		0.4	0.5	0.6	0.7	0.8	0.9	1.0
Children	EVER	2	2	5	5	4	1	1
	STILL	2	3	4	2	3	3	3
Adults	EVER	n/a	n/a	n/a	n/a	6	6	8
	STILL	n/a	n/a	n/a	n/a	5	7	8

The second regression diagnostic was developed from an approximate studentized residual. The residual errors from the LOESS model were divided by standard error (beta) to make their variances approximately constant. These approximately studentized residuals should be approximately normally distributed with a mean of zero and a variance of $\sigma^2 = 1$. To test this assumption, normal probability plots of the residuals were created for each smoothing parameter, combining all the studentized residuals across genders, regions, poverty status, and ages. These normal probability plots are provided in Attachment B, Figures CB-1 to CB-4. The results for the children data indicate little distinction or affect by the selection of a particular smoothing parameter (e.g., see **Figure 5C-1** below), although linearity in the plotted curve is best expressed with smoothing parameters at or above values of 0.6. When considering the adult data sets, again the appropriate value would be 0.9, as Attachment B, Figures 5CB-3 and 5CB-4 supports this conclusion.

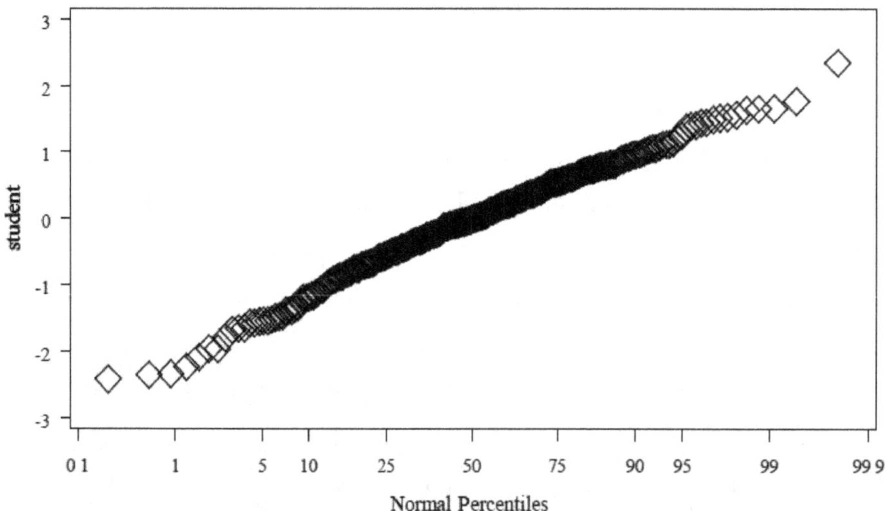

Figure 5C-1. Normal probability plot of studentized residuals generated using logistic model, smoothing set to 0.7, and the children 'EVER' asthmatic data set.

The third regression diagnostic, presented in Attachment B, Figures 5CB-5 to 5CB-8 are plots of the studentized residuals against the smoothed beta values. All the studentized residuals for a given smoothing parameter are plotted together within the same graph. Also plotted is a LOESS smoothed curve fit to the same set of points, with SAS's optimal smoothing parameter choice, to indicate the typical pattern. Ideally there should be no obvious pattern and an average studentized residual close to zero with no regression slope (e.g., see **Figure 5C-2**). For the children data sets, these plots generally indicate no unusual patterns, and the results for smoothing parameters 0.4 through 0.6 indicate a fit LOESS curve closest to the studentized residual equals zero line. When considering the adult data sets, again the appropriate value would be 0.9, as Attachment B, Figures 5CB-7 and 5CB-8 supports this conclusion.

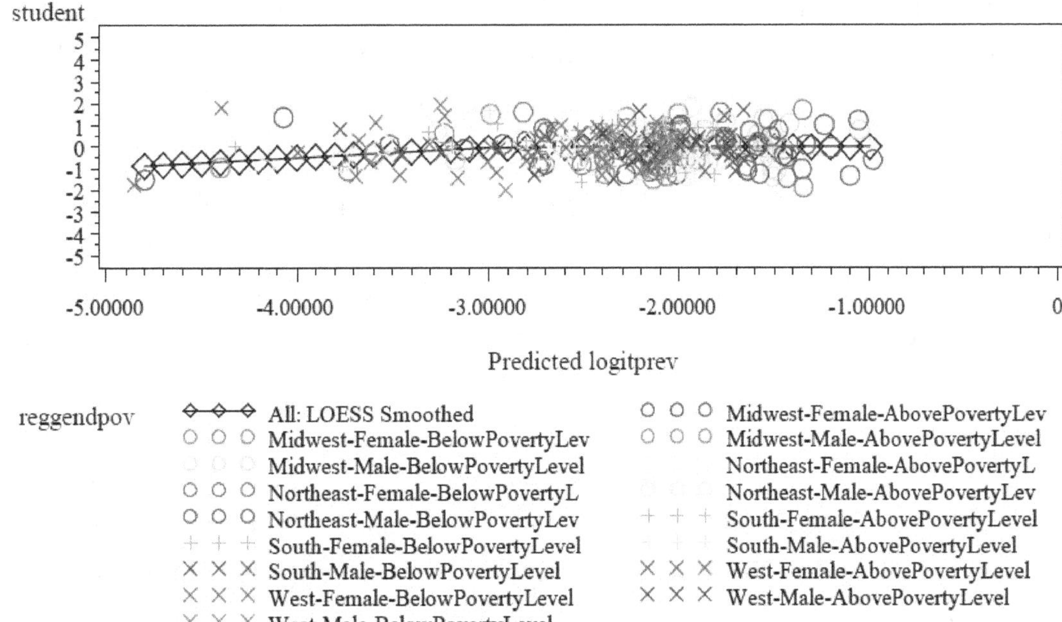

Figure 5C-2. Studentized residuals versus model predicted betas generated using a logistic model and using the children 'EVER' asthmatic data set, with smoothing set to 0.6.

When considering both children asthma prevalence responses evaluated, the residual standard error (estimated values for sigma) suggests the choice of smoothing parameter as 0.6 to 0.8. The normal probability plots of the studentized residuals suggest preference for smoothing at or above 0.6. The plots of residuals against smoothed predictions suggest the choices of 0.4 through 0.6. We therefore chose the final value of 0.6 to use for smoothing the children's asthma prevalence. For the adults, 0.9 was selected for smoothing.

Smoothed asthma prevalence and associated graphical presentation are provided in Attachment C, following a similar format as the unsmoothed data provided in Attachment A.

5C-5 CENSUS TRACT LEVEL POVERTY RATIO DATA SET DESCRIPTION AND PROCESSING

This section describes the approach used to generate census tract level poverty ratios for all US census tracts, stratified by age and age groups where available. The data set generation involved primarily two types of data downloaded from the 2000 US Census, each are described below.

First, individual state level SF3 geographic data ("geo") .uf3 files and associated documentation were downloaded[10] and, following import by SAS (SAS, 2012), were screened for tract level information using the "sumlev" variable equal to '140'. For quality control purposes and ease of matching with the poverty level data, our geo data set retained the following variables: stusab, sumlev, logrecno, state, county, tract, name, latitude, and longitude.

Second, the individual state level SF3 files ("30") were downloaded, retaining the number of persons across the variable "PCT50" for all state "logrecno".[11] The data provided by the PCT50 variable is stratified by age or age groups (ages <5, 5, 6-11, 12-14, 15, 16-17, 18-24, 25-34, 35-44, 45-54, 55-64, 65-74, and ≥75) and income/poverty ratios, given in increments of 0.25. We calculated two new variables for each state logrecno using the number of persons from the PCT50 stratifications; the fraction of those persons having poverty ratios < 1.5 and ≥ 1.5 by summing the appropriate PCT50 variable and dividing by the total number of persons in that age/age group. Finally the poverty ratio data were combined with the above described census tract level geographic data using the "stusab" and "logrecno" variables. The final output was a single file containing relevant tract level poverty probabilities by age groups for all US census tracts (where available).

5C-6 COMBINED CENSUS TRACT LEVEL POVERTY RATIO AND ASTHMA PREVALENCE DATA

Because the prevalence data are stratified by standard US Census defined regions, [12] we first mapped the tract level poverty level data to an appropriate region based on the State. Further, as APEX requires the input data files to be complete, additional processing of the poverty probability file was needed. For where there was missing tract level poverty information,[13] we substituted an age-specific value using the average for the particular county the tract was located within. The frequency of missing data substitution comprised 1.7% of the total poverty probability data set. The two data sets were merged and the final asthma prevalence was calculated using the following weighting scheme:

```
prevalence=round((pov_prob*prev_poor)+((1-pov_prob)*prev_notpoor),0.0001);
```

---

[10] Geographic data were obtained from http://www2.census.gov/census_2000/datasets/Summary_File_3/. Information regarding variable names is given in Figure 2-5 of US Census (2007).

[11] Poverty ratio data were obtained from http://www2.census.gov/census_2000/datasets/Summary_File_3/. Information regarding poverty ratio names variable names is given in chapter 6 of US Census Bureau (2007). We used the variable "PCT50", an income to poverty ratio variable stratified by various ages and age groups and described in chapter 7 of US Census Bureau (2007).

[12] For example, see http://www.cdc.gov/std/stats10/census.htm.

[13] Whether there were no data collected by the Census or whether there were simply no persons in that age group is relatively inconsequential to estimating the asthmatic persons exposed, particularly considering latter case as no persons in that age group would be modeled.

whereas each US census tract value now expresses a tract specific poverty-weighted prevalence, stratified by ages (children 0-17), age groups (adults), and two genders. These final prevalence data are found within the APEX *asthmaprevalence.txt* file.

5C-7 REFERENCES

Cohen, J., and Rosenbaum, A. (2005). Analysis of NHIS Asthma Prevalence Data. Memorandum to John Langstaff by ICF Incorporated. For US EPA Work Assignment 3-08 under EPA contract 68D01052.

SAS. (2012). SAS/STAT 9.2 User's Guide, Second Edition. Available at: http://support.sas.com/documentation/cdl/en/statug/63033/PDF/default/statug.pdf.

US Census Bureau. (2007). 2000 Census of Population and Housing. Summary File 3 (SF3) Technical Documentation, available at: http://www.census.gov/prod/cen2000/doc/sf3.pdf. Individual SF3 files '30' (for income/poverty variables pct50) for each state were downloaded from: http://www2.census.gov/census_2000/datasets/Summary_File_3/.

US EPA. (2007). Ozone Population Exposure Analysis for Selected Urban Areas (July 2007). Office of Air Quality Planning and Standards, Research Triangle Park, NC. EPA-452/R-07-010. Available at: http://epa.gov/ttn/naaqs/standards/ozone/s_o3_cr_td.html.

US EPA. (2008). Risk and Exposure Assessment to Support the Review of the NO_2 Primary National Ambient Air Quality Standard. Report no. EPA-452/R-08-008a. November 2008. Available at: http://www.epa.gov/ttn/naaqs/standards/nox/data/20081121_NO2_REA_final.pdf.

US EPA. (2009). Risk and Exposure Assessment to Support the Review of the SO_2 Primary National Ambient Air Quality Standard. Report no. EPA-452/R-09-007. August 2009. Available at: http://www.epa.gov/ttn/naaqs/standards/so2/data/200908SO2REAFinalReport.pdf.

Appendix 5C, Attachment A

Unsmoothed Asthma Prevalence Tables and Figures

Appendix 5C, Attachment A, Table CA-1. Unsmoothed prevalence for children "EVER" having asthma.								
Smoothed	**Region**	**Gender**	**Poverty Status**	**Age**	**Prevalence**	**SE**	**LowerCl**	**UpperCl**
No	Midwest	Female	Above Poverty	0	0.0018	0.0018	0.0002	0.0129
No	Midwest	Female	Above Poverty	1	0.0387	0.0233	0.0117	0.1208
No	Midwest	Female	Above Poverty	2	0.0367	0.0148	0.0165	0.0797
No	Midwest	Female	Above Poverty	3	0.0395	0.0186	0.0155	0.0972
No	Midwest	Female	Above Poverty	4	0.0815	0.0298	0.0390	0.1624
No	Midwest	Female	Above Poverty	5	0.0885	0.0207	0.0556	0.1382
No	Midwest	Female	Above Poverty	6	0.0438	0.0200	0.0176	0.1046
No	Midwest	Female	Above Poverty	7	0.1374	0.0277	0.0916	0.2010
No	Midwest	Female	Above Poverty	8	0.0820	0.0246	0.0450	0.1450
No	Midwest	Female	Above Poverty	9	0.1027	0.0220	0.0669	0.1545
No	Midwest	Female	Above Poverty	10	0.0995	0.0193	0.0675	0.1442
No	Midwest	Female	Above Poverty	11	0.1129	0.0277	0.0688	0.1797
No	Midwest	Female	Above Poverty	12	0.1752	0.0391	0.1112	0.2652
No	Midwest	Female	Above Poverty	13	0.1331	0.0256	0.0905	0.1916
No	Midwest	Female	Above Poverty	14	0.1944	0.0477	0.1173	0.3049
No	Midwest	Female	Above Poverty	15	0.1383	0.0302	0.0890	0.2086
No	Midwest	Female	Above Poverty	16	0.1731	0.0341	0.1160	0.2502
No	Midwest	Female	Above Poverty	17	0.1311	0.0256	0.0885	0.1898
No	Midwest	Female	Below Poverty	0	0.0564	0.0353	0.0160	0.1799
No	Midwest	Female	Below Poverty	1	0.0585	0.0197	0.0299	0.1112
No	Midwest	Female	Below Poverty	2	0.1256	0.0487	0.0567	0.2552
No	Midwest	Female	Below Poverty	3	0.1127	0.0419	0.0529	0.2240
No	Midwest	Female	Below Poverty	4	0.1746	0.0395	0.1100	0.2658
No	Midwest	Female	Below Poverty	5	0.1584	0.0447	0.0888	0.2664
No	Midwest	Female	Below Poverty	6	0.1229	0.0417	0.0616	0.2301
No	Midwest	Female	Below Poverty	7	0.0867	0.0353	0.0381	0.1851
No	Midwest	Female	Below Poverty	8	0.1523	0.0392	0.0902	0.2456
No	Midwest	Female	Below Poverty	9	0.2070	0.0486	0.1275	0.3182
No	Midwest	Female	Below Poverty	10	0.2293	0.1109	0.0800	0.5043
No	Midwest	Female	Below Poverty	11	0.1359	0.0470	0.0670	0.2562
No	Midwest	Female	Below Poverty	12	0.1501	0.0484	0.0774	0.2710
No	Midwest	Female	Below Poverty	13	0.1527	0.0380	0.0921	0.2427
No	Midwest	Female	Below Poverty	14	0.1197	0.0462	0.0544	0.2431
No	Midwest	Female	Below Poverty	15	0.2103	0.0760	0.0980	0.3949
No	Midwest	Female	Below Poverty	16	0.2054	0.0597	0.1121	0.3462
No	Midwest	Female	Below Poverty	17	0.1844	0.1134	0.0491	0.4976
No	Midwest	Male	Above Poverty	0	0.0061	0.0044	0.0015	0.0247
No	Midwest	Male	Above Poverty	1	0.0258	0.0178	0.0066	0.0957
No	Midwest	Male	Above Poverty	2	0.0848	0.0231	0.0491	0.1426
No	Midwest	Male	Above Poverty	3	0.0996	0.0261	0.0588	0.1636
No	Midwest	Male	Above Poverty	4	0.0876	0.0223	0.0527	0.1423
No	Midwest	Male	Above Poverty	5	0.1593	0.0313	0.1069	0.2306
No	Midwest	Male	Above Poverty	6	0.0977	0.0229	0.0611	0.1527
No	Midwest	Male	Above Poverty	7	0.1793	0.0313	0.1259	0.2489
No	Midwest	Male	Above Poverty	8	0.1503	0.0356	0.0930	0.2340
No	Midwest	Male	Above Poverty	9	0.1418	0.0265	0.0973	0.2021
No	Midwest	Male	Above Poverty	10	0.1569	0.0322	0.1035	0.2306
No	Midwest	Male	Above Poverty	11	0.1717	0.0371	0.1106	0.2568
No	Midwest	Male	Above Poverty	12	0.2054	0.0338	0.1470	0.2795
No	Midwest	Male	Above Poverty	13	0.1846	0.0358	0.1244	0.2650
No	Midwest	Male	Above Poverty	14	0.1671	0.0291	0.1175	0.2322
No	Midwest	Male	Above Poverty	15	0.1454	0.0356	0.0885	0.2297
No	Midwest	Male	Above Poverty	16	0.1557	0.0278	0.1087	0.2182
No	Midwest	Male	Above Poverty	17	0.1320	0.0233	0.0926	0.1848
No	Midwest	Male	Below Poverty	0	0.0293	0.0176	0.0089	0.0922
No	Midwest	Male	Below Poverty	1	0.1051	0.0376	0.0509	0.2047

Appendix 5C, Attachment A, Table CA-1. Unsmoothed prevalence for children "EVER" having asthma.								
Smoothed	**Region**	**Gender**	**Poverty Status**	**Age**	**Prevalence**	**SE**	**LowerCl**	**UpperCl**
No	Midwest	Male	Below Poverty	2	0.1786	0.0652	0.0835	0.3418
No	Midwest	Male	Below Poverty	3	0.2066	0.0513	0.1236	0.3247
No	Midwest	Male	Below Poverty	4	0.2770	0.0638	0.1703	0.4170
No	Midwest	Male	Below Poverty	5	0.2504	0.0499	0.1656	0.3600
No	Midwest	Male	Below Poverty	6	0.2186	0.0447	0.1436	0.3184
No	Midwest	Male	Below Poverty	7	0.2192	0.0456	0.1428	0.3211
No	Midwest	Male	Below Poverty	8	0.2902	0.0649	0.1806	0.4312
No	Midwest	Male	Below Poverty	9	0.1242	0.0437	0.0607	0.2374
No	Midwest	Male	Below Poverty	10	0.2897	0.0639	0.1815	0.4285
No	Midwest	Male	Below Poverty	11	0.2669	0.0613	0.1646	0.4021
No	Midwest	Male	Below Poverty	12	0.2589	0.1050	0.1068	0.5051
No	Midwest	Male	Below Poverty	13	0.2429	0.0693	0.1329	0.4017
No	Midwest	Male	Below Poverty	14	0.1470	0.0490	0.0742	0.2703
No	Midwest	Male	Below Poverty	15	0.1965	0.0509	0.1150	0.3151
No	Midwest	Male	Below Poverty	16	0.1855	0.0611	0.0935	0.3345
No	Midwest	Male	Below Poverty	17	0.3740	0.1042	0.1998	0.5884
No	Northeast	Female	Above Poverty	0	0.0055	0.0054	0.0008	0.0368
No	Northeast	Female	Above Poverty	1	0.0296	0.0164	0.0099	0.0854
No	Northeast	Female	Above Poverty	2	0.0697	0.0252	0.0337	0.1384
No	Northeast	Female	Above Poverty	3	0.0723	0.0250	0.0362	0.1394
No	Northeast	Female	Above Poverty	4	0.1142	0.0254	0.0731	0.1741
No	Northeast	Female	Above Poverty	5	0.1058	0.0296	0.0602	0.1793
No	Northeast	Female	Above Poverty	6	0.0933	0.0254	0.0541	0.1563
No	Northeast	Female	Above Poverty	7	0.1084	0.0251	0.0681	0.1682
No	Northeast	Female	Above Poverty	8	0.0780	0.0221	0.0442	0.1339
No	Northeast	Female	Above Poverty	9	0.1362	0.0374	0.0780	0.2272
No	Northeast	Female	Above Poverty	10	0.0979	0.0298	0.0530	0.1738
No	Northeast	Female	Above Poverty	11	0.1697	0.0382	0.1073	0.2578
No	Northeast	Female	Above Poverty	12	0.0535	0.0229	0.0228	0.1204
No	Northeast	Female	Above Poverty	13	0.0910	0.0273	0.0499	0.1604
No	Northeast	Female	Above Poverty	14	0.1500	0.0207	0.1138	0.1953
No	Northeast	Female	Above Poverty	15	0.1733	0.0355	0.1142	0.2541
No	Northeast	Female	Above Poverty	16	0.1884	0.0510	0.1077	0.3085
No	Northeast	Female	Above Poverty	17	0.1694	0.0395	0.1052	0.2613
No	Northeast	Female	Below Poverty	0	0.0315	0.0251	0.0064	0.1404
No	Northeast	Female	Below Poverty	1	0.1230	0.0576	0.0469	0.2852
No	Northeast	Female	Below Poverty	2	0.0703	0.0277	0.0319	0.1479
No	Northeast	Female	Below Poverty	3	0.1860	0.0555	0.1002	0.3193
No	Northeast	Female	Below Poverty	4	0.1666	0.0598	0.0791	0.3175
No	Northeast	Female	Below Poverty	5	0.2347	0.0636	0.1329	0.3802
No	Northeast	Female	Below Poverty	6	0.0682	0.0250	0.0327	0.1366
No	Northeast	Female	Below Poverty	7	0.0972	0.0362	0.0458	0.1944
No	Northeast	Female	Below Poverty	8	0.2049	0.0604	0.1107	0.3478
No	Northeast	Female	Below Poverty	9	0.1695	0.0698	0.0717	0.3505
No	Northeast	Female	Below Poverty	10	0.0988	0.0440	0.0400	0.2240
No	Northeast	Female	Below Poverty	11	0.2622	0.0734	0.1445	0.4277
No	Northeast	Female	Below Poverty	12	0.1377	0.0525	0.0629	0.2752
No	Northeast	Female	Below Poverty	13	0.3506	0.0762	0.2188	0.5100
No	Northeast	Female	Below Poverty	14	0.1869	0.0537	0.1031	0.3148
No	Northeast	Female	Below Poverty	15	0.1965	0.0534	0.1120	0.3217
No	Northeast	Female	Below Poverty	16	0.1986	0.0470	0.1221	0.3065
No	Northeast	Female	Below Poverty	17	0.1625	0.0602	0.0754	0.3158
No	Northeast	Male	Above Poverty	0	0.0256	0.0130	0.0094	0.0679
No	Northeast	Male	Above Poverty	1	0.0542	0.0231	0.0231	0.1218
No	Northeast	Male	Above Poverty	2	0.0635	0.0220	0.0318	0.1228
No	Northeast	Male	Above Poverty	3	0.0835	0.0232	0.0478	0.1418
No	Northeast	Male	Above Poverty	4	0.1378	0.0329	0.0849	0.2158
No	Northeast	Male	Above Poverty	5	0.1444	0.0357	0.0875	0.2291
No	Northeast	Male	Above Poverty	6	0.2175	0.0482	0.1376	0.3263
No	Northeast	Male	Above Poverty	7	0.2019	0.0343	0.1429	0.2774
No	Northeast	Male	Above Poverty	8	0.1878	0.0373	0.1252	0.2719
No	Northeast	Male	Above Poverty	9	0.1286	0.0342	0.0751	0.2115
No	Northeast	Male	Above Poverty	10	0.1879	0.0278	0.1394	0.2485
No	Northeast	Male	Above Poverty	11	0.2532	0.0420	0.1799	0.3439

Appendix 5C, Attachment A, Table CA-1. Unsmoothed prevalence for children "EVER" having asthma.								
Smoothed	**Region**	**Gender**	**Poverty Status**	**Age**	**Prevalence**	**SE**	**LowerCl**	**UpperCl**
No	Northeast	Male	Above Poverty	12	0.1801	0.0233	0.1388	0.2303
No	Northeast	Male	Above Poverty	13	0.1581	0.0340	0.1022	0.2366
No	Northeast	Male	Above Poverty	14	0.2043	0.0447	0.1303	0.3056
No	Northeast	Male	Above Poverty	15	0.1752	0.0287	0.1257	0.2387
No	Northeast	Male	Above Poverty	16	0.1798	0.0360	0.1195	0.2614
No	Northeast	Male	Above Poverty	17	0.1836	0.0282	0.1346	0.2454
No	Northeast	Male	Below Poverty	0	0.0375	0.0275	0.0087	0.1477
No	Northeast	Male	Below Poverty	1	0.1649	0.0506	0.0877	0.2887
No	Northeast	Male	Below Poverty	2	0.2200	0.0503	0.1371	0.3337
No	Northeast	Male	Below Poverty	3	0.1124	0.0445	0.0501	0.2330
No	Northeast	Male	Below Poverty	4	0.2651	0.0909	0.1262	0.4738
No	Northeast	Male	Below Poverty	5	0.2398	0.0651	0.1355	0.3885
No	Northeast	Male	Below Poverty	6	0.3209	0.0432	0.2427	0.4107
No	Northeast	Male	Below Poverty	7	0.2651	0.0572	0.1686	0.3908
No	Northeast	Male	Below Poverty	8	0.2905	0.0969	0.1401	0.5070
No	Northeast	Male	Below Poverty	9	0.3810	0.0773	0.2446	0.5392
No	Northeast	Male	Below Poverty	10	0.3382	0.1019	0.1732	0.5551
No	Northeast	Male	Below Poverty	11	0.2485	0.0708	0.1359	0.4102
No	Northeast	Male	Below Poverty	12	0.2819	0.0705	0.1656	0.4371
No	Northeast	Male	Below Poverty	13	0.2961	0.0685	0.1808	0.4448
No	Northeast	Male	Below Poverty	14	0.2876	0.0713	0.1695	0.4440
No	Northeast	Male	Below Poverty	15	0.2632	0.0661	0.1548	0.4107
No	Northeast	Male	Below Poverty	16	0.2407	0.0559	0.1483	0.3660
No	Northeast	Male	Below Poverty	17	0.3123	0.0734	0.1885	0.4701
No	South	Female	Above Poverty	0	0.0129	0.0080	0.0038	0.0427
No	South	Female	Above Poverty	1	0.0191	0.0084	0.0080	0.0447
No	South	Female	Above Poverty	2	0.0558	0.0147	0.0330	0.0928
No	South	Female	Above Poverty	3	0.0793	0.0200	0.0479	0.1286
No	South	Female	Above Poverty	4	0.0834	0.0184	0.0537	0.1273
No	South	Female	Above Poverty	5	0.0932	0.0222	0.0579	0.1467
No	South	Female	Above Poverty	6	0.1446	0.0226	0.1057	0.1948
No	South	Female	Above Poverty	7	0.1439	0.0248	0.1017	0.1996
No	South	Female	Above Poverty	8	0.1111	0.0194	0.0784	0.1550
No	South	Female	Above Poverty	9	0.1258	0.0222	0.0883	0.1762
No	South	Female	Above Poverty	10	0.0626	0.0154	0.0383	0.1005
No	South	Female	Above Poverty	11	0.1288	0.0210	0.0928	0.1759
No	South	Female	Above Poverty	12	0.1064	0.0182	0.0756	0.1478
No	South	Female	Above Poverty	13	0.1387	0.0222	0.1006	0.1881
No	South	Female	Above Poverty	14	0.1621	0.0243	0.1198	0.2156
No	South	Female	Above Poverty	15	0.1399	0.0169	0.1100	0.1763
No	South	Female	Above Poverty	16	0.1362	0.0253	0.0938	0.1938
No	South	Female	Above Poverty	17	0.1299	0.0197	0.0959	0.1737
No	South	Female	Below Poverty	0	0.0495	0.0216	0.0207	0.1137
No	South	Female	Below Poverty	1	0.0734	0.0210	0.0415	0.1268
No	South	Female	Below Poverty	2	0.0828	0.0207	0.0503	0.1336
No	South	Female	Below Poverty	3	0.0973	0.0271	0.0556	0.1649
No	South	Female	Below Poverty	4	0.1578	0.0372	0.0976	0.2450
No	South	Female	Below Poverty	5	0.1409	0.0300	0.0917	0.2103
No	South	Female	Below Poverty	6	0.1536	0.0381	0.0927	0.2439
No	South	Female	Below Poverty	7	0.1658	0.0332	0.1104	0.2414
No	South	Female	Below Poverty	8	0.1428	0.0302	0.0931	0.2126
No	South	Female	Below Poverty	9	0.2123	0.0413	0.1425	0.3042
No	South	Female	Below Poverty	10	0.1408	0.0347	0.0855	0.2233
No	South	Female	Below Poverty	11	0.2249	0.0466	0.1467	0.3288
No	South	Female	Below Poverty	12	0.1741	0.0519	0.0941	0.2997
No	South	Female	Below Poverty	13	0.1463	0.0296	0.0972	0.2142
No	South	Female	Below Poverty	14	0.2428	0.0437	0.1675	0.3382
No	South	Female	Below Poverty	15	0.1947	0.0399	0.1280	0.2847
No	South	Female	Below Poverty	16	0.1285	0.0344	0.0747	0.2122
No	South	Female	Below Poverty	17	0.1322	0.0323	0.0807	0.2092
No	South	Male	Above Poverty	0	0.0135	0.0065	0.0052	0.0342
No	South	Male	Above Poverty	1	0.0782	0.0162	0.0517	0.1165
No	South	Male	Above Poverty	2	0.1134	0.0190	0.0811	0.1563
No	South	Male	Above Poverty	3	0.1063	0.0211	0.0714	0.1554

Appendix 5C, Attachment A, Table CA-1. Unsmoothed prevalence for children "EVER" having asthma.								
Smoothed	Region	Gender	Poverty Status	Age	Prevalence	SE	LowerCI	UpperCI
No	South	Male	Above Poverty	4	0.1679	0.0303	0.1165	0.2360
No	South	Male	Above Poverty	5	0.1644	0.0226	0.1247	0.2136
No	South	Male	Above Poverty	6	0.1328	0.0212	0.0964	0.1802
No	South	Male	Above Poverty	7	0.1542	0.0270	0.1083	0.2148
No	South	Male	Above Poverty	8	0.1502	0.0224	0.1114	0.1994
No	South	Male	Above Poverty	9	0.1522	0.0232	0.1121	0.2033
No	South	Male	Above Poverty	10	0.1485	0.0240	0.1073	0.2018
No	South	Male	Above Poverty	11	0.1767	0.0255	0.1322	0.2323
No	South	Male	Above Poverty	12	0.1915	0.0236	0.1495	0.2419
No	South	Male	Above Poverty	13	0.1939	0.0255	0.1487	0.2487
No	South	Male	Above Poverty	14	0.1381	0.0196	0.1039	0.1813
No	South	Male	Above Poverty	15	0.1579	0.0246	0.1154	0.2122
No	South	Male	Above Poverty	16	0.1698	0.0193	0.1352	0.2110
No	South	Male	Above Poverty	17	0.1530	0.0240	0.1117	0.2061
No	South	Male	Below Poverty	0	0.0610	0.0181	0.0338	0.1076
No	South	Male	Below Poverty	1	0.1005	0.0206	0.0667	0.1488
No	South	Male	Below Poverty	2	0.1102	0.0225	0.0732	0.1626
No	South	Male	Below Poverty	3	0.1699	0.0324	0.1154	0.2431
No	South	Male	Below Poverty	4	0.1642	0.0288	0.1152	0.2285
No	South	Male	Below Poverty	5	0.2510	0.0485	0.1682	0.3572
No	South	Male	Below Poverty	6	0.2064	0.0339	0.1477	0.2808
No	South	Male	Below Poverty	7	0.1588	0.0309	0.1072	0.2290
No	South	Male	Below Poverty	8	0.2518	0.0503	0.1663	0.3622
No	South	Male	Below Poverty	9	0.2246	0.0381	0.1588	0.3078
No	South	Male	Below Poverty	10	0.2022	0.0368	0.1394	0.2839
No	South	Male	Below Poverty	11	0.1890	0.0344	0.1305	0.2658
No	South	Male	Below Poverty	12	0.2322	0.0383	0.1656	0.3153
No	South	Male	Below Poverty	13	0.2345	0.0454	0.1573	0.3345
No	South	Male	Below Poverty	14	0.2265	0.0489	0.1448	0.3361
No	South	Male	Below Poverty	15	0.1801	0.0371	0.1183	0.2645
No	South	Male	Below Poverty	16	0.1286	0.0303	0.0799	0.2005
No	South	Male	Below Poverty	17	0.1916	0.0297	0.1399	0.2566
No	West	Female	Above Poverty	0	0.0049	0.0037	0.0011	0.0216
No	West	Female	Above Poverty	1	0.0390	0.0202	0.0139	0.1048
No	West	Female	Above Poverty	2	0.0269	0.0097	0.0132	0.0541
No	West	Female	Above Poverty	3	0.0439	0.0153	0.0219	0.0858
No	West	Female	Above Poverty	4	0.0232	0.0079	0.0118	0.0450
No	West	Female	Above Poverty	5	0.0988	0.0294	0.0544	0.1730
No	West	Female	Above Poverty	6	0.0829	0.0223	0.0484	0.1384
No	West	Female	Above Poverty	7	0.1065	0.0281	0.0627	0.1752
No	West	Female	Above Poverty	8	0.0960	0.0280	0.0534	0.1666
No	West	Female	Above Poverty	9	0.1124	0.0296	0.0662	0.1846
No	West	Female	Above Poverty	10	0.0978	0.0285	0.0545	0.1695
No	West	Female	Above Poverty	11	0.1186	0.0188	0.0864	0.1606
No	West	Female	Above Poverty	12	0.1655	0.0352	0.1074	0.2463
No	West	Female	Above Poverty	13	0.0855	0.0196	0.0542	0.1324
No	West	Female	Above Poverty	14	0.1258	0.0278	0.0806	0.1911
No	West	Female	Above Poverty	15	0.1482	0.0213	0.1111	0.1949
No	West	Female	Above Poverty	16	0.1394	0.0254	0.0967	0.1969
No	West	Female	Above Poverty	17	0.2285	0.0375	0.1632	0.3101
No	West	Female	Below Poverty	0	0.0064	0.0064	0.0009	0.0441
No	West	Female	Below Poverty	1	0.0443	0.0195	0.0185	0.1025
No	West	Female	Below Poverty	2	0.0523	0.0220	0.0226	0.1166
No	West	Female	Below Poverty	3	0.0403	0.0140	0.0202	0.0788
No	West	Female	Below Poverty	4	0.0346	0.0177	0.0126	0.0919
No	West	Female	Below Poverty	5	0.0887	0.0372	0.0380	0.1934
No	West	Female	Below Poverty	6	0.1351	0.0432	0.0703	0.2439
No	West	Female	Below Poverty	7	0.1364	0.0360	0.0798	0.2234
No	West	Female	Below Poverty	8	0.1106	0.0244	0.0711	0.1682
No	West	Female	Below Poverty	9	0.1254	0.0405	0.0650	0.2283
No	West	Female	Below Poverty	10	0.0585	0.0204	0.0292	0.1137
No	West	Female	Below Poverty	11	0.0747	0.0264	0.0368	0.1460
No	West	Female	Below Poverty	12	0.0720	0.0279	0.0331	0.1496
No	West	Female	Below Poverty	13	0.1898	0.0591	0.0993	0.3323

Appendix 5C, Attachment A, Table CA-1. Unsmoothed prevalence for children "EVER" having asthma.								
Smoothed	**Region**	**Gender**	**Poverty Status**	**Age**	**Prevalence**	**SE**	**LowerCl**	**UpperCl**
No	West	Female	Below Poverty	14	0.1431	0.0431	0.0773	0.2495
No	West	Female	Below Poverty	15	0.1168	0.0304	0.0692	0.1906
No	West	Female	Below Poverty	16	0.0814	0.0290	0.0398	0.1593
No	West	Female	Below Poverty	17	0.0637	0.0235	0.0305	0.1285
No	West	Male	Above Poverty	0	0.0000	0.0000	0.0000	0.0000
No	West	Male	Above Poverty	1	0.0244	0.0121	0.0092	0.0635
No	West	Male	Above Poverty	2	0.0517	0.0155	0.0285	0.0920
No	West	Male	Above Poverty	3	0.0601	0.0172	0.0339	0.1041
No	West	Male	Above Poverty	4	0.1698	0.0275	0.1224	0.2307
No	West	Male	Above Poverty	5	0.1236	0.0288	0.0772	0.1918
No	West	Male	Above Poverty	6	0.1376	0.0264	0.0934	0.1980
No	West	Male	Above Poverty	7	0.1288	0.0354	0.0738	0.2152
No	West	Male	Above Poverty	8	0.1018	0.0223	0.0657	0.1547
No	West	Male	Above Poverty	9	0.1884	0.0315	0.1342	0.2579
No	West	Male	Above Poverty	10	0.1604	0.0273	0.1138	0.2215
No	West	Male	Above Poverty	11	0.2121	0.0298	0.1596	0.2762
No	West	Male	Above Poverty	12	0.1833	0.0349	0.1244	0.2618
No	West	Male	Above Poverty	13	0.2105	0.0397	0.1431	0.2987
No	West	Male	Above Poverty	14	0.1475	0.0309	0.0966	0.2187
No	West	Male	Above Poverty	15	0.1641	0.0263	0.1188	0.2224
No	West	Male	Above Poverty	16	0.1958	0.0282	0.1463	0.2569
No	West	Male	Above Poverty	17	0.2113	0.0289	0.1602	0.2733
No	West	Male	Below Poverty	0	0.0135	0.0128	0.0020	0.0832
No	West	Male	Below Poverty	1	0.0812	0.0317	0.0370	0.1691
No	West	Male	Below Poverty	2	0.0417	0.0131	0.0224	0.0765
No	West	Male	Below Poverty	3	0.1182	0.0351	0.0647	0.2061
No	West	Male	Below Poverty	4	0.1349	0.0329	0.0823	0.2131
No	West	Male	Below Poverty	5	0.1562	0.0401	0.0926	0.2514
No	West	Male	Below Poverty	6	0.1853	0.0444	0.1133	0.2883
No	West	Male	Below Poverty	7	0.1484	0.0343	0.0928	0.2288
No	West	Male	Below Poverty	8	0.1549	0.0343	0.0988	0.2346
No	West	Male	Below Poverty	9	0.1275	0.0418	0.0654	0.2338
No	West	Male	Below Poverty	10	0.1742	0.0431	0.1049	0.2751
No	West	Male	Below Poverty	11	0.1909	0.0554	0.1046	0.3227
No	West	Male	Below Poverty	12	0.1678	0.0599	0.0800	0.3185
No	West	Male	Below Poverty	13	0.1793	0.0491	0.1021	0.2959
No	West	Male	Below Poverty	14	0.1919	0.0454	0.1180	0.2966
No	West	Male	Below Poverty	15	0.1410	0.0577	0.0606	0.2946
No	West	Male	Below Poverty	16	0.1863	0.0384	0.1223	0.2734
No	West	Male	Below Poverty	17	0.2030	0.0493	0.1229	0.3165

Appendix 5C, Attachment A, Table CA-2. Unsmoothed prevalence for children "STILL" having asthma.								
Smoothed	Region	Gender	Poverty Status	Age	Prevalence	SE	LowerCl	UpperCl
No	Midwest	Female	Above Poverty	0	0.0018	0.0018	0.0002	0.0129
No	Midwest	Female	Above Poverty	1	0.0387	0.0233	0.0117	0.1208
No	Midwest	Female	Above Poverty	2	0.0302	0.0135	0.0125	0.0715
No	Midwest	Female	Above Poverty	3	0.0395	0.0186	0.0155	0.0972
No	Midwest	Female	Above Poverty	4	0.0531	0.0214	0.0238	0.1142
No	Midwest	Female	Above Poverty	5	0.0617	0.0173	0.0354	0.1055
No	Midwest	Female	Above Poverty	6	0.0386	0.0192	0.0143	0.0999
No	Midwest	Female	Above Poverty	7	0.0801	0.0239	0.0442	0.1411
No	Midwest	Female	Above Poverty	8	0.0492	0.0151	0.0267	0.0888
No	Midwest	Female	Above Poverty	9	0.0789	0.0200	0.0476	0.1280
No	Midwest	Female	Above Poverty	10	0.0625	0.0162	0.0373	0.1029
No	Midwest	Female	Above Poverty	11	0.0856	0.0232	0.0498	0.1433
No	Midwest	Female	Above Poverty	12	0.1269	0.0357	0.0717	0.2145
No	Midwest	Female	Above Poverty	13	0.1089	0.0264	0.0669	0.1724
No	Midwest	Female	Above Poverty	14	0.1580	0.0478	0.0849	0.2751
No	Midwest	Female	Above Poverty	15	0.0863	0.0213	0.0526	0.1382
No	Midwest	Female	Above Poverty	16	0.1300	0.0319	0.0792	0.2062
No	Midwest	Female	Above Poverty	17	0.0989	0.0236	0.0613	0.1556
No	Midwest	Female	Below Poverty	0	0.0564	0.0353	0.0160	0.1799
No	Midwest	Female	Below Poverty	1	0.0486	0.0183	0.0229	0.1000
No	Midwest	Female	Below Poverty	2	0.0959	0.0434	0.0383	0.2206
No	Midwest	Female	Below Poverty	3	0.0697	0.0338	0.0263	0.1723
No	Midwest	Female	Below Poverty	4	0.1697	0.0387	0.1065	0.2594
No	Midwest	Female	Below Poverty	5	0.0819	0.0265	0.0428	0.1512
No	Midwest	Female	Below Poverty	6	0.0809	0.0357	0.0332	0.1840
No	Midwest	Female	Below Poverty	7	0.0680	0.0325	0.0261	0.1661
No	Midwest	Female	Below Poverty	8	0.1257	0.0346	0.0719	0.2105
No	Midwest	Female	Below Poverty	9	0.1394	0.0398	0.0779	0.2369
No	Midwest	Female	Below Poverty	10	0.1871	0.1071	0.0548	0.4777
No	Midwest	Female	Below Poverty	11	0.0726	0.0266	0.0349	0.1451
No	Midwest	Female	Below Poverty	12	0.1101	0.0452	0.0477	0.2340
No	Midwest	Female	Below Poverty	13	0.1258	0.0354	0.0711	0.2130
No	Midwest	Female	Below Poverty	14	0.0999	0.0435	0.0413	0.2226
No	Midwest	Female	Below Poverty	15	0.1648	0.0745	0.0640	0.3629
No	Midwest	Female	Below Poverty	16	0.1647	0.0576	0.0799	0.3094
No	Midwest	Female	Below Poverty	17	0.1747	0.1141	0.0429	0.4997
No	Midwest	Male	Above Poverty	0	0.0061	0.0044	0.0015	0.0247
No	Midwest	Male	Above Poverty	1	0.0214	0.0175	0.0042	0.1008
No	Midwest	Male	Above Poverty	2	0.0752	0.0222	0.0417	0.1319
No	Midwest	Male	Above Poverty	3	0.0692	0.0203	0.0385	0.1213
No	Midwest	Male	Above Poverty	4	0.0527	0.0201	0.0247	0.1090
No	Midwest	Male	Above Poverty	5	0.1293	0.0303	0.0805	0.2011
No	Midwest	Male	Above Poverty	6	0.0710	0.0193	0.0413	0.1193
No	Midwest	Male	Above Poverty	7	0.1369	0.0301	0.0878	0.2072
No	Midwest	Male	Above Poverty	8	0.1047	0.0299	0.0589	0.1793
No	Midwest	Male	Above Poverty	9	0.1096	0.0269	0.0669	0.1745
No	Midwest	Male	Above Poverty	10	0.1004	0.0281	0.0571	0.1704
No	Midwest	Male	Above Poverty	11	0.1340	0.0348	0.0791	0.2179
No	Midwest	Male	Above Poverty	12	0.1093	0.0242	0.0700	0.1665
No	Midwest	Male	Above Poverty	13	0.1029	0.0210	0.0684	0.1520
No	Midwest	Male	Above Poverty	14	0.1230	0.0236	0.0837	0.1771
No	Midwest	Male	Above Poverty	15	0.1007	0.0305	0.0548	0.1780
No	Midwest	Male	Above Poverty	16	0.1141	0.0268	0.0711	0.1780
No	Midwest	Male	Above Poverty	17	0.0644	0.0193	0.0354	0.1143
No	Midwest	Male	Below Poverty	0	0.0274	0.0175	0.0077	0.0925
No	Midwest	Male	Below Poverty	1	0.0892	0.0369	0.0386	0.1927
No	Midwest	Male	Below Poverty	2	0.1786	0.0652	0.0835	0.3418
No	Midwest	Male	Below Poverty	3	0.1620	0.0475	0.0888	0.2772
No	Midwest	Male	Below Poverty	4	0.2557	0.0634	0.1517	0.3974
No	Midwest	Male	Below Poverty	5	0.1914	0.0400	0.1248	0.2821
No	Midwest	Male	Below Poverty	6	0.1432	0.0333	0.0894	0.2215
No	Midwest	Male	Below Poverty	7	0.1788	0.0378	0.1162	0.2649

Appendix 5C, Attachment A, Table CA-2. Unsmoothed prevalence for children "STILL" having asthma.								
Smoothed	**Region**	**Gender**	**Poverty Status**	**Age**	**Prevalence**	**SE**	**LowerCI**	**UpperCI**
No	Midwest	Male	Below Poverty	8	0.2414	0.0604	0.1429	0.3780
No	Midwest	Male	Below Poverty	9	0.1114	0.0404	0.0533	0.2180
No	Midwest	Male	Below Poverty	10	0.2022	0.0624	0.1061	0.3511
No	Midwest	Male	Below Poverty	11	0.1731	0.0406	0.1072	0.2675
No	Midwest	Male	Below Poverty	12	0.2271	0.1064	0.0822	0.4908
No	Midwest	Male	Below Poverty	13	0.1627	0.0591	0.0767	0.3125
No	Midwest	Male	Below Poverty	14	0.0967	0.0413	0.0406	0.2129
No	Midwest	Male	Below Poverty	15	0.1509	0.0506	0.0757	0.2781
No	Midwest	Male	Below Poverty	16	0.1167	0.0490	0.0495	0.2512
No	Midwest	Male	Below Poverty	17	0.3301	0.1005	0.1683	0.5456
No	Northeast	Female	Above Poverty	0	0.0055	0.0054	0.0008	0.0368
No	Northeast	Female	Above Poverty	1	0.0296	0.0164	0.0099	0.0854
No	Northeast	Female	Above Poverty	2	0.0697	0.0252	0.0337	0.1384
No	Northeast	Female	Above Poverty	3	0.0470	0.0158	0.0240	0.0897
No	Northeast	Female	Above Poverty	4	0.0717	0.0199	0.0413	0.1218
No	Northeast	Female	Above Poverty	5	0.0642	0.0196	0.0349	0.1151
No	Northeast	Female	Above Poverty	6	0.0709	0.0254	0.0346	0.1398
No	Northeast	Female	Above Poverty	7	0.0697	0.0180	0.0416	0.1143
No	Northeast	Female	Above Poverty	8	0.0609	0.0209	0.0307	0.1171
No	Northeast	Female	Above Poverty	9	0.0996	0.0334	0.0507	0.1865
No	Northeast	Female	Above Poverty	10	0.0740	0.0260	0.0366	0.1439
No	Northeast	Female	Above Poverty	11	0.1028	0.0305	0.0565	0.1797
No	Northeast	Female	Above Poverty	12	0.0386	0.0187	0.0147	0.0975
No	Northeast	Female	Above Poverty	13	0.0187	0.0095	0.0069	0.0500
No	Northeast	Female	Above Poverty	14	0.0907	0.0181	0.0609	0.1330
No	Northeast	Female	Above Poverty	15	0.1270	0.0344	0.0733	0.2108
No	Northeast	Female	Above Poverty	16	0.0974	0.0267	0.0562	0.1636
No	Northeast	Female	Above Poverty	17	0.1239	0.0375	0.0671	0.2177
No	Northeast	Female	Below Poverty	0	0.0078	0.0078	0.0011	0.0541
No	Northeast	Female	Below Poverty	1	0.1230	0.0576	0.0469	0.2852
No	Northeast	Female	Below Poverty	2	0.0658	0.0272	0.0287	0.1436
No	Northeast	Female	Below Poverty	3	0.1700	0.0576	0.0842	0.3133
No	Northeast	Female	Below Poverty	4	0.1139	0.0456	0.0503	0.2376
No	Northeast	Female	Below Poverty	5	0.2219	0.0583	0.1282	0.3561
No	Northeast	Female	Below Poverty	6	0.0583	0.0290	0.0215	0.1484
No	Northeast	Female	Below Poverty	7	0.0495	0.0252	0.0179	0.1294
No	Northeast	Female	Below Poverty	8	0.0850	0.0368	0.0354	0.1903
No	Northeast	Female	Below Poverty	9	0.0652	0.0294	0.0264	0.1521
No	Northeast	Female	Below Poverty	10	0.0988	0.0440	0.0400	0.2240
No	Northeast	Female	Below Poverty	11	0.2587	0.0734	0.1416	0.4249
No	Northeast	Female	Below Poverty	12	0.0882	0.0426	0.0332	0.2146
No	Northeast	Female	Below Poverty	13	0.3162	0.0739	0.1913	0.4746
No	Northeast	Female	Below Poverty	14	0.1293	0.0372	0.0722	0.2209
No	Northeast	Female	Below Poverty	15	0.1798	0.0479	0.1039	0.2930
No	Northeast	Female	Below Poverty	16	0.1429	0.0381	0.0831	0.2348
No	Northeast	Female	Below Poverty	17	0.1133	0.0426	0.0527	0.2269
No	Northeast	Male	Above Poverty	0	0.0131	0.0101	0.0029	0.0574
No	Northeast	Male	Above Poverty	1	0.0505	0.0227	0.0206	0.1185
No	Northeast	Male	Above Poverty	2	0.0635	0.0220	0.0318	0.1228
No	Northeast	Male	Above Poverty	3	0.0582	0.0216	0.0277	0.1181
No	Northeast	Male	Above Poverty	4	0.1007	0.0281	0.0574	0.1705
No	Northeast	Male	Above Poverty	5	0.1245	0.0318	0.0742	0.2013
No	Northeast	Male	Above Poverty	6	0.1990	0.0511	0.1171	0.3177
No	Northeast	Male	Above Poverty	7	0.1240	0.0274	0.0795	0.1885
No	Northeast	Male	Above Poverty	8	0.1482	0.0321	0.0956	0.2227
No	Northeast	Male	Above Poverty	9	0.0980	0.0321	0.0506	0.1813
No	Northeast	Male	Above Poverty	10	0.0999	0.0216	0.0648	0.1509
No	Northeast	Male	Above Poverty	11	0.1805	0.0342	0.1229	0.2573
No	Northeast	Male	Above Poverty	12	0.1204	0.0211	0.0848	0.1682
No	Northeast	Male	Above Poverty	13	0.0855	0.0237	0.0491	0.1449
No	Northeast	Male	Above Poverty	14	0.1243	0.0351	0.0702	0.2108
No	Northeast	Male	Above Poverty	15	0.1249	0.0247	0.0839	0.1819
No	Northeast	Male	Above Poverty	16	0.1198	0.0283	0.0744	0.1872
No	Northeast	Male	Above Poverty	17	0.0690	0.0173	0.0418	0.1117

Appendix 5C, Attachment A, Table CA-2. Unsmoothed prevalence for children "STILL" having asthma.								
Smoothed	**Region**	**Gender**	**Poverty Status**	**Age**	**Prevalence**	**SE**	**LowerCI**	**UpperCI**
No	Northeast	Male	Below Poverty	0	0.0375	0.0275	0.0087	0.1477
No	Northeast	Male	Below Poverty	1	0.1649	0.0506	0.0877	0.2887
No	Northeast	Male	Below Poverty	2	0.1621	0.0496	0.0864	0.2835
No	Northeast	Male	Below Poverty	3	0.1015	0.0440	0.0420	0.2255
No	Northeast	Male	Below Poverty	4	0.2486	0.0909	0.1131	0.4621
No	Northeast	Male	Below Poverty	5	0.1479	0.0487	0.0753	0.2701
No	Northeast	Male	Below Poverty	6	0.2630	0.0391	0.1939	0.3463
No	Northeast	Male	Below Poverty	7	0.1707	0.0507	0.0926	0.2935
No	Northeast	Male	Below Poverty	8	0.2056	0.0966	0.0751	0.4521
No	Northeast	Male	Below Poverty	9	0.3343	0.0680	0.2162	0.4776
No	Northeast	Male	Below Poverty	10	0.2276	0.0786	0.1093	0.4145
No	Northeast	Male	Below Poverty	11	0.1643	0.0600	0.0770	0.3164
No	Northeast	Male	Below Poverty	12	0.1117	0.0389	0.0552	0.2132
No	Northeast	Male	Below Poverty	13	0.1931	0.0430	0.1223	0.2914
No	Northeast	Male	Below Poverty	14	0.1714	0.0664	0.0764	0.3410
No	Northeast	Male	Below Poverty	15	0.2043	0.0555	0.1162	0.3338
No	Northeast	Male	Below Poverty	16	0.1684	0.0501	0.0912	0.2901
No	Northeast	Male	Below Poverty	17	0.2140	0.0526	0.1286	0.3345
No	South	Female	Above Poverty	0	0.0129	0.0080	0.0038	0.0427
No	South	Female	Above Poverty	1	0.0144	0.0076	0.0051	0.0402
No	South	Female	Above Poverty	2	0.0452	0.0169	0.0215	0.0926
No	South	Female	Above Poverty	3	0.0675	0.0196	0.0379	0.1175
No	South	Female	Above Poverty	4	0.0540	0.0150	0.0311	0.0920
No	South	Female	Above Poverty	5	0.0572	0.0138	0.0354	0.0911
No	South	Female	Above Poverty	6	0.1002	0.0186	0.0692	0.1431
No	South	Female	Above Poverty	7	0.0894	0.0191	0.0584	0.1346
No	South	Female	Above Poverty	8	0.0762	0.0160	0.0502	0.1141
No	South	Female	Above Poverty	9	0.0969	0.0210	0.0627	0.1466
No	South	Female	Above Poverty	10	0.0473	0.0135	0.0269	0.0819
No	South	Female	Above Poverty	11	0.0847	0.0165	0.0576	0.1231
No	South	Female	Above Poverty	12	0.0768	0.0152	0.0518	0.1124
No	South	Female	Above Poverty	13	0.0700	0.0158	0.0447	0.1080
No	South	Female	Above Poverty	14	0.1059	0.0211	0.0711	0.1550
No	South	Female	Above Poverty	15	0.0930	0.0186	0.0624	0.1364
No	South	Female	Above Poverty	16	0.0702	0.0156	0.0451	0.1077
No	South	Female	Above Poverty	17	0.0867	0.0162	0.0597	0.1242
No	South	Female	Below Poverty	0	0.0404	0.0203	0.0149	0.1050
No	South	Female	Below Poverty	1	0.0613	0.0183	0.0338	0.1085
No	South	Female	Below Poverty	2	0.0704	0.0193	0.0408	0.1189
No	South	Female	Below Poverty	3	0.0812	0.0254	0.0434	0.1471
No	South	Female	Below Poverty	4	0.1404	0.0367	0.0826	0.2286
No	South	Female	Below Poverty	5	0.1276	0.0304	0.0789	0.1997
No	South	Female	Below Poverty	6	0.0792	0.0288	0.0381	0.1573
No	South	Female	Below Poverty	7	0.1262	0.0305	0.0775	0.1989
No	South	Female	Below Poverty	8	0.1185	0.0290	0.0724	0.1881
No	South	Female	Below Poverty	9	0.1147	0.0286	0.0694	0.1836
No	South	Female	Below Poverty	10	0.1038	0.0301	0.0579	0.1792
No	South	Female	Below Poverty	11	0.1461	0.0366	0.0879	0.2331
No	South	Female	Below Poverty	12	0.1299	0.0490	0.0600	0.2589
No	South	Female	Below Poverty	13	0.1013	0.0262	0.0602	0.1655
No	South	Female	Below Poverty	14	0.1699	0.0385	0.1071	0.2590
No	South	Female	Below Poverty	15	0.1591	0.0365	0.0998	0.2441
No	South	Female	Below Poverty	16	0.0633	0.0273	0.0267	0.1427
No	South	Female	Below Poverty	17	0.0975	0.0299	0.0526	0.1737
No	South	Male	Above Poverty	0	0.0044	0.0025	0.0014	0.0135
No	South	Male	Above Poverty	1	0.0700	0.0162	0.0442	0.1092
No	South	Male	Above Poverty	2	0.0911	0.0195	0.0595	0.1373
No	South	Male	Above Poverty	3	0.0962	0.0206	0.0627	0.1449
No	South	Male	Above Poverty	4	0.1230	0.0259	0.0805	0.1833
No	South	Male	Above Poverty	5	0.1321	0.0204	0.0970	0.1774
No	South	Male	Above Poverty	6	0.0999	0.0192	0.0681	0.1443
No	South	Male	Above Poverty	7	0.1114	0.0214	0.0758	0.1608
No	South	Male	Above Poverty	8	0.0946	0.0168	0.0664	0.1330
No	South	Male	Above Poverty	9	0.1108	0.0202	0.0770	0.1569

Appendix 5C, Attachment A, Table CA-2. Unsmoothed prevalence for children "STILL" having asthma.								
Smoothed	**Region**	**Gender**	**Poverty Status**	**Age**	**Prevalence**	**SE**	**LowerCI**	**UpperCI**
No	South	Male	Above Poverty	10	0.1010	0.0186	0.0699	0.1438
No	South	Male	Above Poverty	11	0.0946	0.0175	0.0655	0.1348
No	South	Male	Above Poverty	12	0.1340	0.0207	0.0983	0.1801
No	South	Male	Above Poverty	13	0.1122	0.0226	0.0750	0.1646
No	South	Male	Above Poverty	14	0.0713	0.0153	0.0466	0.1077
No	South	Male	Above Poverty	15	0.0899	0.0158	0.0635	0.1260
No	South	Male	Above Poverty	16	0.0871	0.0147	0.0623	0.1206
No	South	Male	Above Poverty	17	0.0700	0.0178	0.0421	0.1141
No	South	Male	Below Poverty	0	0.0477	0.0162	0.0242	0.0916
No	South	Male	Below Poverty	1	0.0859	0.0197	0.0544	0.1330
No	South	Male	Below Poverty	2	0.0820	0.0201	0.0503	0.1309
No	South	Male	Below Poverty	3	0.1434	0.0319	0.0914	0.2178
No	South	Male	Below Poverty	4	0.1320	0.0265	0.0881	0.1931
No	South	Male	Below Poverty	5	0.2314	0.0486	0.1498	0.3397
No	South	Male	Below Poverty	6	0.1395	0.0302	0.0902	0.2097
No	South	Male	Below Poverty	7	0.1207	0.0269	0.0771	0.1840
No	South	Male	Below Poverty	8	0.2064	0.0474	0.1285	0.3145
No	South	Male	Below Poverty	9	0.1364	0.0279	0.0903	0.2009
No	South	Male	Below Poverty	10	0.1473	0.0315	0.0956	0.2203
No	South	Male	Below Poverty	11	0.1390	0.0286	0.0917	0.2051
No	South	Male	Below Poverty	12	0.1673	0.0339	0.1109	0.2445
No	South	Male	Below Poverty	13	0.1684	0.0449	0.0975	0.2752
No	South	Male	Below Poverty	14	0.0936	0.0305	0.0485	0.1729
No	South	Male	Below Poverty	15	0.1379	0.0353	0.0820	0.2226
No	South	Male	Below Poverty	16	0.0816	0.0275	0.0415	0.1544
No	South	Male	Below Poverty	17	0.1057	0.0289	0.0609	0.1772
No	West	Female	Above Poverty	0	0.0013	0.0013	0.0002	0.0095
No	West	Female	Above Poverty	1	0.0353	0.0202	0.0113	0.1045
No	West	Female	Above Poverty	2	0.0159	0.0076	0.0062	0.0401
No	West	Female	Above Poverty	3	0.0284	0.0132	0.0113	0.0695
No	West	Female	Above Poverty	4	0.0183	0.0071	0.0085	0.0389
No	West	Female	Above Poverty	5	0.0689	0.0276	0.0308	0.1468
No	West	Female	Above Poverty	6	0.0477	0.0166	0.0239	0.0928
No	West	Female	Above Poverty	7	0.0469	0.0144	0.0255	0.0846
No	West	Female	Above Poverty	8	0.0756	0.0263	0.0376	0.1459
No	West	Female	Above Poverty	9	0.0686	0.0196	0.0388	0.1185
No	West	Female	Above Poverty	10	0.0791	0.0250	0.0420	0.1440
No	West	Female	Above Poverty	11	0.0763	0.0124	0.0553	0.1043
No	West	Female	Above Poverty	12	0.1023	0.0260	0.0614	0.1655
No	West	Female	Above Poverty	13	0.0571	0.0163	0.0323	0.0989
No	West	Female	Above Poverty	14	0.1012	0.0251	0.0615	0.1622
No	West	Female	Above Poverty	15	0.0923	0.0207	0.0590	0.1416
No	West	Female	Above Poverty	16	0.0787	0.0214	0.0458	0.1322
No	West	Female	Above Poverty	17	0.1303	0.0294	0.0827	0.1993
No	West	Female	Below Poverty	0	0.0064	0.0064	0.0009	0.0441
No	West	Female	Below Poverty	1	0.0443	0.0195	0.0185	0.1025
No	West	Female	Below Poverty	2	0.0249	0.0153	0.0074	0.0805
No	West	Female	Below Poverty	3	0.0372	0.0137	0.0179	0.0756
No	West	Female	Below Poverty	4	0.0114	0.0102	0.0020	0.0638
No	West	Female	Below Poverty	5	0.0491	0.0294	0.0148	0.1506
No	West	Female	Below Poverty	6	0.1016	0.0419	0.0440	0.2174
No	West	Female	Below Poverty	7	0.0908	0.0302	0.0464	0.1698
No	West	Female	Below Poverty	8	0.0874	0.0258	0.0484	0.1529
No	West	Female	Below Poverty	9	0.0839	0.0267	0.0443	0.1532
No	West	Female	Below Poverty	10	0.0275	0.0137	0.0103	0.0715
No	West	Female	Below Poverty	11	0.0339	0.0160	0.0133	0.0839
No	West	Female	Below Poverty	12	0.0551	0.0254	0.0219	0.1315
No	West	Female	Below Poverty	13	0.1028	0.0393	0.0474	0.2089
No	West	Female	Below Poverty	14	0.1312	0.0440	0.0662	0.2435
No	West	Female	Below Poverty	15	0.0630	0.0247	0.0288	0.1324
No	West	Female	Below Poverty	16	0.0758	0.0287	0.0354	0.1546
No	West	Female	Below Poverty	17	0.0328	0.0163	0.0122	0.0850
No	West	Male	Above Poverty	0	0.0000	0.0000	0.0000	0.0000
No	West	Male	Above Poverty	1	0.0039	0.0040	0.0005	0.0289

Appendix 5C, Attachment A, Table CA-2. Unsmoothed prevalence for children "STILL" having asthma.								
Smoothed	Region	Gender	Poverty Status	Age	Prevalence	SE	LowerCl	UpperCl
No	West	Male	Above Poverty	2	0.0305	0.0113	0.0147	0.0623
No	West	Male	Above Poverty	3	0.0384	0.0129	0.0197	0.0735
No	West	Male	Above Poverty	4	0.1363	0.0261	0.0927	0.1960
No	West	Male	Above Poverty	5	0.0933	0.0268	0.0523	0.1608
No	West	Male	Above Poverty	6	0.0803	0.0208	0.0478	0.1317
No	West	Male	Above Poverty	7	0.1014	0.0320	0.0537	0.1834
No	West	Male	Above Poverty	8	0.0537	0.0182	0.0273	0.1029
No	West	Male	Above Poverty	9	0.1120	0.0242	0.0726	0.1689
No	West	Male	Above Poverty	10	0.1202	0.0253	0.0788	0.1791
No	West	Male	Above Poverty	11	0.1333	0.0271	0.0885	0.1959
No	West	Male	Above Poverty	12	0.1258	0.0286	0.0796	0.1934
No	West	Male	Above Poverty	13	0.1039	0.0328	0.0549	0.1879
No	West	Male	Above Poverty	14	0.0873	0.0217	0.0531	0.1404
No	West	Male	Above Poverty	15	0.0881	0.0222	0.0532	0.1425
No	West	Male	Above Poverty	16	0.1066	0.0230	0.0692	0.1607
No	West	Male	Above Poverty	17	0.1364	0.0284	0.0897	0.2021
No	West	Male	Below Poverty	0	0.0135	0.0128	0.0020	0.0832
No	West	Male	Below Poverty	1	0.0812	0.0317	0.0370	0.1691
No	West	Male	Below Poverty	2	0.0308	0.0080	0.0185	0.0510
No	West	Male	Below Poverty	3	0.0944	0.0311	0.0486	0.1755
No	West	Male	Below Poverty	4	0.1056	0.0306	0.0588	0.1822
No	West	Male	Below Poverty	5	0.0856	0.0256	0.0471	0.1508
No	West	Male	Below Poverty	6	0.1277	0.0356	0.0726	0.2149
No	West	Male	Below Poverty	7	0.0943	0.0353	0.0443	0.1897
No	West	Male	Below Poverty	8	0.1282	0.0343	0.0746	0.2115
No	West	Male	Below Poverty	9	0.0883	0.0287	0.0459	0.1632
No	West	Male	Below Poverty	10	0.0697	0.0228	0.0363	0.1298
No	West	Male	Below Poverty	11	0.0954	0.0365	0.0440	0.1947
No	West	Male	Below Poverty	12	0.0759	0.0316	0.0329	0.1655
No	West	Male	Below Poverty	13	0.0600	0.0276	0.0239	0.1427
No	West	Male	Below Poverty	14	0.1457	0.0391	0.0844	0.2398
No	West	Male	Below Poverty	15	0.1099	0.0551	0.0394	0.2713
No	West	Male	Below Poverty	16	0.0957	0.0350	0.0458	0.1894
No	West	Male	Below Poverty	17	0.1136	0.0421	0.0534	0.2254

Appendix 5C, Attachment A, Table CA-3. Unsmoothed prevalence for adults "EVER" having asthma.								
Smoothed	Region	Gender	Poverty Status	Age_grp	Prevalence	SE	LowerCl	UpperCl
No	Midwest	Female	Above Poverty Level	18-24	0.1633	0.0154	0.1353	0.1958
No	Midwest	Female	Above Poverty Level	25-34	0.1347	0.0096	0.1169	0.1547
No	Midwest	Female	Above Poverty Level	35-44	0.1214	0.0084	0.1059	0.1389
No	Midwest	Female	Above Poverty Level	45-54	0.1157	0.0072	0.1022	0.1306
No	Midwest	Female	Above Poverty Level	55-64	0.1360	0.0103	0.1171	0.1575
No	Midwest	Female	Above Poverty Level	65-74	0.1104	0.0107	0.0910	0.1332
No	Midwest	Female	Above Poverty Level	75+	0.0990	0.0095	0.0819	0.1193
No	Midwest	Female	Below Poverty Level	18-24	0.1990	0.0156	0.1701	0.2314
No	Midwest	Female	Below Poverty Level	25-34	0.1896	0.0177	0.1573	0.2268
No	Midwest	Female	Below Poverty Level	35-44	0.1789	0.0209	0.1415	0.2237
No	Midwest	Female	Below Poverty Level	45-54	0.1903	0.0180	0.1576	0.2281
No	Midwest	Female	Below Poverty Level	55-64	0.2760	0.0255	0.2289	0.3285
No	Midwest	Female	Below Poverty Level	65-74	0.1459	0.0205	0.1101	0.1908
No	Midwest	Female	Below Poverty Level	75+	0.1295	0.0202	0.0948	0.1744
No	Midwest	Male	Above Poverty Level	18-24	0.1658	0.0158	0.1371	0.1990
No	Midwest	Male	Above Poverty Level	25-34	0.1254	0.0092	0.1085	0.1446
No	Midwest	Male	Above Poverty Level	35-44	0.0934	0.0083	0.0784	0.1109
No	Midwest	Male	Above Poverty Level	45-54	0.0659	0.0057	0.0555	0.0779
No	Midwest	Male	Above Poverty Level	55-64	0.0856	0.0086	0.0701	0.1040
No	Midwest	Male	Above Poverty Level	65-74	0.0884	0.0106	0.0697	0.1114
No	Midwest	Male	Above Poverty Level	75+	0.0808	0.0110	0.0617	0.1050
No	Midwest	Male	Below Poverty Level	18-24	0.1672	0.0182	0.1345	0.2060
No	Midwest	Male	Below Poverty Level	25-34	0.1103	0.0156	0.0832	0.1447
No	Midwest	Male	Below Poverty Level	35-44	0.0945	0.0191	0.0632	0.1391
No	Midwest	Male	Below Poverty Level	45-54	0.1445	0.0204	0.1089	0.1893
No	Midwest	Male	Below Poverty Level	55-64	0.1623	0.0203	0.1263	0.2061
No	Midwest	Male	Below Poverty Level	65-74	0.1474	0.0307	0.0968	0.2182
No	Midwest	Male	Below Poverty Level	75+	0.0830	0.0217	0.0492	0.1367
No	Northeast	Female	Above Poverty Level	18-24	0.1834	0.0199	0.1476	0.2256
No	Northeast	Female	Above Poverty Level	25-34	0.1375	0.0107	0.1178	0.1598
No	Northeast	Female	Above Poverty Level	35-44	0.1297	0.0109	0.1097	0.1527
No	Northeast	Female	Above Poverty Level	45-54	0.1209	0.0095	0.1034	0.1409
No	Northeast	Female	Above Poverty Level	55-64	0.1306	0.0106	0.1113	0.1528
No	Northeast	Female	Above Poverty Level	65-74	0.1244	0.0130	0.1010	0.1523
No	Northeast	Female	Above Poverty Level	75+	0.0844	0.0101	0.0666	0.1064
No	Northeast	Female	Below Poverty Level	18-24	0.1642	0.0194	0.1296	0.2059
No	Northeast	Female	Below Poverty Level	25-34	0.1726	0.0170	0.1418	0.2084
No	Northeast	Female	Below Poverty Level	35-44	0.1771	0.0172	0.1459	0.2132
No	Northeast	Female	Below Poverty Level	45-54	0.2140	0.0204	0.1767	0.2567
No	Northeast	Female	Below Poverty Level	55-64	0.2174	0.0232	0.1753	0.2664
No	Northeast	Female	Below Poverty Level	65-74	0.1752	0.0186	0.1417	0.2147
No	Northeast	Female	Below Poverty Level	75+	0.0941	0.0132	0.0712	0.1234
No	Northeast	Male	Above Poverty Level	18-24	0.1658	0.0223	0.1265	0.2142
No	Northeast	Male	Above Poverty Level	25-34	0.1262	0.0126	0.1034	0.1531
No	Northeast	Male	Above Poverty Level	35-44	0.0773	0.0094	0.0607	0.0980
No	Northeast	Male	Above Poverty Level	45-54	0.0976	0.0086	0.0820	0.1158
No	Northeast	Male	Above Poverty Level	55-64	0.0911	0.0096	0.0740	0.1117
No	Northeast	Male	Above Poverty Level	65-74	0.0926	0.0128	0.0704	0.1209
No	Northeast	Male	Above Poverty Level	75+	0.0689	0.0127	0.0478	0.0982
No	Northeast	Male	Below Poverty Level	18-24	0.1753	0.0200	0.1395	0.2179
No	Northeast	Male	Below Poverty Level	25-34	0.1255	0.0178	0.0945	0.1648
No	Northeast	Male	Below Poverty Level	35-44	0.1317	0.0244	0.0909	0.1872
No	Northeast	Male	Below Poverty Level	45-54	0.1189	0.0162	0.0906	0.1545
No	Northeast	Male	Below Poverty Level	55-64	0.1681	0.0490	0.0923	0.2865
No	Northeast	Male	Below Poverty Level	65-74	0.1383	0.0313	0.0875	0.2118
No	Northeast	Male	Below Poverty Level	75+	0.0943	0.0265	0.0536	0.1606
No	South	Female	Above Poverty Level	18-24	0.1501	0.0121	0.1279	0.1754
No	South	Female	Above Poverty Level	25-34	0.1290	0.0084	0.1134	0.1464
No	South	Female	Above Poverty Level	35-44	0.1050	0.0074	0.0914	0.1205
No	South	Female	Above Poverty Level	45-54	0.1163	0.0060	0.1051	0.1285
No	South	Female	Above Poverty Level	55-64	0.1279	0.0087	0.1119	0.1459
No	South	Female	Above Poverty Level	65-74	0.1231	0.0102	0.1044	0.1446

Appendix 5C, Attachment A, Table CA-3. Unsmoothed prevalence for adults "EVER" having asthma.								
Smoothed	**Region**	**Gender**	**Poverty Status**	**Age grp**	**Prevalence**	**SE**	**LowerCl**	**UpperCl**
No	South	Female	Above Poverty Level	75+	0.0939	0.0092	0.0773	0.1136
No	South	Female	Below Poverty Level	18-24	0.1511	0.0133	0.1269	0.1790
No	South	Female	Below Poverty Level	25-34	0.1336	0.0087	0.1175	0.1515
No	South	Female	Below Poverty Level	35-44	0.1452	0.0125	0.1224	0.1714
No	South	Female	Below Poverty Level	45-54	0.1622	0.0128	0.1386	0.1889
No	South	Female	Below Poverty Level	55-64	0.2039	0.0179	0.1711	0.2413
No	South	Female	Below Poverty Level	65-74	0.1616	0.0163	0.1321	0.1962
No	South	Female	Below Poverty Level	75+	0.1127	0.0133	0.0891	0.1415
No	South	Male	Above Poverty Level	18-24	0.1438	0.0100	0.1253	0.1645
No	South	Male	Above Poverty Level	25-34	0.1095	0.0078	0.0952	0.1258
No	South	Male	Above Poverty Level	35-44	0.0890	0.0066	0.0769	0.1027
No	South	Male	Above Poverty Level	45-54	0.0704	0.0051	0.0610	0.0811
No	South	Male	Above Poverty Level	55-64	0.0782	0.0071	0.0654	0.0932
No	South	Male	Above Poverty Level	65-74	0.0789	0.0078	0.0649	0.0956
No	South	Male	Above Poverty Level	75+	0.0893	0.0111	0.0698	0.1135
No	South	Male	Below Poverty Level	18-24	0.1473	0.0152	0.1199	0.1797
No	South	Male	Below Poverty Level	25-34	0.0914	0.0122	0.0701	0.1184
No	South	Male	Below Poverty Level	35-44	0.0972	0.0139	0.0732	0.1280
No	South	Male	Below Poverty Level	45-54	0.1062	0.0138	0.0821	0.1363
No	South	Male	Below Poverty Level	55-64	0.1068	0.0156	0.0799	0.1414
No	South	Male	Below Poverty Level	65-74	0.0966	0.0149	0.0710	0.1301
No	South	Male	Below Poverty Level	75+	0.0702	0.0130	0.0486	0.1004
No	West	Female	Above Poverty Level	18-24	0.1595	0.0150	0.1323	0.1911
No	West	Female	Above Poverty Level	25-34	0.1387	0.0096	0.1209	0.1586
No	West	Female	Above Poverty Level	35-44	0.1368	0.0109	0.1168	0.1595
No	West	Female	Above Poverty Level	45-54	0.1431	0.0092	0.1261	0.1621
No	West	Female	Above Poverty Level	55-64	0.1478	0.0094	0.1303	0.1671
No	West	Female	Above Poverty Level	65-74	0.1541	0.0130	0.1302	0.1813
No	West	Female	Above Poverty Level	75+	0.1231	0.0117	0.1020	0.1479
No	West	Female	Below Poverty Level	18-24	0.1522	0.0184	0.1195	0.1920
No	West	Female	Below Poverty Level	25-34	0.1191	0.0118	0.0978	0.1441
No	West	Female	Below Poverty Level	35-44	0.1466	0.0182	0.1145	0.1859
No	West	Female	Below Poverty Level	45-54	0.1874	0.0219	0.1483	0.2341
No	West	Female	Below Poverty Level	55-64	0.1747	0.0181	0.1419	0.2131
No	West	Female	Below Poverty Level	65-74	0.1318	0.0179	0.1005	0.1709
No	West	Female	Below Poverty Level	75+	0.1370	0.0198	0.1027	0.1806
No	West	Male	Above Poverty Level	18-24	0.1499	0.0188	0.1167	0.1905
No	West	Male	Above Poverty Level	25-34	0.1304	0.0107	0.1108	0.1527
No	West	Male	Above Poverty Level	35-44	0.0984	0.0080	0.0837	0.1153
No	West	Male	Above Poverty Level	45-54	0.0944	0.0081	0.0796	0.1116
No	West	Male	Above Poverty Level	55-64	0.0917	0.0075	0.0780	0.1076
No	West	Male	Above Poverty Level	65-74	0.1168	0.0126	0.0943	0.1438
No	West	Male	Above Poverty Level	75+	0.1208	0.0160	0.0928	0.1558
No	West	Male	Below Poverty Level	18-24	0.1589	0.0222	0.1201	0.2073
No	West	Male	Below Poverty Level	25-34	0.0846	0.0128	0.0626	0.1133
No	West	Male	Below Poverty Level	35-44	0.0760	0.0135	0.0535	0.1069
No	West	Male	Below Poverty Level	45-54	0.1422	0.0214	0.1052	0.1894
No	West	Male	Below Poverty Level	55-64	0.0979	0.0176	0.0684	0.1381
No	West	Male	Below Poverty Level	65-74	0.1349	0.0323	0.0831	0.2116
No	West	Male	Below Poverty Level	75+	0.0937	0.0194	0.0620	0.1393

Appendix 5C, Attachment A, Table CA-4. Unsmoothed prevalence for adults "STILL" having asthma.								
Smoothed	Region	Gender	Poverty Status	Age_grp	Prevalence	SE	LowerCl	UpperCl
No	Midwest	Female	Above Poverty Level	18-24	0.1062	0.0133	0.0828	0.1354
No	Midwest	Female	Above Poverty Level	25-34	0.0859	0.0090	0.0699	0.1052
No	Midwest	Female	Above Poverty Level	35-44	0.0859	0.0081	0.0713	0.1031
No	Midwest	Female	Above Poverty Level	45-44	0.0858	0.0061	0.0746	0.0986
No	Midwest	Female	Above Poverty Level	55-64	0.0996	0.0090	0.0832	0.1188
No	Midwest	Female	Above Poverty Level	65-74	0.0755	0.0083	0.0608	0.0934
No	Midwest	Female	Above Poverty Level	75+	0.0643	0.0073	0.0514	0.0802
No	Midwest	Female	Below Poverty Level	18-24	0.1306	0.0144	0.1049	0.1614
No	Midwest	Female	Below Poverty Level	25-34	0.1329	0.0143	0.1073	0.1634
No	Midwest	Female	Below Poverty Level	35-44	0.1354	0.0187	0.1027	0.1764
No	Midwest	Female	Below Poverty Level	45-44	0.1398	0.0166	0.1102	0.1757
No	Midwest	Female	Below Poverty Level	55-64	0.2110	0.0221	0.1709	0.2575
No	Midwest	Female	Below Poverty Level	65-74	0.1190	0.0180	0.0879	0.1590
No	Midwest	Female	Below Poverty Level	75+	0.1029	0.0183	0.0722	0.1448
No	Midwest	Male	Above Poverty Level	18-24	0.0790	0.0125	0.0577	0.1071
No	Midwest	Male	Above Poverty Level	25-34	0.0599	0.0066	0.0482	0.0743
No	Midwest	Male	Above Poverty Level	35-44	0.0486	0.0063	0.0377	0.0625
No	Midwest	Male	Above Poverty Level	45-44	0.0447	0.0049	0.0360	0.0554
No	Midwest	Male	Above Poverty Level	55-64	0.0555	0.0059	0.0450	0.0683
No	Midwest	Male	Above Poverty Level	65-74	0.0524	0.0076	0.0394	0.0694
No	Midwest	Male	Above Poverty Level	75+	0.0477	0.0088	0.0331	0.0682
No	Midwest	Male	Below Poverty Level	18-24	0.0938	0.0143	0.0693	0.1258
No	Midwest	Male	Below Poverty Level	25-34	0.0572	0.0137	0.0355	0.0908
No	Midwest	Male	Below Poverty Level	35-44	0.0731	0.0162	0.0470	0.1119
No	Midwest	Male	Below Poverty Level	45-44	0.0969	0.0208	0.0630	0.1461
No	Midwest	Male	Below Poverty Level	55-64	0.1350	0.0205	0.0997	0.1804
No	Midwest	Male	Below Poverty Level	65-74	0.1349	0.0294	0.0869	0.2035
No	Midwest	Male	Below Poverty Level	75+	0.0643	0.0213	0.0332	0.1208
No	Northeast	Female	Above Poverty Level	18-24	0.1123	0.0148	0.0864	0.1447
No	Northeast	Female	Above Poverty Level	25-34	0.0917	0.0102	0.0735	0.1138
No	Northeast	Female	Above Poverty Level	35-44	0.0944	0.0092	0.0778	0.1141
No	Northeast	Female	Above Poverty Level	45-44	0.0858	0.0080	0.0714	0.1029
No	Northeast	Female	Above Poverty Level	55-64	0.0945	0.0086	0.0790	0.1127
No	Northeast	Female	Above Poverty Level	65-74	0.0898	0.0106	0.0711	0.1128
No	Northeast	Female	Above Poverty Level	75+	0.0706	0.0098	0.0537	0.0924
No	Northeast	Female	Below Poverty Level	18-24	0.1232	0.0182	0.0918	0.1634
No	Northeast	Female	Below Poverty Level	25-34	0.1180	0.0147	0.0921	0.1499
No	Northeast	Female	Below Poverty Level	35-44	0.1265	0.0138	0.1018	0.1560
No	Northeast	Female	Below Poverty Level	45-44	0.1745	0.0185	0.1412	0.2137
No	Northeast	Female	Below Poverty Level	55-64	0.1744	0.0211	0.1369	0.2196
No	Northeast	Female	Below Poverty Level	65-74	0.1388	0.0148	0.1123	0.1704
No	Northeast	Female	Below Poverty Level	75+	0.0488	0.0088	0.0341	0.0693
No	Northeast	Male	Above Poverty Level	18-24	0.0888	0.0161	0.0620	0.1257
No	Northeast	Male	Above Poverty Level	25-34	0.0655	0.0093	0.0495	0.0862
No	Northeast	Male	Above Poverty Level	35-44	0.0409	0.0061	0.0304	0.0547
No	Northeast	Male	Above Poverty Level	45-44	0.0564	0.0078	0.0429	0.0738
No	Northeast	Male	Above Poverty Level	55-64	0.0469	0.0085	0.0328	0.0667
No	Northeast	Male	Above Poverty Level	65-74	0.0641	0.0105	0.0463	0.0880
No	Northeast	Male	Above Poverty Level	75+	0.0527	0.0110	0.0348	0.0789
No	Northeast	Male	Below Poverty Level	18-24	0.0780	0.0129	0.0562	0.1075
No	Northeast	Male	Below Poverty Level	25-34	0.0847	0.0171	0.0566	0.1248
No	Northeast	Male	Below Poverty Level	35-44	0.0795	0.0212	0.0467	0.1322
No	Northeast	Male	Below Poverty Level	45-44	0.0798	0.0196	0.0489	0.1275
No	Northeast	Male	Below Poverty Level	55-64	0.1322	0.0492	0.0617	0.2608
No	Northeast	Male	Below Poverty Level	65-74	0.1055	0.0296	0.0600	0.1789
No	Northeast	Male	Below Poverty Level	75+	0.0758	0.0247	0.0395	0.1406
No	South	Female	Above Poverty Level	18-24	0.0893	0.0090	0.0732	0.1086
No	South	Female	Above Poverty Level	25-34	0.0731	0.0064	0.0615	0.0866
No	South	Female	Above Poverty Level	35-44	0.0689	0.0051	0.0595	0.0797
No	South	Female	Above Poverty Level	45-44	0.0716	0.0049	0.0626	0.0818
No	South	Female	Above Poverty Level	55-64	0.0865	0.0064	0.0747	0.1000
No	South	Female	Above Poverty Level	65-74	0.0914	0.0090	0.0753	0.1105

Appendix 5C, Attachment A, Table CA-4. Unsmoothed prevalence for adults "STILL" having asthma.								
Smoothed	**Region**	**Gender**	**Poverty Status**	**Age grp**	**Prevalence**	**SE**	**LowerCI**	**UpperCI**
No	South	Female	Above Poverty Level	75+	0.0599	0.0072	0.0473	0.0756
No	South	Female	Below Poverty Level	18-24	0.0996	0.0119	0.0786	0.1254
No	South	Female	Below Poverty Level	25-34	0.0867	0.0079	0.0725	0.1035
No	South	Female	Below Poverty Level	35-44	0.1152	0.0113	0.0948	0.1393
No	South	Female	Below Poverty Level	45-44	0.1369	0.0123	0.1144	0.1629
No	South	Female	Below Poverty Level	55-64	0.1780	0.0173	0.1467	0.2144
No	South	Female	Below Poverty Level	65-74	0.1303	0.0152	0.1033	0.1631
No	South	Female	Below Poverty Level	75+	0.0895	0.0118	0.0689	0.1154
No	South	Male	Above Poverty Level	18-24	0.0608	0.0079	0.0471	0.0782
No	South	Male	Above Poverty Level	25-34	0.0471	0.0053	0.0377	0.0587
No	South	Male	Above Poverty Level	35-44	0.0451	0.0048	0.0365	0.0556
No	South	Male	Above Poverty Level	45-44	0.0359	0.0040	0.0288	0.0446
No	South	Male	Above Poverty Level	55-64	0.0413	0.0055	0.0317	0.0535
No	South	Male	Above Poverty Level	65-74	0.0441	0.0057	0.0342	0.0567
No	South	Male	Above Poverty Level	75+	0.0636	0.0097	0.0470	0.0855
No	South	Male	Below Poverty Level	18-24	0.0617	0.0086	0.0468	0.0810
No	South	Male	Below Poverty Level	25-34	0.0344	0.0064	0.0239	0.0494
No	South	Male	Below Poverty Level	35-44	0.0488	0.0109	0.0314	0.0751
No	South	Male	Below Poverty Level	45-44	0.0800	0.0131	0.0579	0.1097
No	South	Male	Below Poverty Level	55-64	0.0676	0.0122	0.0473	0.0957
No	South	Male	Below Poverty Level	65-74	0.0687	0.0129	0.0473	0.0987
No	South	Male	Below Poverty Level	75+	0.0331	0.0083	0.0202	0.0539
No	West	Female	Above Poverty Level	18-24	0.0908	0.0143	0.0663	0.1231
No	West	Female	Above Poverty Level	25-34	0.0819	0.0070	0.0691	0.0968
No	West	Female	Above Poverty Level	35-44	0.0994	0.0090	0.0830	0.1186
No	West	Female	Above Poverty Level	45-44	0.0937	0.0095	0.0766	0.1141
No	West	Female	Above Poverty Level	55-64	0.1013	0.0087	0.0854	0.1197
No	West	Female	Above Poverty Level	65-74	0.1103	0.0114	0.0898	0.1347
No	West	Female	Above Poverty Level	75+	0.0783	0.0092	0.0621	0.0982
No	West	Female	Below Poverty Level	18-24	0.0901	0.0135	0.0669	0.1202
No	West	Female	Below Poverty Level	25-34	0.0861	0.0111	0.0667	0.1105
No	West	Female	Below Poverty Level	35-44	0.1081	0.0143	0.0831	0.1394
No	West	Female	Below Poverty Level	45-44	0.1391	0.0179	0.1075	0.1781
No	West	Female	Below Poverty Level	55-64	0.1293	0.0164	0.1005	0.1648
No	West	Female	Below Poverty Level	65-74	0.1053	0.0166	0.0770	0.1425
No	West	Female	Below Poverty Level	75+	0.1061	0.0162	0.0782	0.1424
No	West	Male	Above Poverty Level	18-24	0.0620	0.0104	0.0445	0.0858
No	West	Male	Above Poverty Level	25-34	0.0528	0.0068	0.0410	0.0679
No	West	Male	Above Poverty Level	35-44	0.0582	0.0061	0.0473	0.0715
No	West	Male	Above Poverty Level	45-44	0.0499	0.0065	0.0386	0.0642
No	West	Male	Above Poverty Level	55-64	0.0542	0.0072	0.0416	0.0702
No	West	Male	Above Poverty Level	65-74	0.0756	0.0102	0.0579	0.0982
No	West	Male	Above Poverty Level	75+	0.0711	0.0133	0.0491	0.1019
No	West	Male	Below Poverty Level	18-24	0.0741	0.0132	0.0520	0.1046
No	West	Male	Below Poverty Level	25-34	0.0457	0.0097	0.0301	0.0689
No	West	Male	Below Poverty Level	35-44	0.0344	0.0089	0.0207	0.0568
No	West	Male	Below Poverty Level	45-44	0.1119	0.0198	0.0786	0.1570
No	West	Male	Below Poverty Level	55-64	0.0528	0.0137	0.0316	0.0870
No	West	Male	Below Poverty Level	65-74	0.1159	0.0336	0.0644	0.1996
No	West	Male	Below Poverty Level	75+	0.0442	0.0131	0.0246	0.0781

Figure 1. Raw asthma 'EVER' prevalence rates and confidence intervals
region=Midwest pov_rat=Above Poverty Level

Figure 1. Raw asthma 'EVER' prevalence rates and confidence intervals
region=Northeast pov_rat=Above Poverty Level

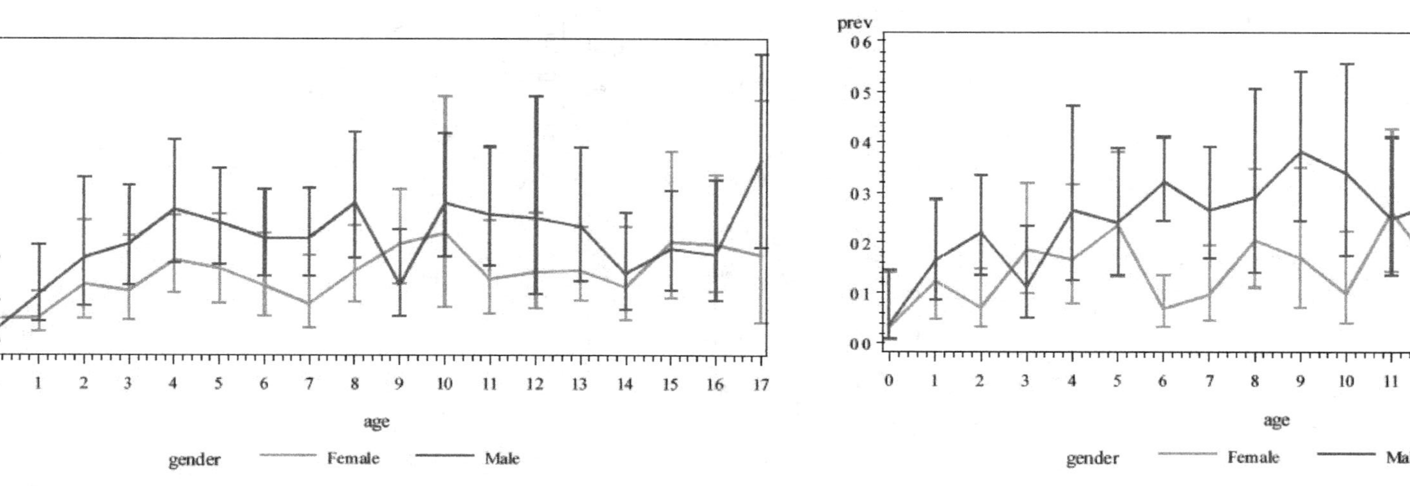

Figure 1. Raw asthma 'EVER' prevalence rates and confidence intervals
region=Midwest pov_rat=Below Poverty Level

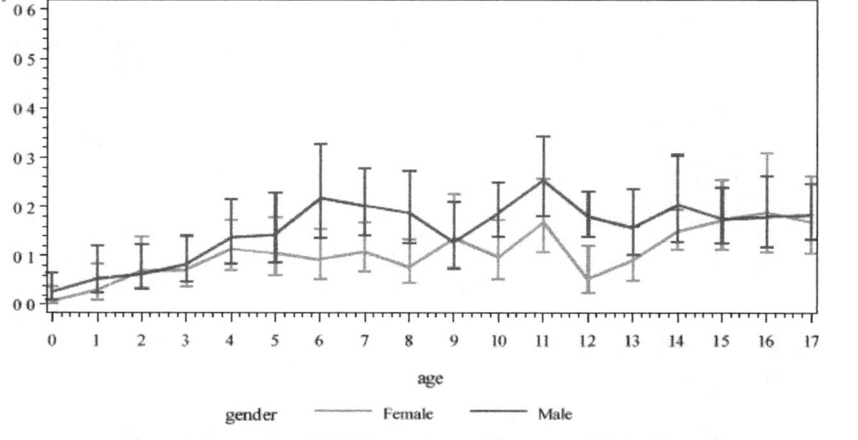

Figure 1. Raw asthma 'EVER' prevalence rates and confidence intervals
region=Northeast pov_rat=Below Poverty Level

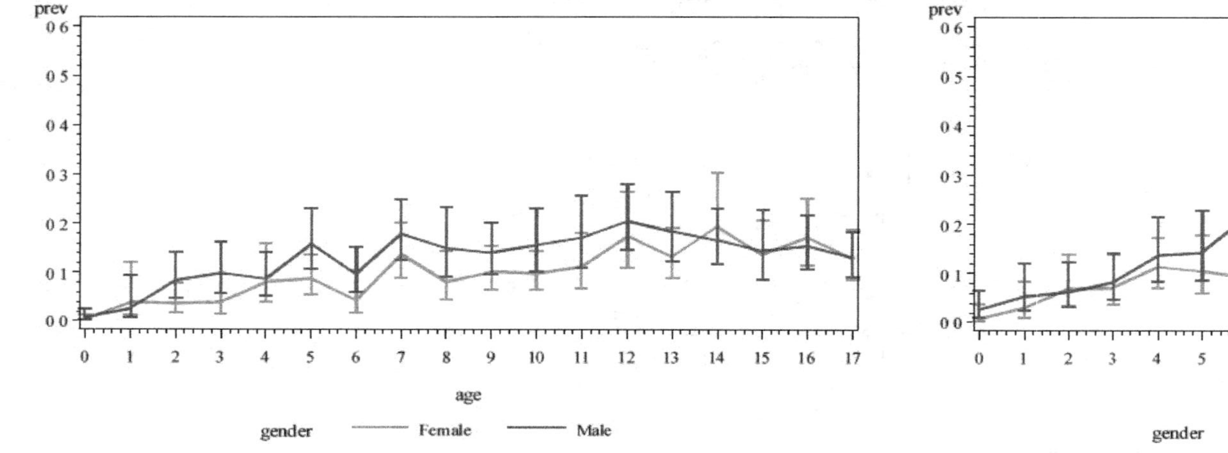

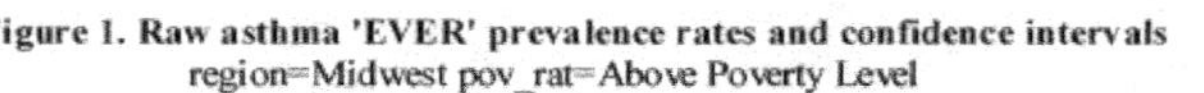

Appendix 5C, Attachment A, Figure CA-1. Unsmoothed prevalence and confidence intervals for children 'EVER' having asthma.

Figure 1. Raw asthma 'EVER' prevalence rates and confidence intervals
region=South pov_rat=Above Poverty Level

Figure 1. Raw asthma 'EVER' prevalence rates and confidence intervals
region=West pov_rat=Above Poverty Level

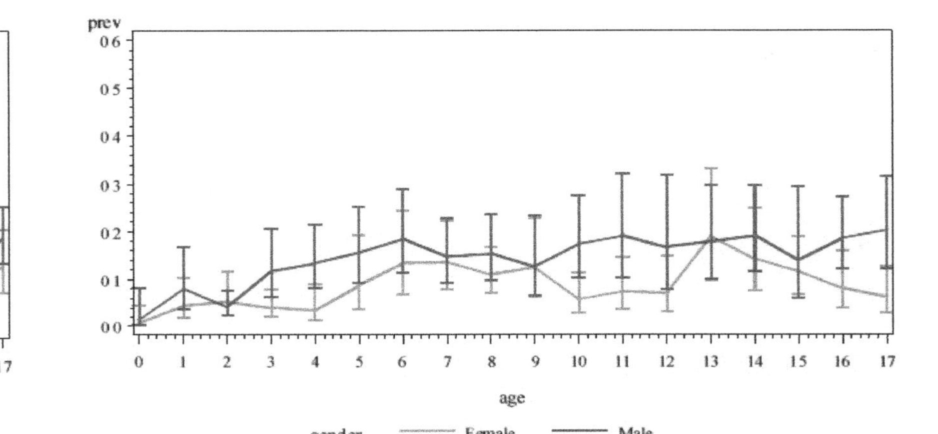

Figure 1. Raw asthma 'EVER' prevalence rates and confidence intervals
region=South pov_rat=Below Poverty Level

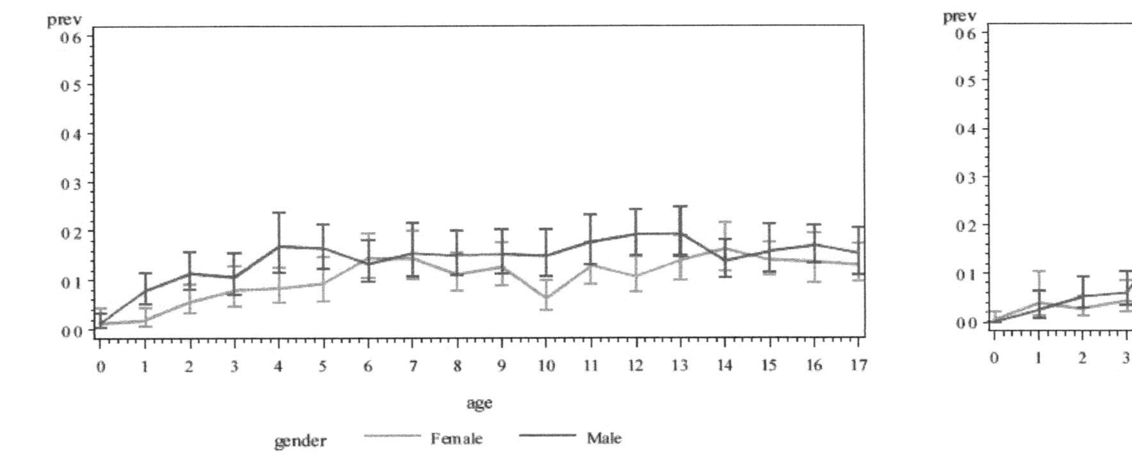

Figure 1. Raw asthma 'EVER' prevalence rates and confidence intervals
region=West pov_rat=Below Poverty Level

Appendix 5C, Attachment A, Figure CA-1, cont. Unsmoothed prevalence and confidence intervals for children 'EVER' having asthma.

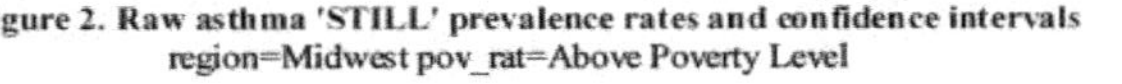

Figure 2. Raw asthma 'STILL' prevalence rates and confidence intervals
region=Midwest pov_rat=Above Poverty Level

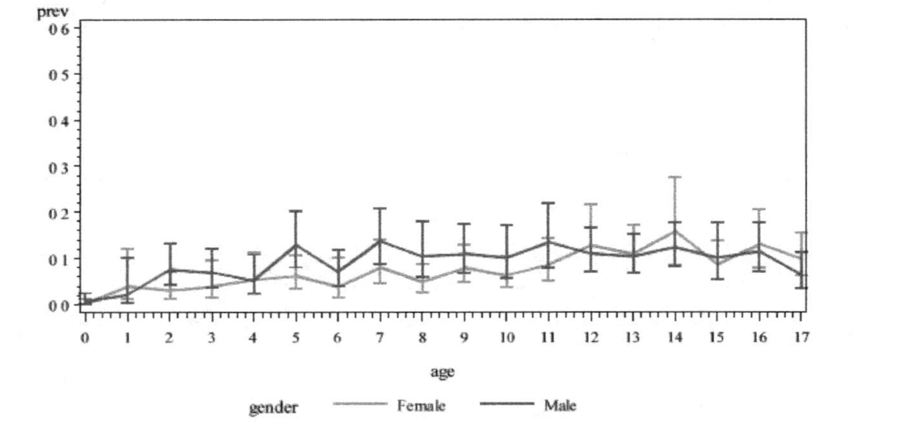

Figure 2. Raw asthma 'STILL' prevalence rates and confidence intervals
region=Midwest pov_rat=Below Poverty Level

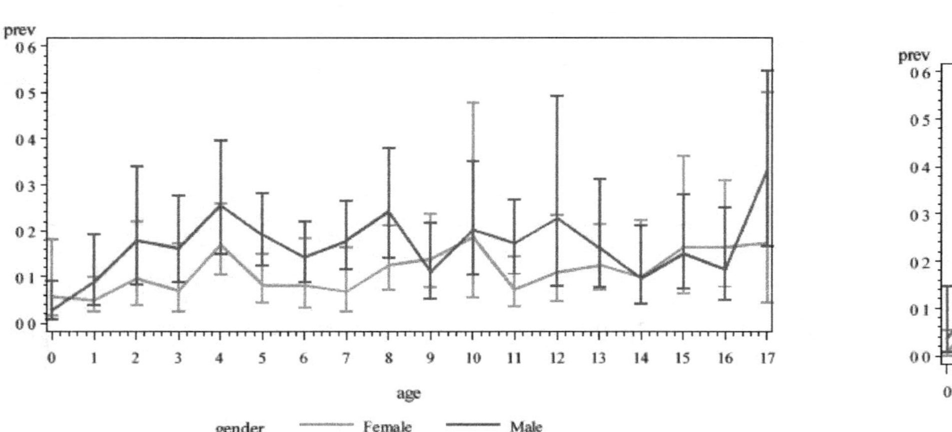

Figure 2. Raw asthma 'STILL' prevalence rates and confidence intervals
region=Northeast pov_rat=Above Poverty Level

gender Female Male

Figure 2. Raw asthma 'STILL' prevalence rates and confidence intervals
region=Northeast pov_rat=Below Poverty Level

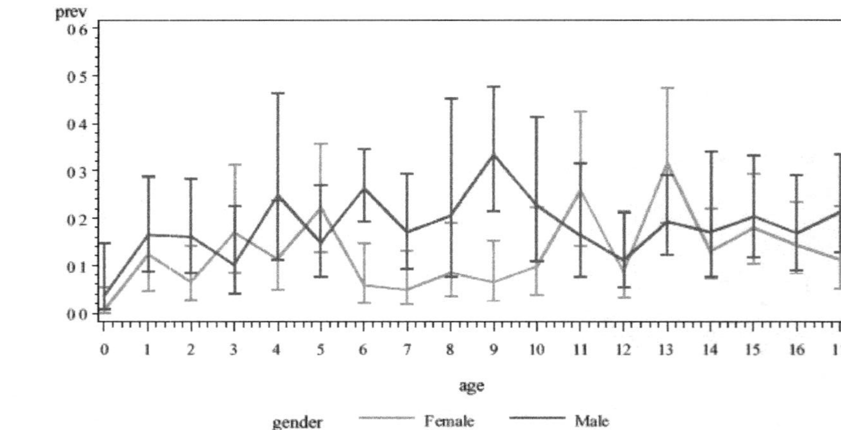

Appendix 5C, Attachment A, Figure CA-2. Unsmoothed prevalence and confidence intervals for children 'STILL' having asthma.

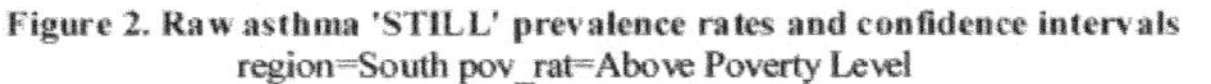

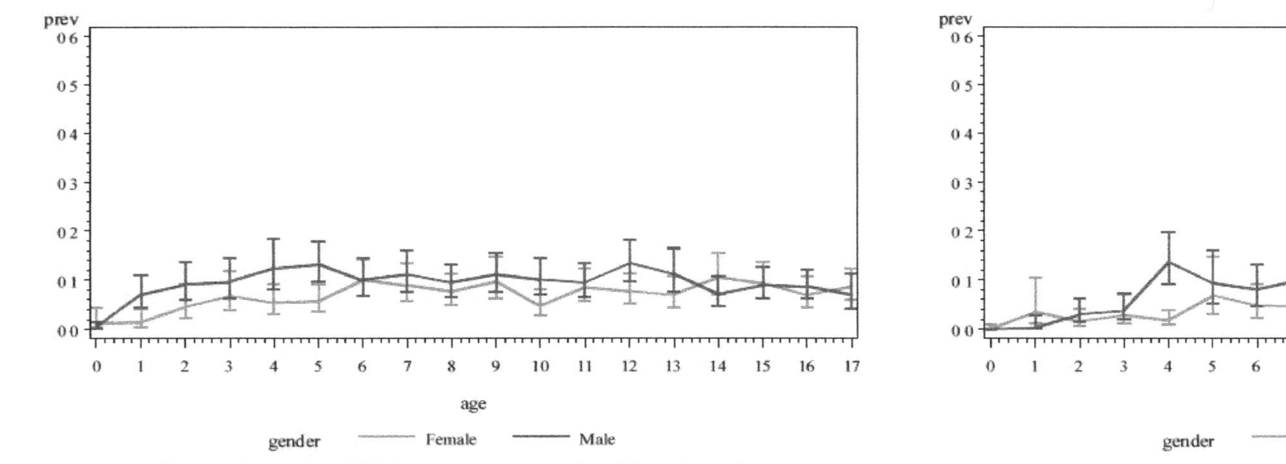

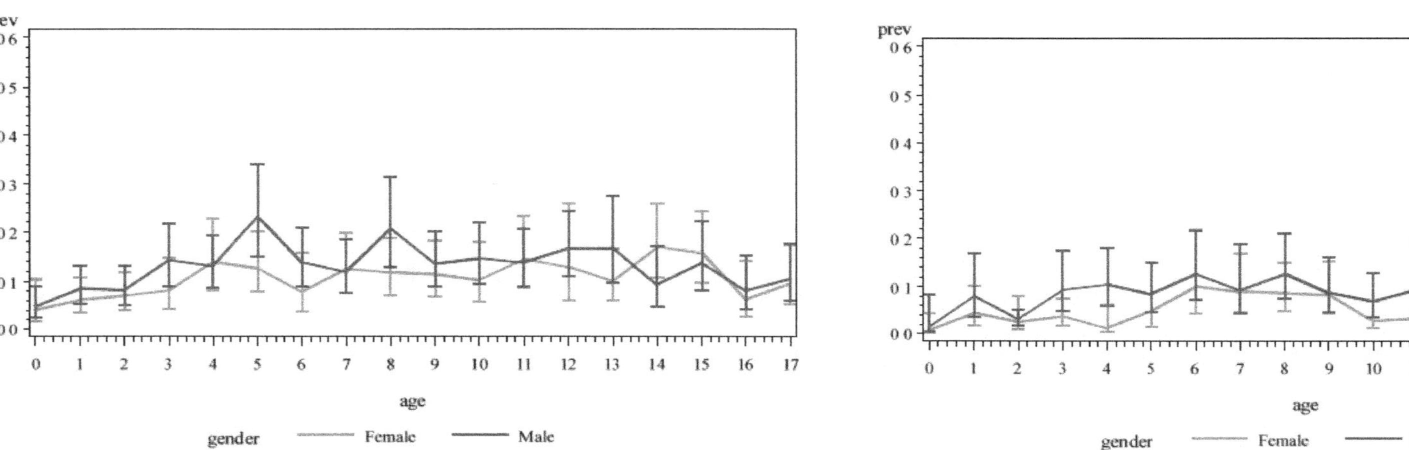

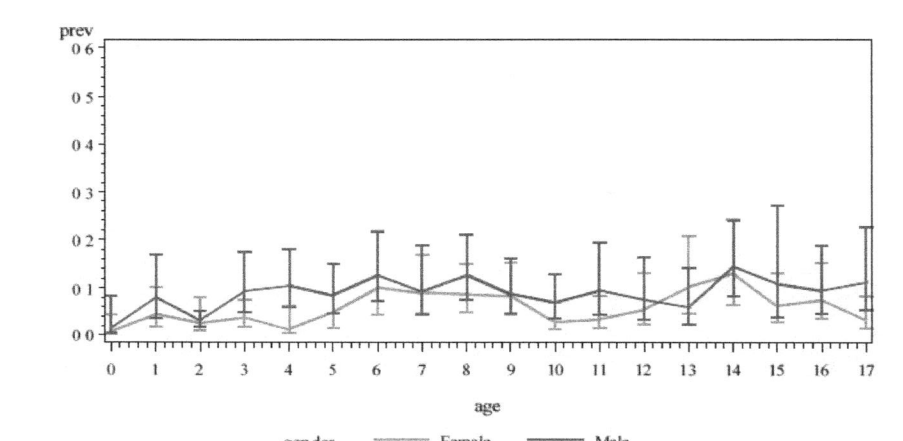

Appendix 5C, Attachment A, Figure CA-2, cont. Unsmoothed prevalence and confidence intervals for children 'STILL' having asthma.

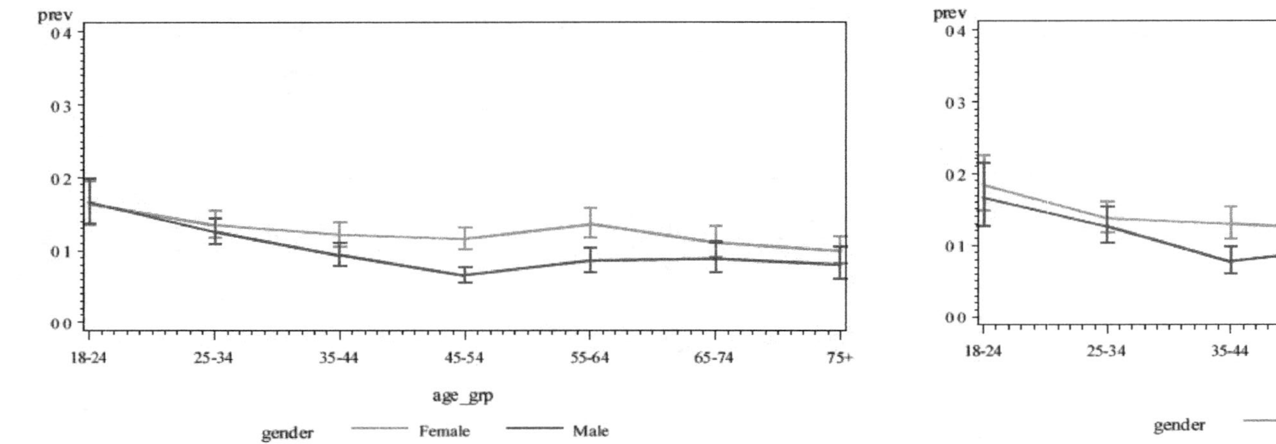

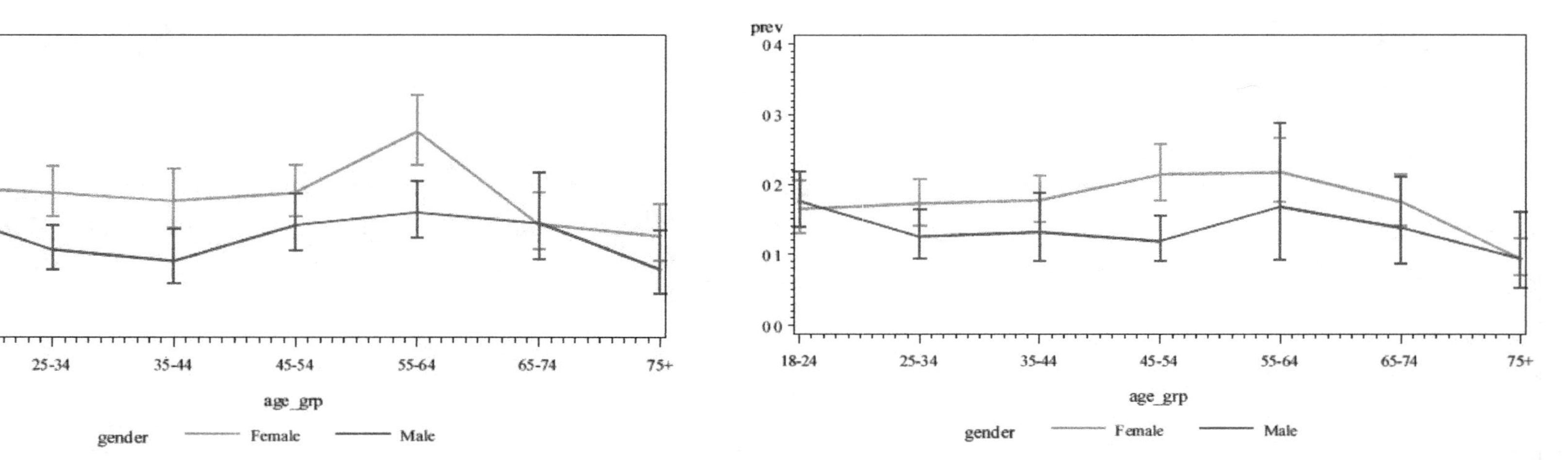

Appendix 5C, Attachment A, Figure CA-3. Unsmoothed prevalence and confidence intervals for adults 'EVER' having asthma.

Figure 3. Raw adult asthma 'EVER' prevalence rates and confidence intervals
region=South pov_rat=Above Poverty Level

Figure 3. Raw adult asthma 'EVER' prevalence rates and confidence intervals
region=South pov_rat=Below Poverty Level

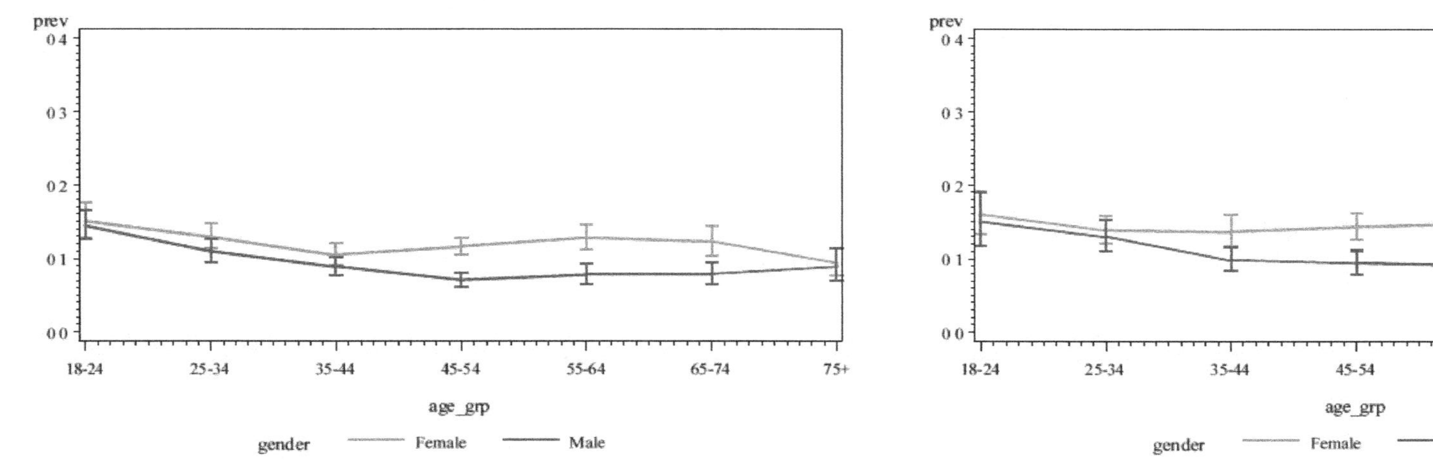

Figure 3. Raw adult asthma 'EVER' prevalence rates and confidence intervals
region=West pov_rat=Above Poverty Level

Figure 3. Raw adult asthma 'EVER' prevalence rates and confidence intervals
region=West pov_rat=Below Poverty Level

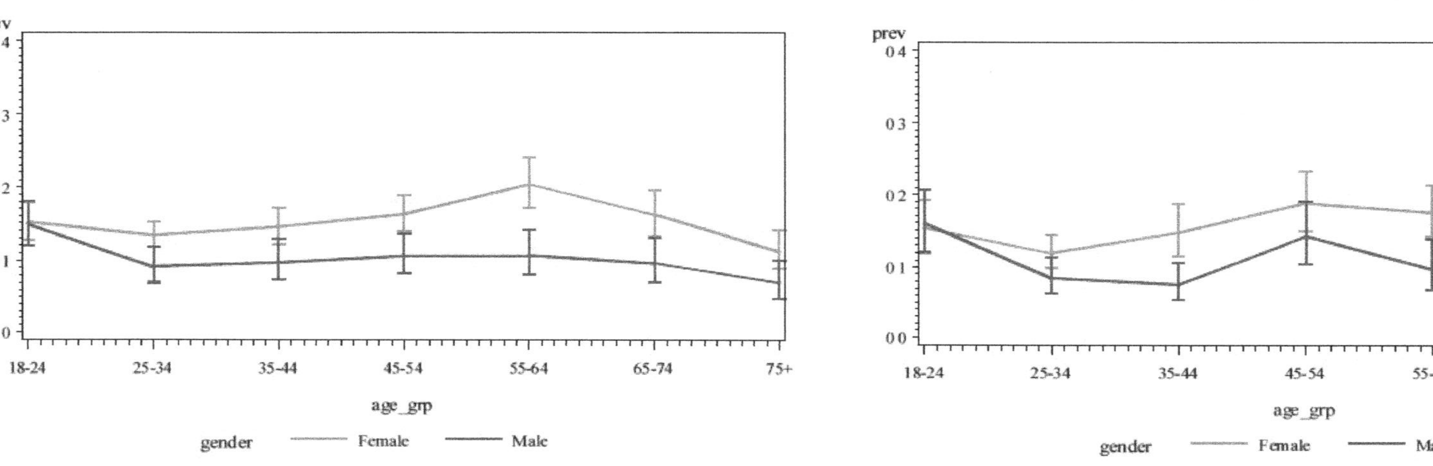

Appendix 5C, Attachment A, Figure CA-3, cont. Unsmoothed prevalence and confidence intervals for adults 'EVER' having asthma.

Figure 4. Raw adult asthma 'STILL' prevalence rates and confidence intervals
region=Midwest pov_rat=Above Poverty Level

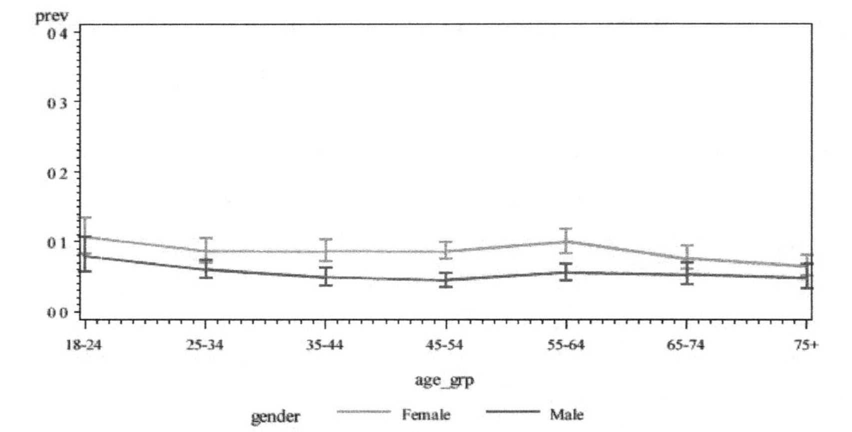

Figure 4. Raw adult asthma 'STILL' prevalence rates and confidence intervals
region=Northeast pov_rat=Above Poverty Level

prev
0 4
0 3
0 2
0 1
0 0
18-24 25-34 35-44 45-54 55-64 65-74 75+
age_grp
gender Female Male

Figure 4. Raw adult asthma 'STILL' prevalence rates and confidence intervals
region=Midwest pov_rat=Below Poverty Level

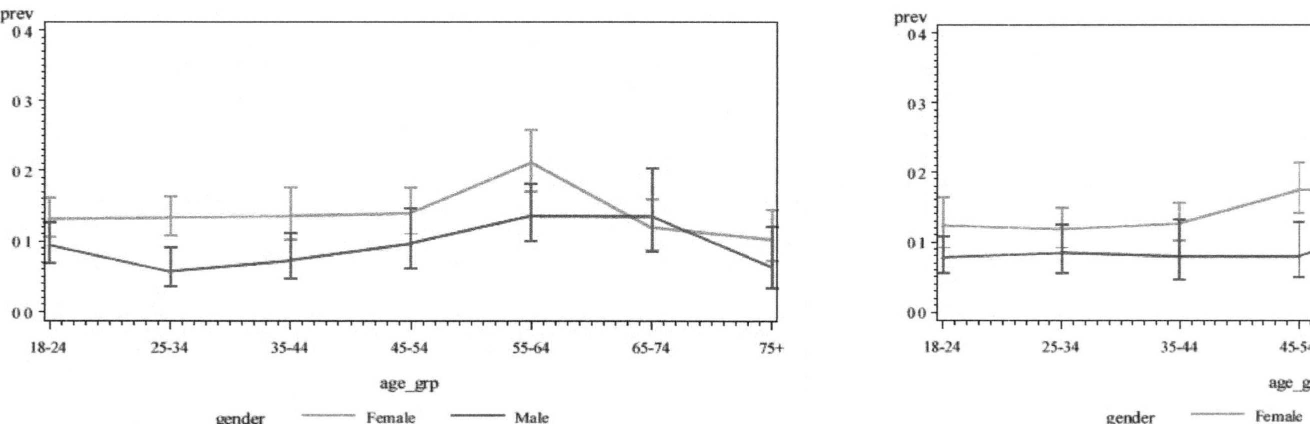

Figure 4. Raw adult asthma 'STILL' prevalence rates and confidence intervals
region=Northeast pov_rat=Below Poverty Level

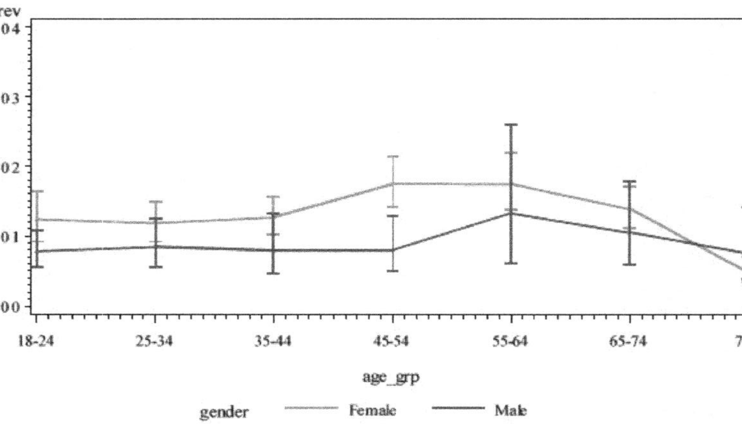

Appendix 5C, Attachment A, Figure CA-4. Unsmoothed prevalence and confidence intervals for adults 'STILL' having asthma.

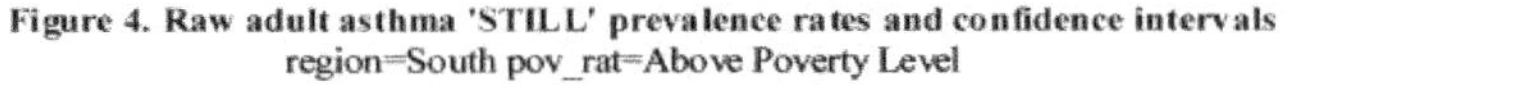

Figure 4. Raw adult asthma 'STILL' prevalence rates and confidence intervals
region=South pov_rat=Above Poverty Level

Figure 4. Raw adult asthma 'STILL' prevalence rates and confidence intervals
region=West pov_rat=Above Poverty Level

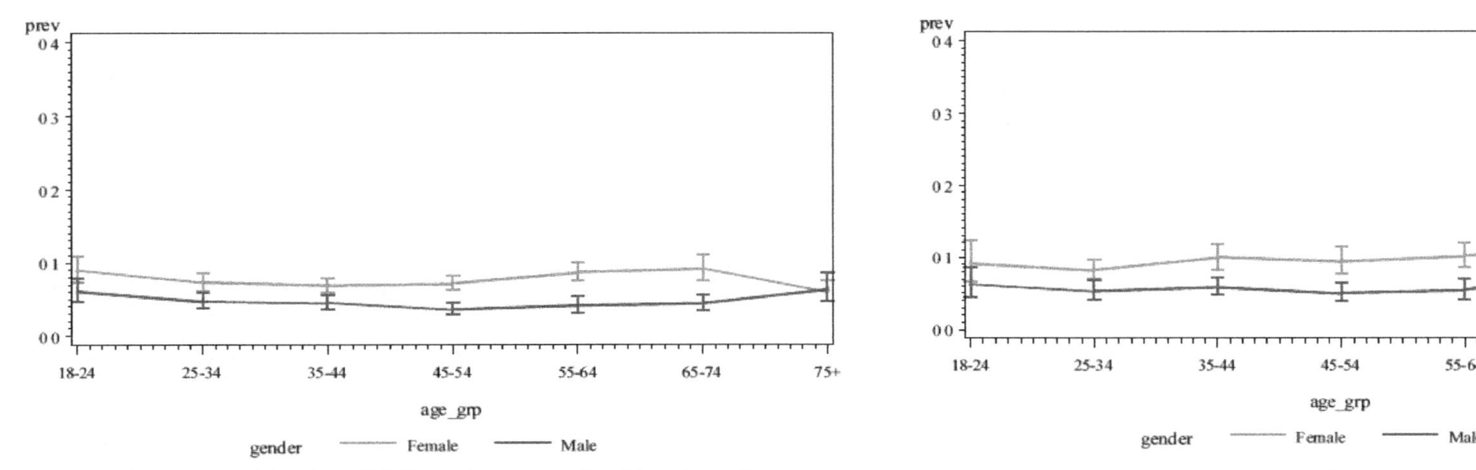

Figure 4. Raw adult asthma 'STILL' prevalence rates and confidence intervals
region=South pov_rat=Below Poverty Level

Figure 4. Raw adult asthma 'STILL' prevalence rates and confidence intervals
region=West pov_rat=Below Poverty Level

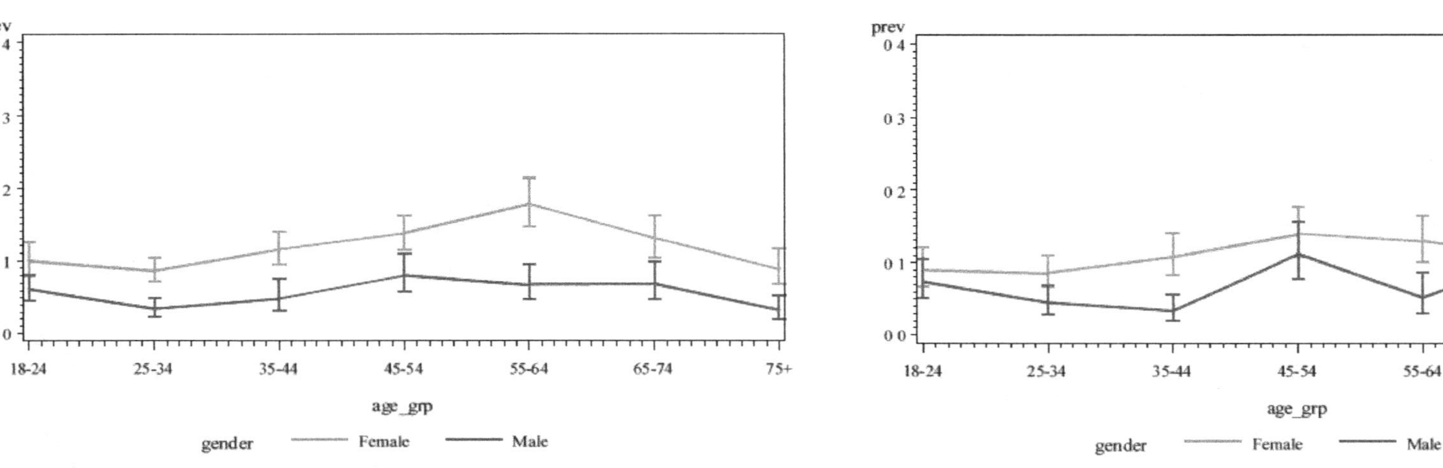

Appendix 5C, Attachment A, Figure CA-4, cont. Unsmoothed prevalence and confidence intervals for adults 'STILL' having asthma.

Appendix 5C, Attachment B
Logistic Model Fit Tables and Figures

Appendix 5C, Attachment B, Table 5CB-1. Alternative logistic models for estimating child asthma prevalence using the "EVER" asthma response variable and goodness of fit test results.

Description	Stratification Variable	-2 log likelihood	DF
1. logit(prob) = linear in age	1. none	288740115.1	2
1. logit(prob) = linear in age	2. gender	287062346.4	4
1. logit(prob) = linear in age	3. region	288120804.1	8
1. logit(prob) = linear in age	4. poverty	287385013.1	4
1. logit(prob) = linear in age	5. region, gender	286367652.6	16
1. logit(prob) = linear in age	6. region, poverty	286283543.6	16
1. logit(prob) = linear in age	7. gender, poverty	285696164.7	8
1. logit(prob) = linear in age	8. region, gender, poverty	284477928.1	32
2. logit(prob) = quadratic in age	1. none	286862135.1	3
2. logit(prob) = quadratic in age	2. gender	285098650.6	6
2. logit(prob) = quadratic in age	3. region	286207721.5	12
2. logit(prob) = quadratic in age	4. poverty	285352164	6
2. logit(prob) = quadratic in age	5. region, gender	284330346.1	24
2. logit(prob) = quadratic in age	6. region, poverty	284182547.5	24
2. logit(prob) = quadratic in age	7. gender, poverty	283587631.7	12
2. logit(prob) = quadratic in age	8. region, gender, poverty	282241318.6	48
3. logit(prob) = cubic in age	1. none	286227019.6	4
3. logit(prob) = cubic in age	2. gender	284470413	8
3. logit(prob) = cubic in age	3. region	285546716.1	16
3. logit(prob) = cubic in age	4. poverty	284688169.9	8
3. logit(prob) = cubic in age	5. region, gender	283662673.5	32
3. logit(prob) = cubic in age	6. region, poverty	283404487.5	32
3. logit(prob) = cubic in age	7. gender, poverty	282890785.3	16
3. logit(prob) = cubic in age	8. region, gender, poverty	281407414.3	64
4. logit(prob) = f(age)	1. none	285821686.2	18
4. logit(prob) = f(age)	2. gender	283843266.2	36
4. logit(prob) = f(age)	3. region	284761522.8	72
4. logit(prob) = f(age)	4. poverty	284045849.2	36
4. logit(prob) = f(age)	5. region, gender	282099156.1	144
4. logit(prob) = f(age)	6. region, poverty	281929968.5	144
4. logit(prob) = f(age)	7. gender, poverty	281963915.7	72
4. logit(prob) = f(age)	8. region, gender, poverty	278655423.1	288

Appendix 5C, Attachment B, Table 5CB-2. Alternative logistic models for estimating child asthma prevalence using the "STILL" asthma response variable and goodness of fit test results.

Description	Stratification Variable	-2 log likelihood	DF
1. logit(prob) = linear in age	1. none	181557347.7	2
1. logit(prob) = linear in age	2. gender	180677544.6	4
1. logit(prob) = linear in age	3. region	180947344.2	8
1. logit(prob) = linear in age	4. poverty	180502490.5	4
1. logit(prob) = linear in age	5. region, gender	179996184.8	16
1. logit(prob) = linear in age	6. region, poverty	179517528	16
1. logit(prob) = linear in age	7. gender, poverty	179637601.4	8
1. logit(prob) = linear in age	8. region, gender, poverty	178567573.9	32
2. logit(prob) = quadratic in age	1. none	180752073.1	3
2. logit(prob) = quadratic in age	2. gender	179771977.6	6
2. logit(prob) = quadratic in age	3. region	180088080.5	12
2. logit(prob) = quadratic in age	4. poverty	179611530.4	6
2. logit(prob) = quadratic in age	5. region, gender	179004935.6	24
2. logit(prob) = quadratic in age	6. region, poverty	178519078.1	24
2. logit(prob) = quadratic in age	7. gender, poverty	178640744.8	12
2. logit(prob) = quadratic in age	8. region, gender, poverty	177414967.2	48
3. logit(prob) = cubic in age	1. none	180247874.1	4
3. logit(prob) = cubic in age	2. gender	179235170	8
3. logit(prob) = cubic in age	3. region	179583725.1	16

Appendix 5C, Attachment B, Table 5CB-2. Alternative logistic models for estimating child asthma prevalence using the "STILL" asthma response variable and goodness of fit test results.			
Description	**Stratification Variable**	**-2 log likelihood**	**DF**
3. logit(prob) = cubic in age	4. poverty	179067549.2	8
3. logit(prob) = cubic in age	5. region, gender	178407915.7	32
3. logit(prob) = cubic in age	6. region, poverty	177897359.3	32
3. logit(prob) = cubic in age	7. gender, poverty	178029240	16
3. logit(prob) = cubic in age	8. region, gender, poverty	176642073.7	64
4. logit(prob) = f(age)	1. none	179972765.3	18
4. logit(prob) = f(age)	2. gender	178918713.8	36
4. logit(prob) = f(age)	3. region	178852704.9	72
4. logit(prob) = f(age)	4. poverty	178599743.4	36
4. logit(prob) = f(age)	5. region, gender	177075815.4	144
4. logit(prob) = f(age)	6. region, poverty	176418872.7	144
4. logit(prob) = f(age)	7. gender, poverty	177422457.4	72
4. logit(prob) = f(age)	8. region, gender, poverty	173888684.9	288

Appendix 5C, Attachment B, Table 5CB- 3. Alternative logistic models for estimating adult asthma prevalence using the "EVER" asthma response variable and goodness of fit test results.			
Description	**Stratification Variable**	**-2 log likelihood**	**DF**
4. logit(prob) = f(age_grp)	1. none	825494282	7
4. logit(prob) = f(age_grp)	2. gender	821614711.2	14
4. logit(prob) = f(age_grp)	3. region	824598583.4	28
4. logit(prob) = f(age_grp)	4. poverty	823443004.3	14
4. logit(prob) = f(age_grp)	5. region, gender	820520390.7	56
4. logit(prob) = f(age_grp)	6. region, poverty	821958349.1	56
4. logit(prob) = f(age_grp)	7. gender, poverty	819560679.9	28
4. logit(prob) = f(age_grp)	8. region, gender, poverty	817723710	112

Appendix 5C, Attachment B, Table 5CB-4. Alternative logistic models for estimating adult asthma prevalence using the "STILL" asthma response variable and goodness of fit test results.			
Description	**Stratification Variable**	**-2 log likelihood**	**DF**
4. logit(prob) = f(age_grp)	1. none	600538044.1	7
4. logit(prob) = f(age_grp)	2. gender	594277797.3	14
4. logit(prob) = f(age_grp)	3. region	599561222.3	28
4. logit(prob) = f(age_grp)	4. poverty	597511872.6	14
4. logit(prob) = f(age_grp)	5. region, gender	593112157.6	56
4. logit(prob) = f(age_grp)	6. region, poverty	596008068.6	56
4. logit(prob) = f(age_grp)	7. gender, poverty	591394271.8	28
4. logit(prob) = f(age_grp)	8. region, gender, poverty	589398969.5	112

Appendix 5C, Attachment B, Table 5CB-5. Effect on residual standard error by varying LOESS smoothing parameter while fitting children "EVER" having asthma data set.

Region	Gender	Poverty Ratio	Smoothing Parameter	Residual Standard Error
South	Female	Above Poverty Level	0.5	0.999919
Northeast	Female	Above Poverty Level	0.7	1.00088
South	Male	Above Poverty Level	0.6	1.003839
Midwest	Male	Above Poverty Level	0.9	1.00548
Midwest	Male	Below Poverty Level	0.8	1.010889
South	Female	Above Poverty Level	0.8	1.012178
South	Male	Above Poverty Level	0.5	0.982885
Midwest	Male	Above Poverty Level	1	1.023284
West	Female	Below Poverty Level	0.7	0.973279
South	Female	Above Poverty Level	0.7	0.97298
Midwest	Female	Above Poverty Level	0.7	1.028007
Midwest	Male	Above Poverty Level	0.4	0.970948
Midwest	Male	Above Poverty Level	0.8	0.965591
Midwest	Female	Above Poverty Level	0.6	1.038233
Northeast	Female	Above Poverty Level	0.4	0.961444
South	Male	Above Poverty Level	0.7	1.040867
South	Female	Above Poverty Level	0.6	0.954946
Midwest	Female	Above Poverty Level	0.8	1.045107
West	Male	Above Poverty Level	0.6	1.052418
Northeast	Female	Above Poverty Level	0.6	0.946315
South	Female	Below Poverty Level	0.5	0.945525
Northeast	Female	Above Poverty Level	0.8	1.054556
Midwest	Male	Above Poverty Level	0.7	0.940657
Northeast	Female	Above Poverty Level	0.5	0.940383
Midwest	Male	Below Poverty Level	0.9	1.063971
West	Female	Below Poverty Level	0.8	1.066819
West	Male	Above Poverty Level	0.5	1.067075
South	Female	Above Poverty Level	0.9	1.067923
South	Female	Below Poverty Level	0.4	0.930104
Midwest	Male	Below Poverty Level	0.7	0.929292
Midwest	Female	Above Poverty Level	0.9	1.072631
South	Male	Below Poverty Level	0.6	0.927161
Northeast	Female	Above Poverty Level	0.9	1.074984
Midwest	Male	Above Poverty Level	0.5	0.917969
South	Male	Below Poverty Level	0.7	0.912266
South	Female	Above Poverty Level	0.4	1.089646
Midwest	Male	Above Poverty Level	0.6	0.90827
Midwest	Male	Below Poverty Level	0.4	0.906073
Midwest	Male	Below Poverty Level	1	1.094737
Midwest	Female	Above Poverty Level	0.5	1.096459
South	Male	Above Poverty Level	0.8	1.099725
South	Male	Below Poverty Level	0.5	0.898228
Northeast	Female	Above Poverty Level	1	1.101884
South	Male	Below Poverty Level	1	0.896985
Midwest	Female	Above Poverty Level	1	1.103976
West	Male	Below Poverty Level	0.4	0.894137
South	Male	Below Poverty Level	0.8	0.893364
South	Female	Below Poverty Level	0.6	0.891551
South	Male	Below Poverty Level	0.9	0.890138
West	Female	Below Poverty Level	0.9	1.111538
South	Male	Above Poverty Level	0.4	0.885511
West	Male	Above Poverty Level	0.4	1.115223
South	Female	Below Poverty Level	0.7	0.86999
Northeast	Male	Below Poverty Level	0.6	0.86934
Midwest	Male	Below Poverty Level	0.6	0.86245
Midwest	Male	Below Poverty Level	0.5	0.857982
South	Female	Below Poverty Level	0.8	0.857778
Northeast	Male	Below Poverty Level	0.5	0.857592

Appendix 5C, Attachment B, Table 5CB-5. Effect on residual standard error by varying LOESS smoothing parameter while fitting children "EVER" having asthma data set.				
Region	**Gender**	**Poverty Ratio**	**Smoothing Parameter**	**Residual Standard Error**
West	Female	Below Poverty Level	0.6	0.852664
West	Female	Below Poverty Level	1	1.147894
South	Female	Below Poverty Level	1	0.849143
South	Female	Below Poverty Level	0.9	0.847567
Northeast	Male	Below Poverty Level	0.7	0.844668
West	Male	Above Poverty Level	0.7	1.163749
West	Female	Above Poverty Level	0.9	1.163943
West	Female	Above Poverty Level	0.8	1.166005
South	Male	Below Poverty Level	0.4	0.826195
West	Female	Above Poverty Level	0.7	1.174564
West	Female	Above Poverty Level	1	1.178045
South	Male	Above Poverty Level	0.9	1.178803
Northeast	Male	Below Poverty Level	0.8	0.820245
South	Female	Above Poverty Level	1	1.182254
West	Female	Above Poverty Level	0.6	1.187757
Northeast	Male	Below Poverty Level	1	0.811815
West	Female	Below Poverty Level	0.5	0.808706
Northeast	Male	Below Poverty Level	0.9	0.805685
West	Male	Below Poverty Level	1	0.804743
Midwest	Female	Below Poverty Level	1	0.799988
Northeast	Male	Above Poverty Level	1	0.799128
Northeast	Male	Above Poverty Level	0.7	0.798212
Midwest	Female	Above Poverty Level	0.4	1.20612
West	Male	Below Poverty Level	0.5	0.793132
Midwest	Female	Below Poverty Level	0.9	0.788082
Northeast	Male	Above Poverty Level	0.6	0.78547
South	Male	Above Poverty Level	1	1.216423
Northeast	Male	Above Poverty Level	0.8	0.78144
West	Male	Below Poverty Level	0.9	0.780843
Northeast	Male	Above Poverty Level	0.9	0.779772
West	Female	Above Poverty Level	0.5	1.224495
Northeast	Male	Below Poverty Level	0.4	0.769037
West	Male	Below Poverty Level	0.6	0.763027
West	Female	Below Poverty Level	0.4	0.762134
Midwest	Female	Below Poverty Level	0.8	0.758775
West	Male	Below Poverty Level	0.8	0.756848
West	Male	Below Poverty Level	0.7	0.752592
Northeast	Male	Above Poverty Level	0.5	0.729776
West	Male	Above Poverty Level	0.8	1.284153
Northeast	Female	Below Poverty Level	0.8	1.292845
Northeast	Female	Below Poverty Level	0.7	1.296274
Northeast	Female	Below Poverty Level	0.9	1.308752
Northeast	Female	Below Poverty Level	0.6	1.309671
Midwest	Female	Below Poverty Level	0.7	0.688366
Northeast	Female	Below Poverty Level	0.5	1.314991
West	Female	Above Poverty Level	0.4	1.31595
Northeast	Female	Below Poverty Level	1	1.327129
West	Male	Above Poverty Level	0.9	1.35931
Northeast	Female	Below Poverty Level	0.4	1.37577
Northeast	Male	Above Poverty Level	0.4	0.618785
Midwest	Female	Below Poverty Level	0.6	0.607758
West	Male	Above Poverty Level	1	1.395061
Midwest	Female	Below Poverty Level	0.5	0.541466
Midwest	Female	Below Poverty Level	0.4	0.522325

Appendix 5C, Attachment B, Table 5CB-6. Effect on residual standard error by varying LOESS smoothing parameter while fitting children "STILL" having asthma data set.				
Region	**Gender**	**Poverty Ratio**	**Smoothing Parameter**	**Residual Standard Error**
South	Female	Above Poverty Level	1	1.000117
Northeast	Male	Above Poverty Level	0.9	1.000909
Northeast	Male	Below Poverty Level	0.7	1.000993
Northeast	Male	Below Poverty Level	0.9	0.997502
Northeast	Male	Below Poverty Level	0.4	0.997275
Midwest	Male	Above Poverty Level	0.7	0.996943
Midwest	Male	Above Poverty Level	0.8	0.996544
Midwest	Female	Above Poverty Level	1	1.003498
Midwest	Male	Above Poverty Level	0.6	0.995815
Northeast	Male	Below Poverty Level	0.8	0.995723
South	Male	Below Poverty Level	0.6	1.007198
Midwest	Female	Above Poverty Level	0.5	0.99235
Northeast	Male	Above Poverty Level	0.5	1.008536
South	Female	Above Poverty Level	0.4	0.99041
Northeast	Male	Below Poverty Level	0.6	1.009859
Northeast	Male	Below Poverty Level	0.5	1.01048
Northeast	Male	Above Poverty Level	0.8	1.011028
Midwest	Male	Above Poverty Level	0.9	1.011038
South	Female	Above Poverty Level	0.6	1.013156
Northeast	Male	Above Poverty Level	1	1.01445
Northeast	Male	Below Poverty Level	1	1.016505
Midwest	Male	Above Poverty Level	0.5	1.01692
Midwest	Female	Above Poverty Level	0.9	0.979917
Midwest	Male	Above Poverty Level	1	1.020707
Northeast	Male	Above Poverty Level	0.7	1.021388
Midwest	Male	Below Poverty Level	0.7	0.977074
South	Female	Above Poverty Level	0.7	0.976479
Northeast	Male	Above Poverty Level	0.6	1.024042
South	Male	Below Poverty Level	1	0.975784
West	Male	Below Poverty Level	0.9	1.025093
South	Male	Below Poverty Level	0.5	1.026184
South	Male	Below Poverty Level	0.7	0.971057
South	Female	Above Poverty Level	0.8	0.965833
South	Female	Above Poverty Level	0.9	0.965238
West	Male	Below Poverty Level	0.8	1.03481
South	Male	Below Poverty Level	0.9	0.964953
West	Female	Below Poverty Level	0.7	1.036384
West	Female	Above Poverty Level	1	1.040924
South	Male	Below Poverty Level	0.8	0.957162
West	Female	Above Poverty Level	0.9	1.044522
Midwest	Male	Below Poverty Level	0.8	1.04601
West	Male	Below Poverty Level	0.7	1.04802
West	Male	Below Poverty Level	1	1.050309
Midwest	Female	Above Poverty Level	0.8	0.946142
Northeast	Female	Above Poverty Level	0.4	0.94543
West	Female	Above Poverty Level	0.8	1.055218
Midwest	Female	Above Poverty Level	0.6	0.938888
West	Male	Above Poverty Level	0.7	1.063545
South	Female	Above Poverty Level	0.5	1.063816
Midwest	Female	Above Poverty Level	0.7	0.931681
West	Male	Below Poverty Level	0.6	1.079146
Midwest	Male	Above Poverty Level	0.4	1.080605
Northeast	Female	Above Poverty Level	0.5	1.083479
West	Female	Above Poverty Level	0.7	1.084472
Midwest	Male	Below Poverty Level	0.9	1.084476
Midwest	Female	Below Poverty Level	0.9	0.914962
Midwest	Female	Below Poverty Level	1	0.913089

Appendix 5C, Attachment B, Table 5CB-6. Effect on residual standard error by varying LOESS smoothing parameter while fitting children "STILL" having asthma data set.				
Region	**Gender**	**Poverty Ratio**	**Smoothing Parameter**	**Residual Standard Error**
South	Male	Below Poverty Level	0.4	1.087093
Midwest	Female	Below Poverty Level	0.8	0.912722
West	Female	Below Poverty Level	0.6	0.912605
Midwest	Male	Below Poverty Level	0.6	0.907737
Midwest	Male	Below Poverty Level	1	1.103127
Northeast	Female	Above Poverty Level	0.6	1.103286
South	Male	Above Poverty Level	0.4	1.112998
Midwest	Male	Below Poverty Level	0.5	0.878223
West	Female	Above Poverty Level	0.6	1.124127
Midwest	Female	Below Poverty Level	0.7	0.875579
Northeast	Male	Above Poverty Level	0.4	0.874469
West	Female	Below Poverty Level	0.5	0.873529
West	Male	Below Poverty Level	0.5	1.127032
South	Female	Below Poverty Level	0.6	0.87206
Midwest	Male	Below Poverty Level	0.4	0.869726
Midwest	Female	Above Poverty Level	0.4	1.135372
West	Female	Below Poverty Level	0.8	1.136048
South	Female	Below Poverty Level	1	0.863066
Northeast	Female	Above Poverty Level	0.7	1.140006
South	Female	Below Poverty Level	0.5	0.858107
Northeast	Female	Above Poverty Level	0.9	1.147352
Northeast	Female	Above Poverty Level	1	1.148471
West	Male	Below Poverty Level	0.4	1.152015
Northeast	Female	Above Poverty Level	0.8	1.153553
West	Male	Above Poverty Level	0.4	0.845979
South	Female	Below Poverty Level	0.7	0.842335
West	Male	Above Poverty Level	0.6	0.8413
South	Female	Below Poverty Level	0.9	0.841106
West	Female	Above Poverty Level	0.5	1.166931
South	Female	Below Poverty Level	0.8	0.830955
West	Female	Below Poverty Level	0.4	0.826586
West	Female	Below Poverty Level	0.9	1.183444
West	Male	Above Poverty Level	0.5	0.815615
Midwest	Female	Below Poverty Level	0.6	0.802622
West	Female	Below Poverty Level	1	1.20757
Midwest	Female	Below Poverty Level	0.4	0.78769
South	Male	Above Poverty Level	0.5	1.214019
South	Male	Above Poverty Level	0.6	1.216661
South	Female	Below Poverty Level	0.4	0.781555
South	Male	Above Poverty Level	0.7	1.242272
West	Female	Above Poverty Level	0.4	1.252141
West	Male	Above Poverty Level	0.8	1.254244
Midwest	Female	Below Poverty Level	0.5	0.742493
South	Male	Above Poverty Level	0.8	1.294055
Northeast	Female	Below Poverty Level	0.7	1.32003
Northeast	Female	Below Poverty Level	0.6	1.355219
West	Male	Above Poverty Level	0.9	1.356792
South	Male	Above Poverty Level	0.9	1.365737
Northeast	Female	Below Poverty Level	0.8	1.39015
West	Male	Above Poverty Level	1	1.405599
South	Male	Above Poverty Level	1	1.408469
Northeast	Female	Below Poverty Level	0.5	1.431367
Northeast	Female	Below Poverty Level	0.9	1.503674
Northeast	Female	Below Poverty Level	1	1.574778
Northeast	Female	Below Poverty Level	0.4	1.605

Appendix 5C, Attachment B, Table 5CB-7. Effect on residual standard error by varying LOESS smoothing parameter while fitting adults "EVER" having asthma data set.				
Region	**Gender**	**Poverty Ratio**	**Smoothing Parameter**	**Residual Standard Error**
Midwest	Female	Above Poverty Level	1	0.983356
South	Female	Below Poverty Level	1	1.040607
West	Female	Below Poverty Level	0.9	1.044712
West	Male	Above Poverty Level	0.8	0.937658
South	Female	Above Poverty Level	1	1.06598
Midwest	Female	Above Poverty Level	0.9	0.911278
West	Male	Below Poverty Level	0.8	1.095844
West	Female	Below Poverty Level	0.8	0.893319
West	Male	Above Poverty Level	0.9	0.886119
Northeast	Female	Above Poverty Level	1	0.875056
West	Male	Above Poverty Level	1	0.858542
Midwest	Female	Above Poverty Level	0.8	0.843191
Northeast	Male	Above Poverty Level	0.8	1.177547
South	Male	Below Poverty Level	1	0.813689
Midwest	Male	Above Poverty Level	0.9	1.190978
Midwest	Male	Below Poverty Level	1	0.785268
South	Female	Above Poverty Level	0.9	0.77381
Northeast	Male	Above Poverty Level	1	1.241548
South	Female	Above Poverty Level	0.8	0.751726
South	Female	Below Poverty Level	0.9	0.747912
South	Female	Below Poverty Level	0.8	0.740577
Northeast	Male	Below Poverty Level	1	0.732859
West	Female	Below Poverty Level	1	1.275049
South	Male	Above Poverty Level	0.9	0.708509
South	Male	Above Poverty Level	1	0.706944
Northeast	Female	Above Poverty Level	0.9	0.699107
Northeast	Male	Above Poverty Level	0.9	1.301543
Northeast	Male	Below Poverty Level	0.9	0.677309
West	Female	Above Poverty Level	1	0.669638
Northeast	Female	Below Poverty Level	1	0.662619
Northeast	Male	Below Poverty Level	0.8	0.646318
South	Male	Below Poverty Level	0.9	0.64328
Midwest	Male	Above Poverty Level	1	1.395026
West	Female	Above Poverty Level	0.8	0.597305
South	Male	Below Poverty Level	0.8	0.58427
West	Female	Above Poverty Level	0.9	0.567466
Northeast	Female	Above Poverty Level	0.8	0.528031
Midwest	Male	Below Poverty Level	0.9	0.49517
West	Male	Below Poverty Level	0.9	1.523816
West	Male	Below Poverty Level	1	1.537805
South	Male	Above Poverty Level	0.8	0.400237
Northeast	Female	Below Poverty Level	0.9	0.394894
Northeast	Female	Below Poverty Level	0.8	0.362058
Midwest	Male	Below Poverty Level	0.8	0.306085
Midwest	Male	Above Poverty Level	0.8	0.169594
Midwest	Female	Below Poverty Level	1	1.910643
Midwest	Female	Below Poverty Level	0.9	1.920542
Midwest	Female	Below Poverty Level	0.8	2.249162

Appendix 5C, Attachment B, Table 5CB-8. Effect on residual standard error by varying LOESS smoothing parameter while fitting adults "STILL" having asthma data set.				
Region	**Gender**	**Poverty Ratio**	**Smoothing Parameter**	**Residual Standard Error**
South	Male	Below Poverty Level	0.8	1.015193
West	Female	Above Poverty Level	0.8	1.045714
West	Female	Above Poverty Level	0.9	1.051807
West	Female	Above Poverty Level	1	1.061488
West	Male	Above Poverty Level	1	0.92928
West	Male	Above Poverty Level	0.8	0.925921
West	Male	Above Poverty Level	0.9	0.915895
South	Female	Below Poverty Level	0.9	1.097531
Midwest	Female	Above Poverty Level	1	0.89825
Northeast	Female	Below Poverty Level	1	1.102905
Midwest	Female	Above Poverty Level	0.9	0.876146
South	Female	Below Poverty Level	0.8	1.128781
Midwest	Female	Above Poverty Level	0.8	0.870507
South	Female	Above Poverty Level	1	1.130393
South	Female	Above Poverty Level	0.9	0.835583
West	Female	Below Poverty Level	1	0.825684
South	Male	Below Poverty Level	0.9	1.192655
Midwest	Male	Below Poverty Level	1	0.788217
Northeast	Female	Below Poverty Level	0.9	0.786205
Northeast	Male	Above Poverty Level	1	1.21537
South	Female	Below Poverty Level	1	1.23752
South	Male	Above Poverty Level	0.9	0.748499
South	Male	Above Poverty Level	0.8	0.717121
West	Female	Below Poverty Level	0.9	0.670751
South	Male	Above Poverty Level	1	0.664236
Northeast	Female	Below Poverty Level	0.8	0.65848
Northeast	Female	Above Poverty Level	1	0.653985
Midwest	Male	Above Poverty Level	1	0.650735
Northeast	Female	Above Poverty Level	0.9	0.630298
Northeast	Male	Above Poverty Level	0.9	1.370134
Northeast	Male	Above Poverty Level	0.8	1.375365
Midwest	Male	Below Poverty Level	0.9	0.620174
South	Male	Below Poverty Level	1	1.400273
Northeast	Male	Below Poverty Level	1	0.581032
South	Female	Above Poverty Level	0.8	0.568428
Midwest	Male	Above Poverty Level	0.9	0.508247
Midwest	Male	Below Poverty Level	0.8	0.503315
Northeast	Female	Above Poverty Level	0.8	0.478186
West	Female	Below Poverty Level	0.8	0.464598
Northeast	Male	Below Poverty Level	0.9	0.453855
Northeast	Male	Below Poverty Level	0.8	0.396203
Midwest	Female	Below Poverty Level	1	1.616706
Midwest	Female	Below Poverty Level	0.9	1.636938
Midwest	Male	Above Poverty Level	0.8	0.295923
Midwest	Female	Below Poverty Level	0.8	1.883863
West	Male	Below Poverty Level	0.8	2.16547
West	Male	Below Poverty Level	1	2.200364
West	Male	Below Poverty Level	0.9	2.396381

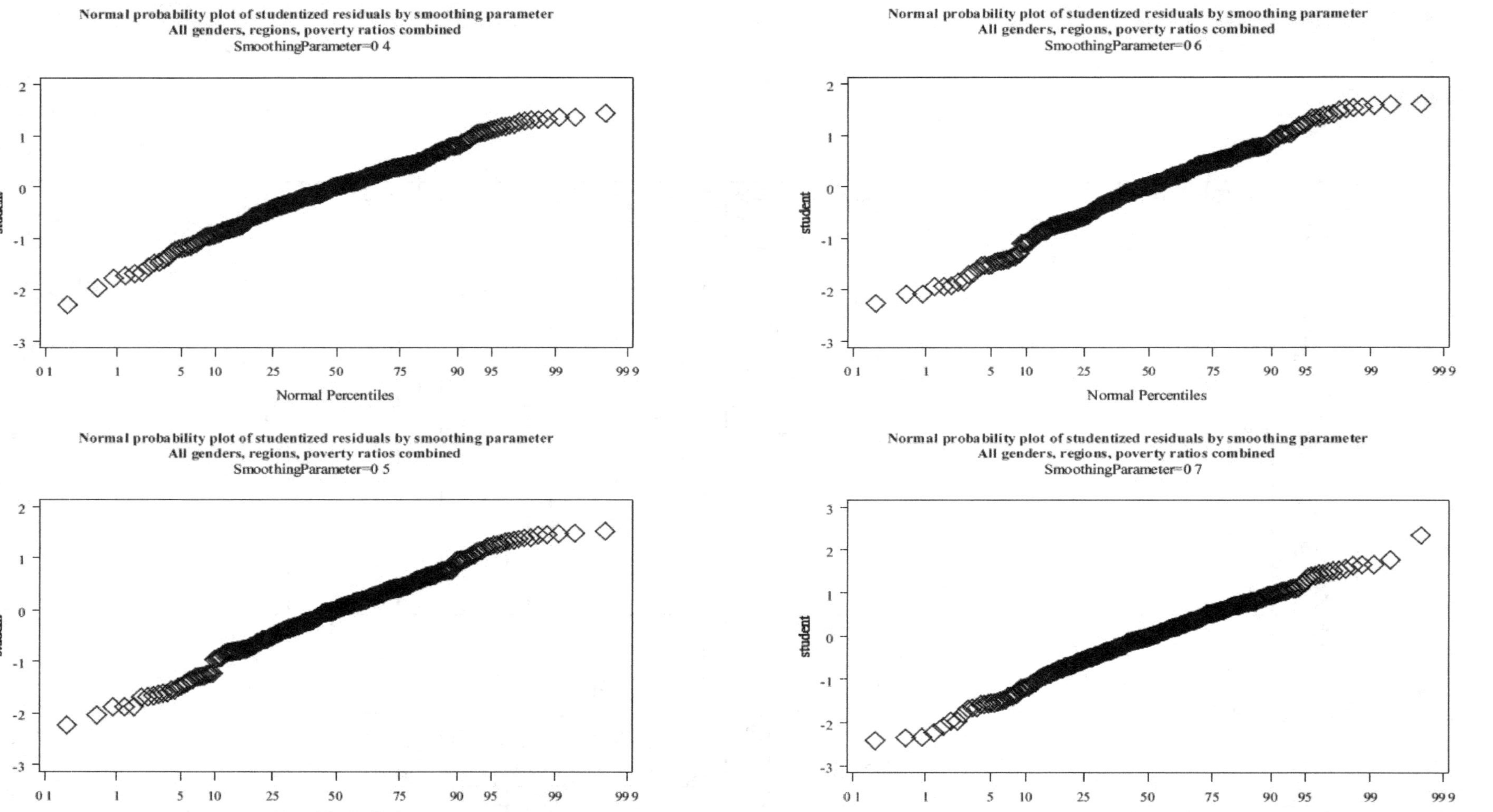

Appendix 5C, Attachment B, Figure 5CB-1. Normal probability plots of studentized residuals generated using logistic model and children 'EVER' asthmatic data set.

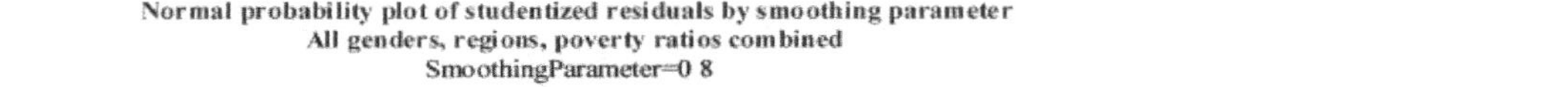

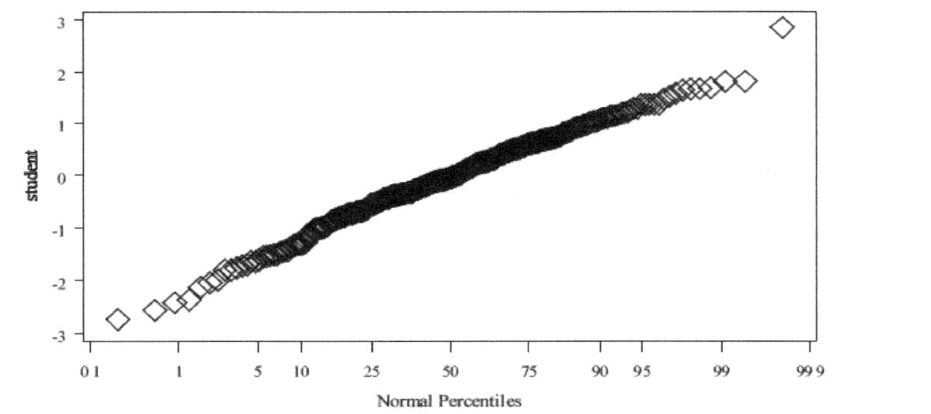

Normal probability plot of studentized residuals by smoothing parameter
All genders, regions, poverty ratios combined
SmoothingParameter=0 9

student

Normal Percentiles

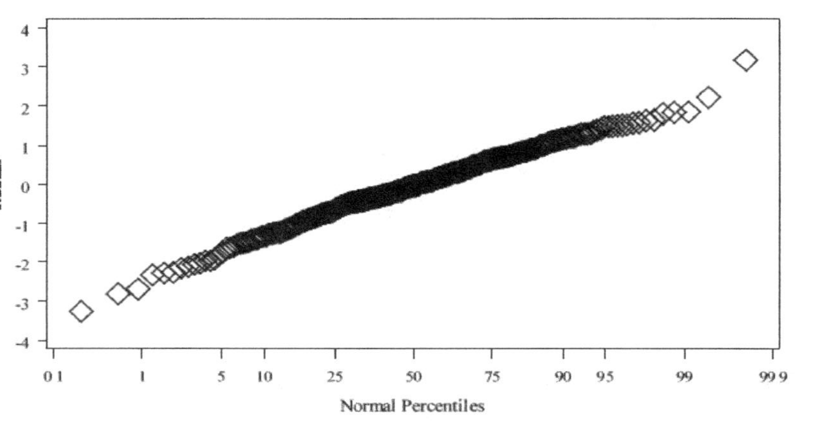

Appendix 5C, Attachment B, Figure 5CB-1, cont. Normal probability plots of studentized residuals generated using logistic model and children 'EVER' asthmatic data set.

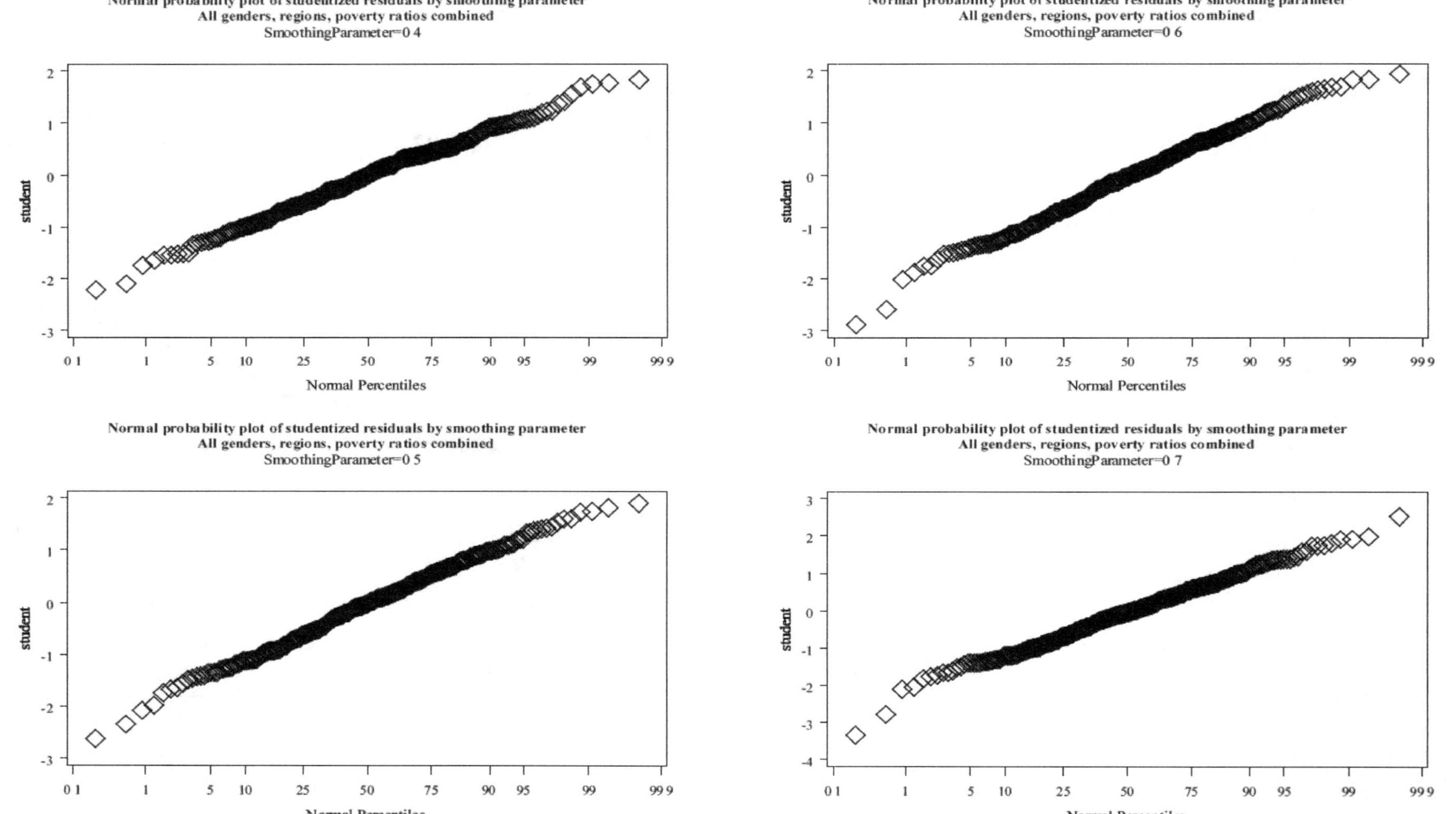

Appendix 5C, Attachment B, Figure 5CB-2. Normal probability plots of studentized residuals generated using logistic model and children 'STILL' asthmatic data set.

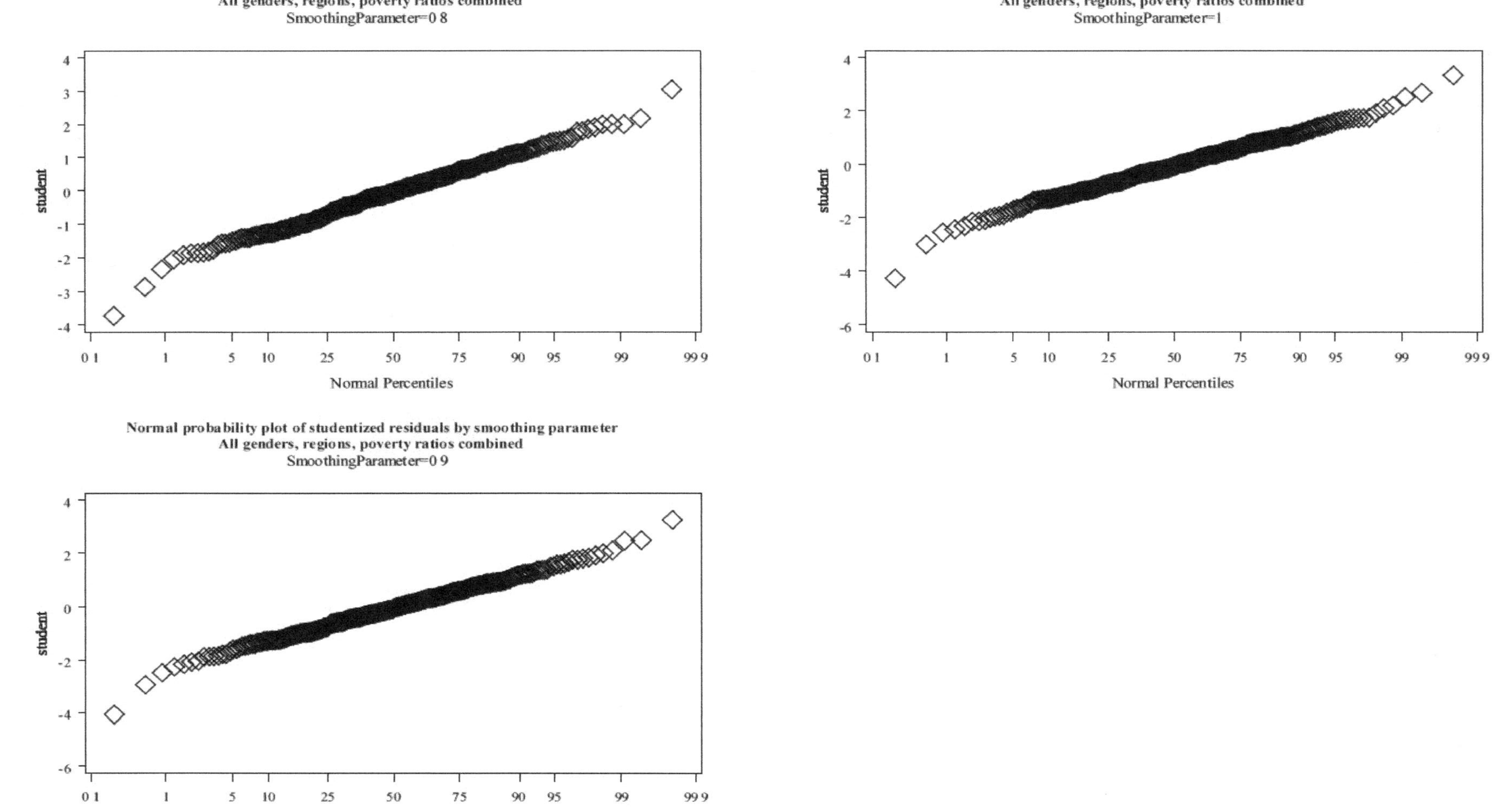

Appendix 5C, Attachment B, Figure 5CB-2, cont. Normal probability plots of studentized residuals generated using logistic model and children 'STILL' asthmatic data set.

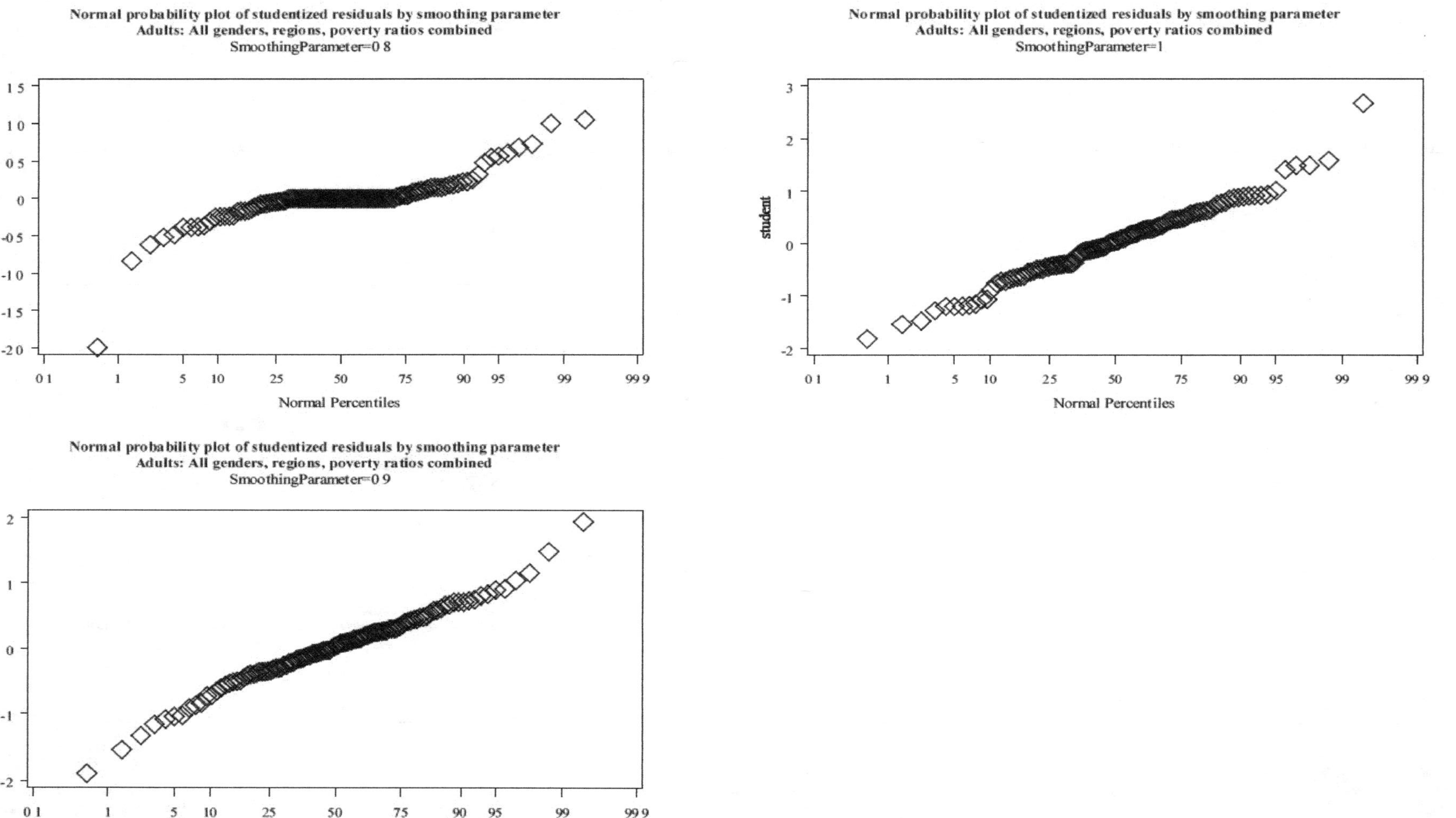

Appendix 5C, Attachment B, Figure 5CB-3. Normal probability plots of studentized residuals generated using logistic model and adult 'EVER' asthmatic data set.

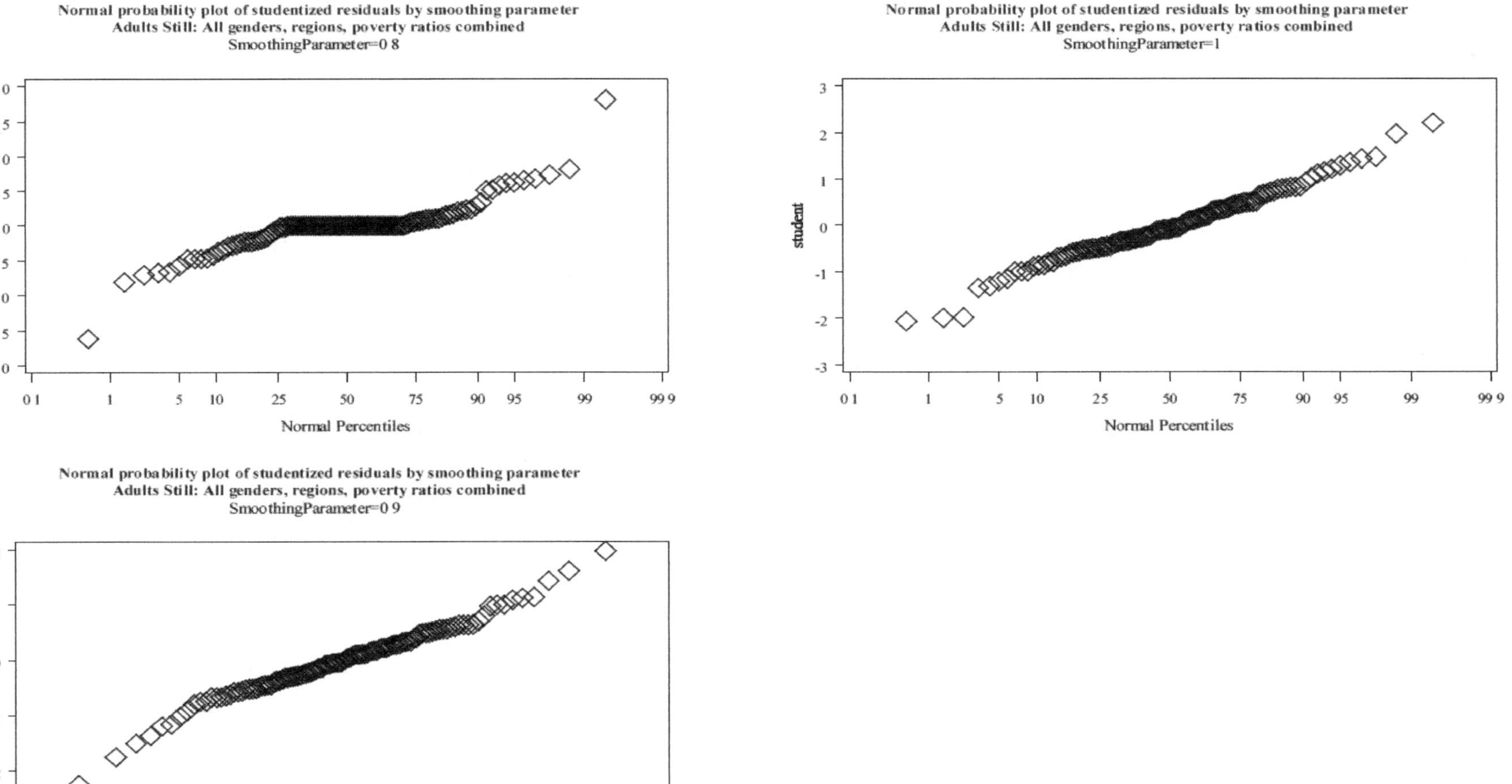

Appendix 5C, Attachment B, Figure 5CB-4. Normal probability plots of studentized residuals generated using logistic model and adult 'STILL' asthmatic data set.

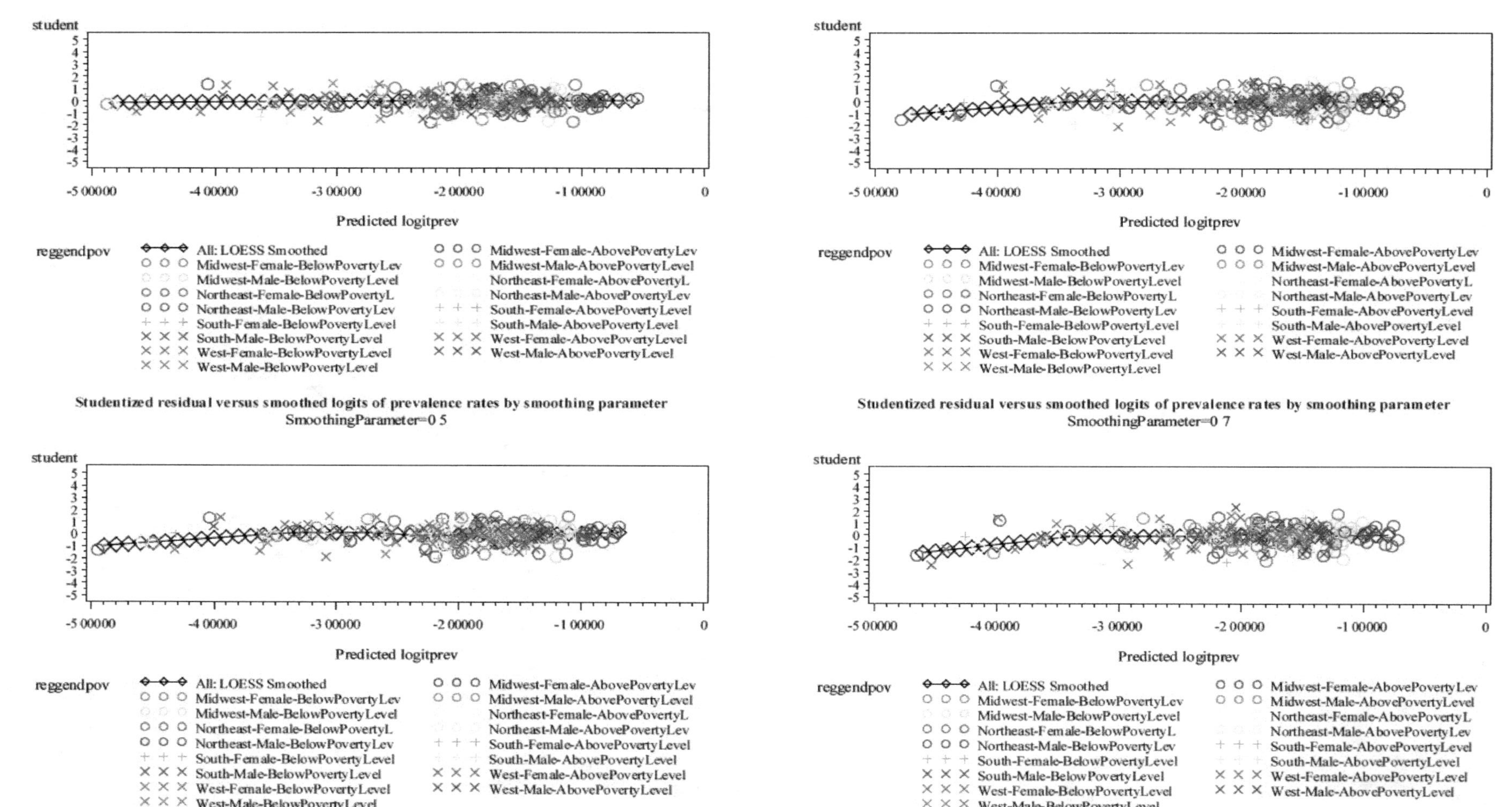

Appendix 5C, Attachment B, Figure 5CB-5. Studentized residuals generated using logistic model versus model predicted betas and the child 'EVER' asthmatic data set.

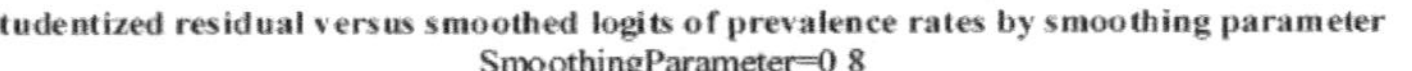

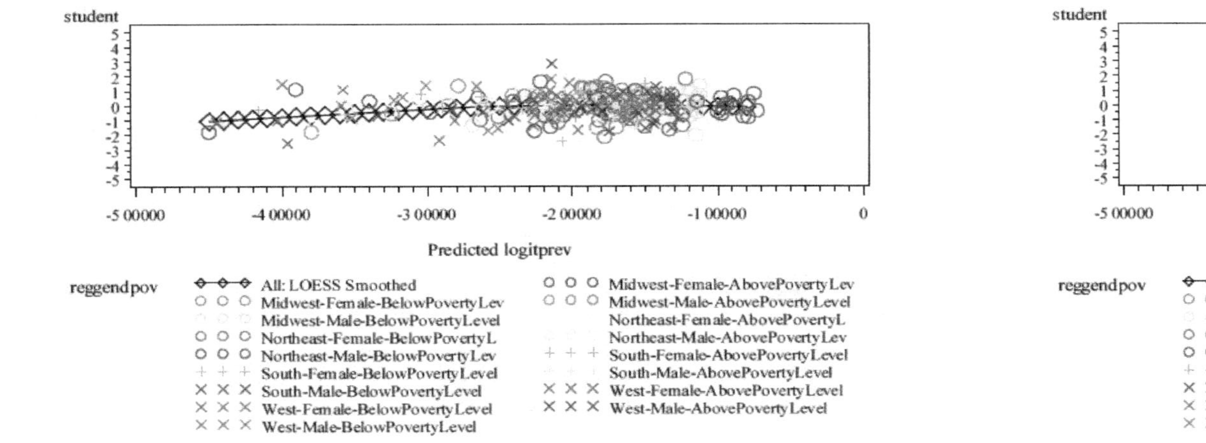

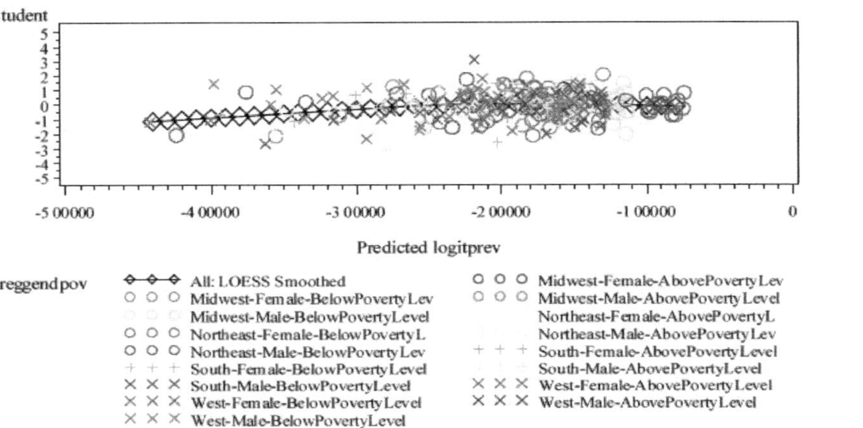

Studentized residual versus smoothed logits of prevalence rates by smoothing parameter
SmoothingParameter=1

student

-5 00000 -4 00000 -3 00000 -2 00000 -1 00000 0

Predicted logitprev

reggendpov
All: LOESS Smoothed
Midwest-Female-BelowPovertyLev
Midwest-Male-BelowPovertyLevel
Northeast-Female-BelowPovertyL
Northeast-Male-BelowPovertyLev
South-Female-BelowPovertyLevel
South-Male-BelowPovertyLevel
West-Female-BelowPovertyLevel
West-Male-BelowPovertyLevel
Midwest-Female-AbovePovertyLev
Midwest-Male-AbovePovertyLevel
Northeast-Female-AbovePovertyL
Northeast-Male-AbovePovertyLev
South-Female-AbovePovertyLevel
South-Male-AbovePovertyLevel
West-Female-AbovePovertyLevel
West-Male-AbovePovertyLevel

Appendix 5C, Attachment B, Figure 5CB-5, cont. Studentized residuals generated using logistic model versus model predicted betas and the child 'EVER' asthmatic data set.

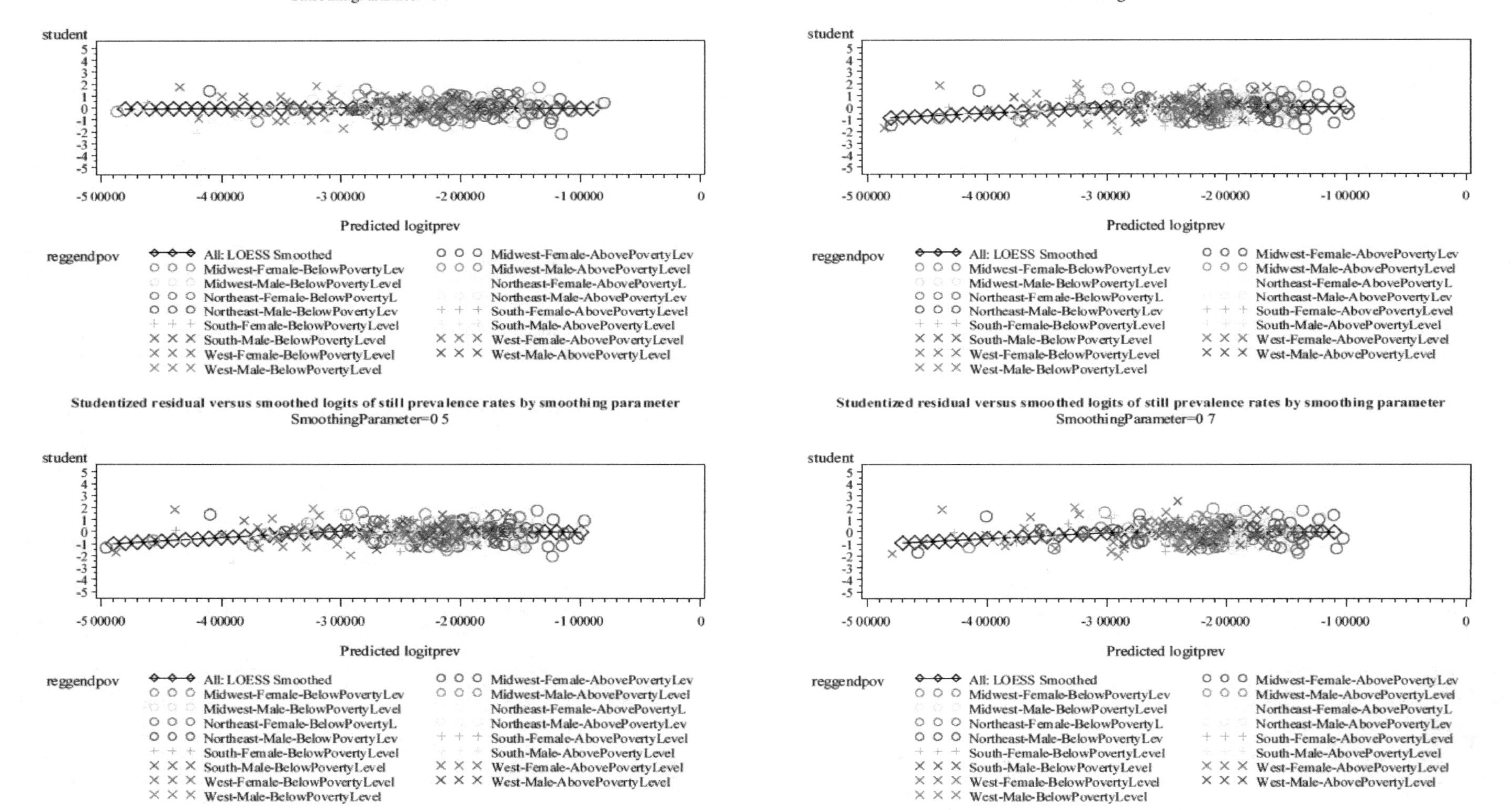

Appendix 5C, Attachment B, Figure 5CB-6. Studentized residuals generated using logistic model versus model predicted betas and the child 'STILL' asthmatic data set.

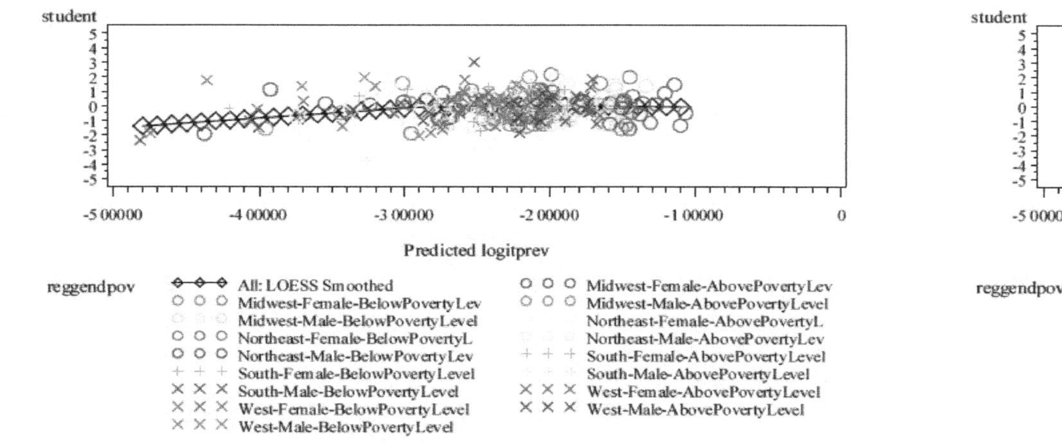

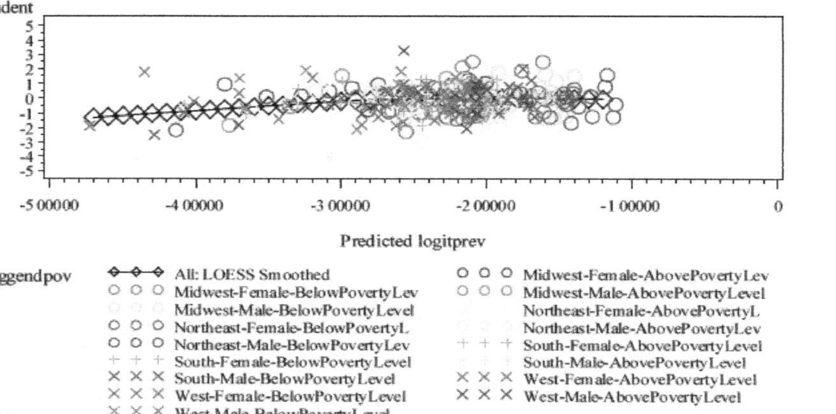

Appendix 5C, Attachment B, Figure 5CB-6, cont. Studentized residuals generated using logistic model versus model predicted betas using child 'STILL' asthmatic data set.

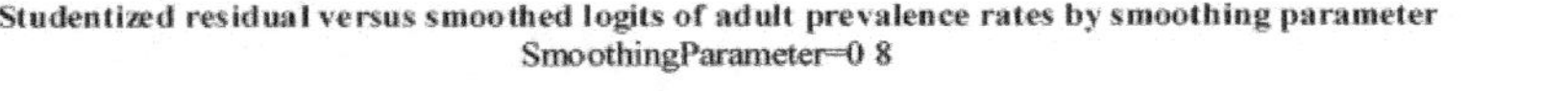

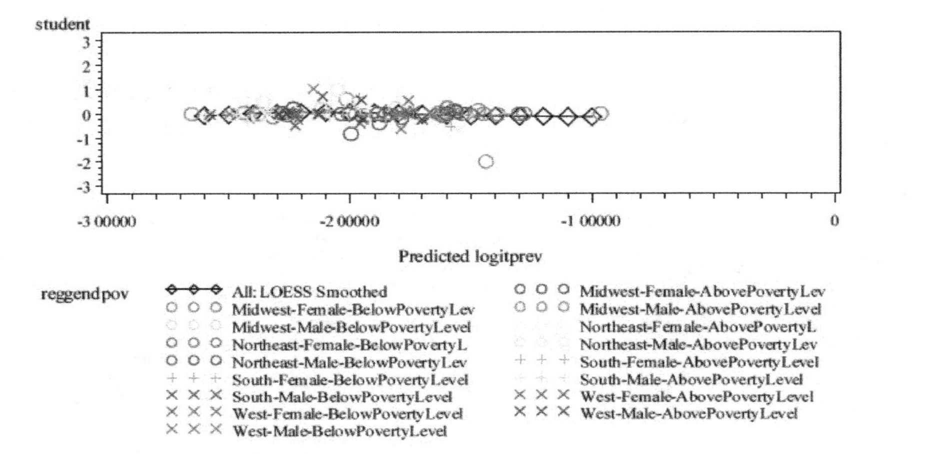

Studentized residual versus smoothed logits of adult prevalence rates by smoothing parameter
SmoothingParameter=1

student
3
2
1
0
-1
-2
-3
-3 00000
-2 00000
-1 00000
0
Predicted logitprev
reggendpov
All: LOESS Smoothed
Midwest-Female-BelowPovertyLev
Midwest-Male-BelowPovertyLevel
Northeast-Female-BelowPovertyL
Northeast-Male-BelowPovertyLev
South-Female-BelowPovertyLevel
South-Male-BelowPovertyLevel
West-Female-BelowPovertyLevel
West-Male-BelowPovertyLevel
Midwest-Female-AbovePovertyLev
Midwest-Male-AbovePovertyLevel
Northeast-Female-AbovePovertyL
Northeast-Male-AbovePovertyLev
South-Female-AbovePovertyLevel
South-Male-AbovePovertyLevel
West-Female-AbovePovertyLevel
West-Male-AbovePovertyLevel

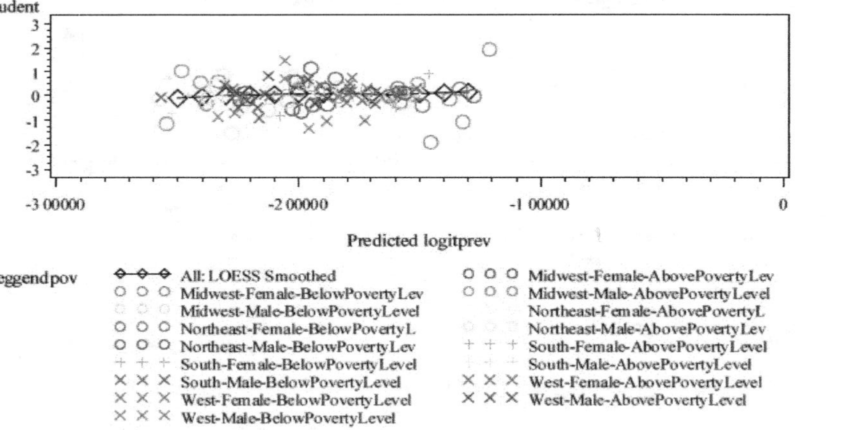

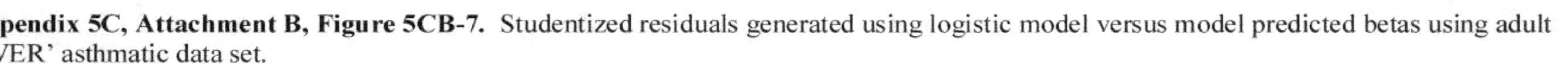

Appendix 5C, Attachment B, Figure 5CB-7. Studentized residuals generated using logistic model versus model predicted betas using adult 'EVER' asthmatic data set.

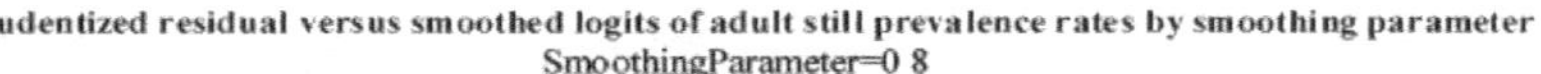

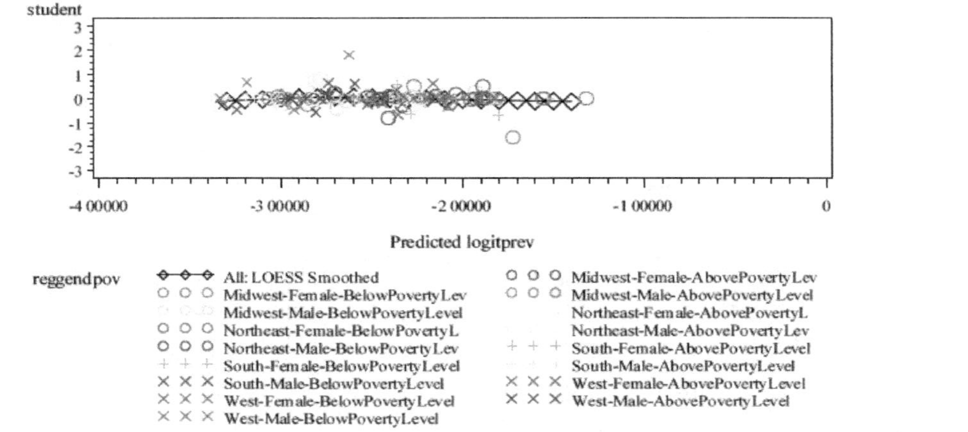

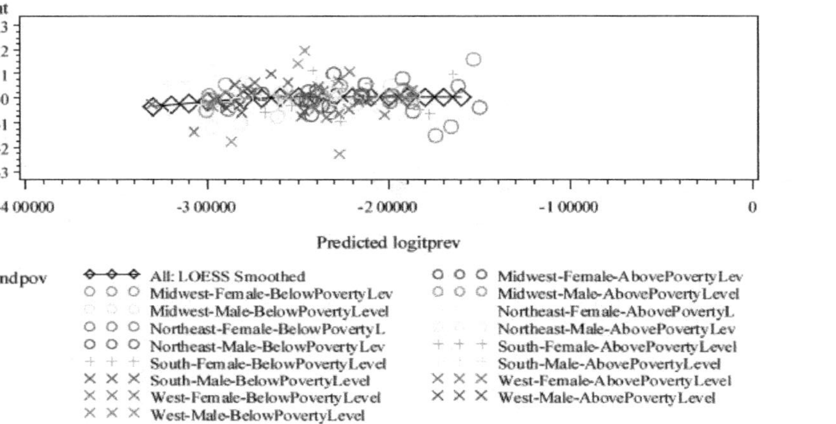

Studentized residual versus smoothed logits of adult still prevalence rates by smoothing parameter
SmoothingParameter=1

student
3
2
1
0
-1
-2
-3
-4 00000
-3 00000
-2 00000
-1 00000
0
Predicted logitprev
reggendpov
All: LOESS Smoothed
Midwest-Female-BelowPovertyLev
Midwest-Male-BelowPovertyLevel
Northeast-Female-BelowPovertyL
Northeast-Male-BelowPovertyLev
South-Female-BelowPovertyLevel
South-Male-BelowPovertyLevel
West-Female-BelowPovertyLevel
West-Male-BelowPovertyLevel
Midwest-Female-AbovePovertyLev
Midwest-Male-AbovePovertyLevel
Northeast-Female-AbovePovertyL
Northeast-Male-AbovePovertyLev
South-Female-AbovePovertyLevel
South-Male-AbovePovertyLevel
West-Female-AbovePovertyLevel
West-Male-AbovePovertyLevel

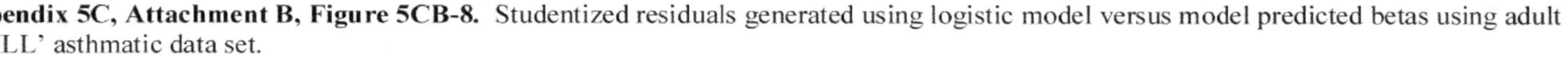

Appendix 5C, Attachment B, Figure 5CB-8. Studentized residuals generated using logistic model versus model predicted betas using adult 'STILL' asthmatic data set.

Appendix 5C, Attachment C
Smoothed Asthma Prevalence Tables and Figures

Appendix 5C, Attachment C, Table 5CC-1. Smoothed prevalence for children "EVER" having asthma.								
Smoothed	**Region**	**Gender**	**Poverty Status**	**Age**	**Prevalence**	**SE**	**LowerCI**	**UpperCI**
Yes	Midwest	Female	Above Poverty Level	0	0.0083	0.0050	0.0022	0.0310
Yes	Midwest	Female	Above Poverty Level	1	0.0179	0.0066	0.0079	0.0397
Yes	Midwest	Female	Above Poverty Level	2	0.0327	0.0076	0.0195	0.0541
Yes	Midwest	Female	Above Poverty Level	3	0.0509	0.0096	0.0336	0.0766
Yes	Midwest	Female	Above Poverty Level	4	0.0671	0.0122	0.0448	0.0993
Yes	Midwest	Female	Above Poverty Level	5	0.0854	0.0134	0.0602	0.1198
Yes	Midwest	Female	Above Poverty Level	6	0.0995	0.0141	0.0725	0.1351
Yes	Midwest	Female	Above Poverty Level	7	0.1041	0.0145	0.0765	0.1403
Yes	Midwest	Female	Above Poverty Level	8	0.1024	0.0132	0.0769	0.1352
Yes	Midwest	Female	Above Poverty Level	9	0.1020	0.0121	0.0784	0.1317
Yes	Midwest	Female	Above Poverty Level	10	0.1055	0.0127	0.0806	0.1369
Yes	Midwest	Female	Above Poverty Level	11	0.1192	0.0137	0.0922	0.1527
Yes	Midwest	Female	Above Poverty Level	12	0.1390	0.0163	0.1070	0.1787
Yes	Midwest	Female	Above Poverty Level	13	0.1529	0.0176	0.1182	0.1956
Yes	Midwest	Female	Above Poverty Level	14	0.1603	0.0176	0.1254	0.2026
Yes	Midwest	Female	Above Poverty Level	15	0.1597	0.0160	0.1277	0.1979
Yes	Midwest	Female	Above Poverty Level	16	0.1517	0.0161	0.1197	0.1903
Yes	Midwest	Female	Above Poverty Level	17	0.1374	0.0229	0.0945	0.1956
Yes	Midwest	Female	Below Poverty Level	0	0.0413	0.0168	0.0167	0.0985
Yes	Midwest	Female	Below Poverty Level	1	0.0706	0.0168	0.0416	0.1174
Yes	Midwest	Female	Below Poverty Level	2	0.1047	0.0173	0.0724	0.1491
Yes	Midwest	Female	Below Poverty Level	3	0.1356	0.0208	0.0962	0.1879
Yes	Midwest	Female	Below Poverty Level	4	0.1553	0.0237	0.1100	0.2146
Yes	Midwest	Female	Below Poverty Level	5	0.1488	0.0229	0.1053	0.2062
Yes	Midwest	Female	Below Poverty Level	6	0.1327	0.0228	0.0902	0.1910
Yes	Midwest	Female	Below Poverty Level	7	0.1341	0.0224	0.0920	0.1912
Yes	Midwest	Female	Below Poverty Level	8	0.1535	0.0239	0.1080	0.2136
Yes	Midwest	Female	Below Poverty Level	9	0.1729	0.0270	0.1215	0.2401
Yes	Midwest	Female	Below Poverty Level	10	0.1861	0.0311	0.1272	0.2640
Yes	Midwest	Female	Below Poverty Level	11	0.1691	0.0300	0.1131	0.2451
Yes	Midwest	Female	Below Poverty Level	12	0.1470	0.0247	0.1006	0.2097
Yes	Midwest	Female	Below Poverty Level	13	0.1439	0.0239	0.0990	0.2045
Yes	Midwest	Female	Below Poverty Level	14	0.1541	0.0244	0.1078	0.2156
Yes	Midwest	Female	Below Poverty Level	15	0.1707	0.0275	0.1186	0.2395
Yes	Midwest	Female	Below Poverty Level	16	0.1962	0.0427	0.1187	0.3065
Yes	Midwest	Female	Below Poverty Level	17	0.2323	0.0813	0.1002	0.4512
Yes	Midwest	Male	Above Poverty Level	0	0.0133	0.0066	0.0045	0.0391
Yes	Midwest	Male	Above Poverty Level	1	0.0313	0.0091	0.0164	0.0588
Yes	Midwest	Male	Above Poverty Level	2	0.0585	0.0102	0.0398	0.0851
Yes	Midwest	Male	Above Poverty Level	3	0.0898	0.0121	0.0666	0.1200
Yes	Midwest	Male	Above Poverty Level	4	0.1111	0.0145	0.0831	0.1471
Yes	Midwest	Male	Above Poverty Level	5	0.1256	0.0149	0.0964	0.1621
Yes	Midwest	Male	Above Poverty Level	6	0.1411	0.0158	0.1100	0.1793
Yes	Midwest	Male	Above Poverty Level	7	0.1496	0.0164	0.1171	0.1892
Yes	Midwest	Male	Above Poverty Level	8	0.1502	0.0161	0.1182	0.1891
Yes	Midwest	Male	Above Poverty Level	9	0.1542	0.0166	0.1211	0.1942
Yes	Midwest	Male	Above Poverty Level	10	0.1627	0.0173	0.1283	0.2041
Yes	Midwest	Male	Above Poverty Level	11	0.1760	0.0181	0.1397	0.2193
Yes	Midwest	Male	Above Poverty Level	12	0.1876	0.0186	0.1501	0.2319
Yes	Midwest	Male	Above Poverty Level	13	0.1847	0.0181	0.1483	0.2277
Yes	Midwest	Male	Above Poverty Level	14	0.1764	0.0170	0.1422	0.2167
Yes	Midwest	Male	Above Poverty Level	15	0.1641	0.0149	0.1341	0.1994
Yes	Midwest	Male	Above Poverty Level	16	0.1487	0.0144	0.1198	0.1833
Yes	Midwest	Male	Above Poverty Level	17	0.1318	0.0201	0.0937	0.1823
Yes	Midwest	Male	Below Poverty Level	0	0.0429	0.0176	0.0173	0.1026
Yes	Midwest	Male	Below Poverty Level	1	0.0908	0.0214	0.0536	0.1498
Yes	Midwest	Male	Below Poverty Level	2	0.1530	0.0235	0.1084	0.2118

Appendix 5C, Attachment C, Table 5CC-1. Smoothed prevalence for children "EVER" having asthma.								
Smoothed	Region	Gender	Poverty Status	Age	Prevalence	SE	LowerCI	UpperCI
Yes	Midwest	Male	Below Poverty Level	3	0.2110	0.0277	0.1566	0.2780
Yes	Midwest	Male	Below Poverty Level	4	0.2428	0.0303	0.1828	0.3150
Yes	Midwest	Male	Below Poverty Level	5	0.2458	0.0285	0.1888	0.3133
Yes	Midwest	Male	Below Poverty Level	6	0.2393	0.0270	0.1853	0.3033
Yes	Midwest	Male	Below Poverty Level	7	0.2261	0.0268	0.1729	0.2900
Yes	Midwest	Male	Below Poverty Level	8	0.2225	0.0290	0.1655	0.2924
Yes	Midwest	Male	Below Poverty Level	9	0.2354	0.0311	0.1741	0.3101
Yes	Midwest	Male	Below Poverty Level	10	0.2499	0.0339	0.1831	0.3311
Yes	Midwest	Male	Below Poverty Level	11	0.2553	0.0357	0.1852	0.3409
Yes	Midwest	Male	Below Poverty Level	12	0.2512	0.0377	0.1779	0.3423
Yes	Midwest	Male	Below Poverty Level	13	0.2149	0.0355	0.1473	0.3025
Yes	Midwest	Male	Below Poverty Level	14	0.1941	0.0308	0.1353	0.2703
Yes	Midwest	Male	Below Poverty Level	15	0.2027	0.0292	0.1462	0.2741
Yes	Midwest	Male	Below Poverty Level	16	0.2364	0.0390	0.1617	0.3320
Yes	Midwest	Male	Below Poverty Level	17	0.3045	0.0768	0.1652	0.4921
Yes	Northeast	Female	Above Poverty Level	0	0.0115	0.0066	0.0032	0.0402
Yes	Northeast	Female	Above Poverty Level	1	0.0278	0.0095	0.0131	0.0583
Yes	Northeast	Female	Above Poverty Level	2	0.0533	0.0108	0.0340	0.0827
Yes	Northeast	Female	Above Poverty Level	3	0.0823	0.0127	0.0584	0.1150
Yes	Northeast	Female	Above Poverty Level	4	0.1027	0.0152	0.0737	0.1413
Yes	Northeast	Female	Above Poverty Level	5	0.1066	0.0150	0.0777	0.1445
Yes	Northeast	Female	Above Poverty Level	6	0.1023	0.0143	0.0749	0.1383
Yes	Northeast	Female	Above Poverty Level	7	0.0979	0.0137	0.0715	0.1325
Yes	Northeast	Female	Above Poverty Level	8	0.1010	0.0144	0.0734	0.1375
Yes	Northeast	Female	Above Poverty Level	9	0.1146	0.0166	0.0828	0.1566
Yes	Northeast	Female	Above Poverty Level	10	0.1179	0.0171	0.0852	0.1611
Yes	Northeast	Female	Above Poverty Level	11	0.1170	0.0175	0.0836	0.1615
Yes	Northeast	Female	Above Poverty Level	12	0.1154	0.0164	0.0838	0.1568
Yes	Northeast	Female	Above Poverty Level	13	0.1246	0.0148	0.0955	0.1611
Yes	Northeast	Female	Above Poverty Level	14	0.1405	0.0148	0.1109	0.1765
Yes	Northeast	Female	Above Poverty Level	15	0.1551	0.0152	0.1245	0.1916
Yes	Northeast	Female	Above Poverty Level	16	0.1714	0.0209	0.1302	0.2223
Yes	Northeast	Female	Above Poverty Level	17	0.1883	0.0376	0.1189	0.2851
Yes	Northeast	Female	Below Poverty Level	0	0.0394	0.0211	0.0119	0.1222
Yes	Northeast	Female	Below Poverty Level	1	0.0754	0.0229	0.0383	0.1433
Yes	Northeast	Female	Below Poverty Level	2	0.1188	0.0229	0.0770	0.1789
Yes	Northeast	Female	Below Poverty Level	3	0.1539	0.0265	0.1043	0.2214
Yes	Northeast	Female	Below Poverty Level	4	0.1684	0.0295	0.1131	0.2432
Yes	Northeast	Female	Below Poverty Level	5	0.1503	0.0269	0.1003	0.2193
Yes	Northeast	Female	Below Poverty Level	6	0.1355	0.0245	0.0902	0.1987
Yes	Northeast	Female	Below Poverty Level	7	0.1263	0.0231	0.0836	0.1862
Yes	Northeast	Female	Below Poverty Level	8	0.1322	0.0257	0.0853	0.1993
Yes	Northeast	Female	Below Poverty Level	9	0.1583	0.0301	0.1029	0.2358
Yes	Northeast	Female	Below Poverty Level	10	0.1818	0.0342	0.1183	0.2689
Yes	Northeast	Female	Below Poverty Level	11	0.2030	0.0358	0.1355	0.2926
Yes	Northeast	Female	Below Poverty Level	12	0.2293	0.0359	0.1600	0.3172
Yes	Northeast	Female	Below Poverty Level	13	0.2437	0.0366	0.1726	0.3323
Yes	Northeast	Female	Below Poverty Level	14	0.2368	0.0335	0.1713	0.3179
Yes	Northeast	Female	Below Poverty Level	15	0.2188	0.0286	0.1625	0.2879
Yes	Northeast	Female	Below Poverty Level	16	0.1906	0.0298	0.1335	0.2645
Yes	Northeast	Female	Below Poverty Level	17	0.1572	0.0443	0.0822	0.2796
Yes	Northeast	Male	Above Poverty Level	0	0.0279	0.0107	0.0119	0.0639
Yes	Northeast	Male	Above Poverty Level	1	0.0444	0.0103	0.0265	0.0733
Yes	Northeast	Male	Above Poverty Level	2	0.0668	0.0106	0.0470	0.0940
Yes	Northeast	Male	Above Poverty Level	3	0.0948	0.0134	0.0692	0.1284
Yes	Northeast	Male	Above Poverty Level	4	0.1269	0.0174	0.0933	0.1702
Yes	Northeast	Male	Above Poverty Level	5	0.1665	0.0209	0.1257	0.2173
Yes	Northeast	Male	Above Poverty Level	6	0.1891	0.0207	0.1478	0.2387
Yes	Northeast	Male	Above Poverty Level	7	0.1901	0.0204	0.1494	0.2389
Yes	Northeast	Male	Above Poverty Level	8	0.1858	0.0189	0.1479	0.2307
Yes	Northeast	Male	Above Poverty Level	9	0.1873	0.0189	0.1494	0.2322
Yes	Northeast	Male	Above Poverty Level	10	0.1908	0.0180	0.1545	0.2333
Yes	Northeast	Male	Above Poverty Level	11	0.1926	0.0163	0.1595	0.2307
Yes	Northeast	Male	Above Poverty Level	12	0.1934	0.0168	0.1592	0.2329

Appendix 5C, Attachment C, Table 5CC-1. Smoothed prevalence for children "EVER" having asthma.								
Smoothed	Region	Gender	Poverty Status	Age	Prevalence	SE	LowerCl	UpperCl
Yes	Northeast	Male	Above Poverty Level	13	0.1847	0.0172	0.1499	0.2253
Yes	Northeast	Male	Above Poverty Level	14	0.1797	0.0168	0.1458	0.2195
Yes	Northeast	Male	Above Poverty Level	15	0.1781	0.0156	0.1465	0.2149
Yes	Northeast	Male	Above Poverty Level	16	0.1795	0.0162	0.1467	0.2178
Yes	Northeast	Male	Above Poverty Level	17	0.1838	0.0251	0.1350	0.2452
Yes	Northeast	Male	Below Poverty Level	0	0.0946	0.0396	0.0365	0.2240
Yes	Northeast	Male	Below Poverty Level	1	0.1345	0.0296	0.0817	0.2134
Yes	Northeast	Male	Below Poverty Level	2	0.1759	0.0264	0.1251	0.2416
Yes	Northeast	Male	Below Poverty Level	3	0.2132	0.0326	0.1503	0.2932
Yes	Northeast	Male	Below Poverty Level	4	0.2353	0.0361	0.1653	0.3236
Yes	Northeast	Male	Below Poverty Level	5	0.2638	0.0316	0.2004	0.3388
Yes	Northeast	Male	Below Poverty Level	6	0.2909	0.0305	0.2287	0.3621
Yes	Northeast	Male	Below Poverty Level	7	0.3169	0.0339	0.2475	0.3954
Yes	Northeast	Male	Below Poverty Level	8	0.3272	0.0405	0.2451	0.4214
Yes	Northeast	Male	Below Poverty Level	9	0.3238	0.0439	0.2356	0.4265
Yes	Northeast	Male	Below Poverty Level	10	0.3163	0.0429	0.2304	0.4169
Yes	Northeast	Male	Below Poverty Level	11	0.3022	0.0412	0.2199	0.3995
Yes	Northeast	Male	Below Poverty Level	12	0.2846	0.0388	0.2074	0.3769
Yes	Northeast	Male	Below Poverty Level	13	0.2779	0.0367	0.2048	0.3651
Yes	Northeast	Male	Below Poverty Level	14	0.2702	0.0343	0.2016	0.3518
Yes	Northeast	Male	Below Poverty Level	15	0.2698	0.0316	0.2062	0.3445
Yes	Northeast	Male	Below Poverty Level	16	0.2745	0.0349	0.2048	0.3573
Yes	Northeast	Male	Below Poverty Level	17	0.2843	0.0575	0.1760	0.4250
Yes	South	Female	Above Poverty Level	0	0.0137	0.0056	0.0056	0.0334
Yes	South	Female	Above Poverty Level	1	0.0266	0.0064	0.0156	0.0450
Yes	South	Female	Above Poverty Level	2	0.0453	0.0068	0.0325	0.0629
Yes	South	Female	Above Poverty Level	3	0.0687	0.0086	0.0522	0.0901
Yes	South	Female	Above Poverty Level	4	0.0928	0.0112	0.0710	0.1203
Yes	South	Female	Above Poverty Level	5	0.1142	0.0123	0.0900	0.1439
Yes	South	Female	Above Poverty Level	6	0.1298	0.0128	0.1042	0.1605
Yes	South	Female	Above Poverty Level	7	0.1333	0.0123	0.1085	0.1627
Yes	South	Female	Above Poverty Level	8	0.1231	0.0117	0.0996	0.1512
Yes	South	Female	Above Poverty Level	9	0.1095	0.0109	0.0877	0.1359
Yes	South	Female	Above Poverty Level	10	0.1033	0.0102	0.0830	0.1279
Yes	South	Female	Above Poverty Level	11	0.1086	0.0103	0.0881	0.1332
Yes	South	Female	Above Poverty Level	12	0.1212	0.0110	0.0991	0.1475
Yes	South	Female	Above Poverty Level	13	0.1368	0.0113	0.1138	0.1635
Yes	South	Female	Above Poverty Level	14	0.1437	0.0111	0.1210	0.1699
Yes	South	Female	Above Poverty Level	15	0.1448	0.0104	0.1235	0.1690
Yes	South	Female	Above Poverty Level	16	0.1395	0.0113	0.1166	0.1661
Yes	South	Female	Above Poverty Level	17	0.1283	0.0172	0.0952	0.1709
Yes	South	Female	Below Poverty Level	0	0.0496	0.0153	0.0250	0.0962
Yes	South	Female	Below Poverty Level	1	0.0682	0.0123	0.0458	0.1004
Yes	South	Female	Below Poverty Level	2	0.0893	0.0116	0.0670	0.1181
Yes	South	Female	Below Poverty Level	3	0.1111	0.0141	0.0838	0.1459
Yes	South	Female	Below Poverty Level	4	0.1319	0.0171	0.0987	0.1740
Yes	South	Female	Below Poverty Level	5	0.1473	0.0181	0.1120	0.1914
Yes	South	Female	Below Poverty Level	6	0.1553	0.0183	0.1193	0.1997
Yes	South	Female	Below Poverty Level	7	0.1592	0.0183	0.1231	0.2035
Yes	South	Female	Below Poverty Level	8	0.1650	0.0188	0.1277	0.2104
Yes	South	Female	Below Poverty Level	9	0.1766	0.0198	0.1374	0.2241
Yes	South	Female	Below Poverty Level	10	0.1825	0.0216	0.1398	0.2347
Yes	South	Female	Below Poverty Level	11	0.1805	0.0219	0.1373	0.2336
Yes	South	Female	Below Poverty Level	12	0.1837	0.0221	0.1401	0.2371
Yes	South	Female	Below Poverty Level	13	0.1932	0.0218	0.1499	0.2453
Yes	South	Female	Below Poverty Level	14	0.1891	0.0202	0.1487	0.2374
Yes	South	Female	Below Poverty Level	15	0.1760	0.0181	0.1398	0.2192
Yes	South	Female	Below Poverty Level	16	0.1560	0.0195	0.1178	0.2037
Yes	South	Female	Below Poverty Level	17	0.1298	0.0271	0.0810	0.2015
Yes	South	Male	Above Poverty Level	0	0.0335	0.0089	0.0186	0.0596
Yes	South	Male	Above Poverty Level	1	0.0629	0.0093	0.0453	0.0867
Yes	South	Male	Above Poverty Level	2	0.0985	0.0094	0.0797	0.1212
Yes	South	Male	Above Poverty Level	3	0.1306	0.0116	0.1073	0.1581
Yes	South	Male	Above Poverty Level	4	0.1472	0.0133	0.1204	0.1787

Appendix 5C, Attachment C, Table 5CC-1. Smoothed prevalence for children "EVER" having asthma.								
Smoothed	Region	Gender	Poverty Status	Age	Prevalence	SE	LowerCl	UpperCl
Yes	South	Male	Above Poverty Level	5	0.1523	0.0130	0.1259	0.1831
Yes	South	Male	Above Poverty Level	6	0.1539	0.0128	0.1278	0.1842
Yes	South	Male	Above Poverty Level	7	0.1485	0.0125	0.1231	0.1782
Yes	South	Male	Above Poverty Level	8	0.1461	0.0123	0.1212	0.1752
Yes	South	Male	Above Poverty Level	9	0.1517	0.0124	0.1265	0.1810
Yes	South	Male	Above Poverty Level	10	0.1639	0.0129	0.1375	0.1943
Yes	South	Male	Above Poverty Level	11	0.1772	0.0134	0.1496	0.2085
Yes	South	Male	Above Poverty Level	12	0.1794	0.0128	0.1530	0.2093
Yes	South	Male	Above Poverty Level	13	0.1752	0.0127	0.1491	0.2049
Yes	South	Male	Above Poverty Level	14	0.1705	0.0120	0.1458	0.1984
Yes	South	Male	Above Poverty Level	15	0.1652	0.0108	0.1428	0.1902
Yes	South	Male	Above Poverty Level	16	0.1600	0.0118	0.1358	0.1876
Yes	South	Male	Above Poverty Level	17	0.1562	0.0190	0.1189	0.2026
Yes	South	Male	Below Poverty Level	0	0.0629	0.0140	0.0383	0.1016
Yes	South	Male	Below Poverty Level	1	0.0922	0.0118	0.0694	0.1215
Yes	South	Male	Below Poverty Level	2	0.1253	0.0123	0.1008	0.1547
Yes	South	Male	Below Poverty Level	3	0.1578	0.0156	0.1265	0.1951
Yes	South	Male	Below Poverty Level	4	0.1852	0.0186	0.1479	0.2294
Yes	South	Male	Below Poverty Level	5	0.1975	0.0190	0.1592	0.2424
Yes	South	Male	Below Poverty Level	6	0.2038	0.0198	0.1639	0.2506
Yes	South	Male	Below Poverty Level	7	0.2087	0.0204	0.1675	0.2570
Yes	South	Male	Below Poverty Level	8	0.2078	0.0203	0.1669	0.2558
Yes	South	Male	Below Poverty Level	9	0.2080	0.0206	0.1664	0.2567
Yes	South	Male	Below Poverty Level	10	0.2122	0.0203	0.1711	0.2601
Yes	South	Male	Below Poverty Level	11	0.2137	0.0202	0.1727	0.2612
Yes	South	Male	Below Poverty Level	12	0.2192	0.0214	0.1759	0.2698
Yes	South	Male	Below Poverty Level	13	0.2199	0.0220	0.1755	0.2718
Yes	South	Male	Below Poverty Level	14	0.2059	0.0209	0.1639	0.2554
Yes	South	Male	Below Poverty Level	15	0.1946	0.0186	0.1571	0.2385
Yes	South	Male	Below Poverty Level	16	0.1827	0.0177	0.1471	0.2246
Yes	South	Male	Below Poverty Level	17	0.1709	0.0246	0.1235	0.2317
Yes	West	Female	Above Poverty Level	0	0.0131	0.0067	0.0042	0.0400
Yes	West	Female	Above Poverty Level	1	0.0188	0.0057	0.0096	0.0365
Yes	West	Female	Above Poverty Level	2	0.0264	0.0053	0.0171	0.0407
Yes	West	Female	Above Poverty Level	3	0.0361	0.0064	0.0245	0.0531
Yes	West	Female	Above Poverty Level	4	0.0469	0.0083	0.0317	0.0689
Yes	West	Female	Above Poverty Level	5	0.0647	0.0105	0.0451	0.0919
Yes	West	Female	Above Poverty Level	6	0.0857	0.0130	0.0611	0.1189
Yes	West	Female	Above Poverty Level	7	0.1008	0.0144	0.0733	0.1372
Yes	West	Female	Above Poverty Level	8	0.1032	0.0151	0.0746	0.1412
Yes	West	Female	Above Poverty Level	9	0.1063	0.0144	0.0786	0.1424
Yes	West	Female	Above Poverty Level	10	0.1166	0.0140	0.0893	0.1509
Yes	West	Female	Above Poverty Level	11	0.1181	0.0129	0.0927	0.1494
Yes	West	Female	Above Poverty Level	12	0.1196	0.0131	0.0938	0.1513
Yes	West	Female	Above Poverty Level	13	0.1202	0.0130	0.0945	0.1519
Yes	West	Female	Above Poverty Level	14	0.1241	0.0127	0.0987	0.1548
Yes	West	Female	Above Poverty Level	15	0.1389	0.0125	0.1136	0.1687
Yes	West	Female	Above Poverty Level	16	0.1665	0.0152	0.1358	0.2025
Yes	West	Female	Above Poverty Level	17	0.2118	0.0305	0.1525	0.2864
Yes	West	Female	Below Poverty Level	0	0.0250	0.0138	0.0073	0.0819
Yes	West	Female	Below Poverty Level	1	0.0309	0.0099	0.0152	0.0618
Yes	West	Female	Below Poverty Level	2	0.0387	0.0082	0.0243	0.0612
Yes	West	Female	Below Poverty Level	3	0.0488	0.0099	0.0312	0.0757
Yes	West	Female	Below Poverty Level	4	0.0602	0.0129	0.0374	0.0955
Yes	West	Female	Below Poverty Level	5	0.0843	0.0169	0.0538	0.1296
Yes	West	Female	Below Poverty Level	6	0.1143	0.0197	0.0776	0.1652
Yes	West	Female	Below Poverty Level	7	0.1295	0.0191	0.0930	0.1775
Yes	West	Female	Below Poverty Level	8	0.1195	0.0175	0.0861	0.1636
Yes	West	Female	Below Poverty Level	9	0.0950	0.0151	0.0666	0.1338
Yes	West	Female	Below Poverty Level	10	0.0786	0.0139	0.0530	0.1150
Yes	West	Female	Below Poverty Level	11	0.0812	0.0150	0.0537	0.1209
Yes	West	Female	Below Poverty Level	12	0.0979	0.0179	0.0651	0.1447
Yes	West	Female	Below Poverty Level	13	0.1278	0.0221	0.0866	0.1848
Yes	West	Female	Below Poverty Level	14	0.1324	0.0211	0.0925	0.1859

Appendix 5C, Attachment C, Table 5CC-1. Smoothed prevalence for children "EVER" having asthma.								
Smoothed	**Region**	**Gender**	**Poverty Status**	**Age**	**Prevalence**	**SE**	**LowerCl**	**UpperCl**
Yes	West	Female	Below Poverty Level	15	0.1188	0.0176	0.0853	0.1631
Yes	West	Female	Below Poverty Level	16	0.0917	0.0164	0.0615	0.1347
Yes	West	Female	Below Poverty Level	17	0.0600	0.0186	0.0300	0.1163
Yes	West	Male	Above Poverty Level	0	0.0057	0.0035	0.0014	0.0229
Yes	West	Male	Above Poverty Level	1	0.0191	0.0067	0.0084	0.0428
Yes	West	Male	Above Poverty Level	2	0.0479	0.0092	0.0306	0.0743
Yes	West	Male	Above Poverty Level	3	0.0903	0.0114	0.0673	0.1201
Yes	West	Male	Above Poverty Level	4	0.1300	0.0149	0.0993	0.1685
Yes	West	Male	Above Poverty Level	5	0.1437	0.0158	0.1110	0.1842
Yes	West	Male	Above Poverty Level	6	0.1374	0.0157	0.1050	0.1779
Yes	West	Male	Above Poverty Level	7	0.1290	0.0148	0.0985	0.1671
Yes	West	Male	Above Poverty Level	8	0.1365	0.0148	0.1058	0.1743
Yes	West	Male	Above Poverty Level	9	0.1560	0.0154	0.1236	0.1950
Yes	West	Male	Above Poverty Level	10	0.1794	0.0160	0.1454	0.2193
Yes	West	Male	Above Poverty Level	11	0.1980	0.0175	0.1608	0.2413
Yes	West	Male	Above Poverty Level	12	0.1948	0.0180	0.1566	0.2396
Yes	West	Male	Above Poverty Level	13	0.1818	0.0175	0.1449	0.2256
Yes	West	Male	Above Poverty Level	14	0.1771	0.0164	0.1423	0.2183
Yes	West	Male	Above Poverty Level	15	0.1801	0.0148	0.1484	0.2167
Yes	West	Male	Above Poverty Level	16	0.1897	0.0149	0.1577	0.2264
Yes	West	Male	Above Poverty Level	17	0.2081	0.0248	0.1567	0.2709
Yes	West	Male	Below Poverty Level	0	0.0258	0.0126	0.0087	0.0738
Yes	West	Male	Below Poverty Level	1	0.0442	0.0124	0.0237	0.0812
Yes	West	Male	Below Poverty Level	2	0.0700	0.0119	0.0479	0.1013
Yes	West	Male	Below Poverty Level	3	0.1005	0.0144	0.0729	0.1370
Yes	West	Male	Below Poverty Level	4	0.1323	0.0190	0.0959	0.1799
Yes	West	Male	Below Poverty Level	5	0.1609	0.0218	0.1186	0.2147
Yes	West	Male	Below Poverty Level	6	0.1663	0.0213	0.1247	0.2184
Yes	West	Male	Below Poverty Level	7	0.1582	0.0205	0.1182	0.2086
Yes	West	Male	Below Poverty Level	8	0.1536	0.0204	0.1140	0.2040
Yes	West	Male	Below Poverty Level	9	0.1543	0.0214	0.1128	0.2075
Yes	West	Male	Below Poverty Level	10	0.1630	0.0240	0.1168	0.2228
Yes	West	Male	Below Poverty Level	11	0.1746	0.0270	0.1230	0.2420
Yes	West	Male	Below Poverty Level	12	0.1828	0.0270	0.1306	0.2498
Yes	West	Male	Below Poverty Level	13	0.1809	0.0276	0.1280	0.2495
Yes	West	Male	Below Poverty Level	14	0.1800	0.0259	0.1298	0.2440
Yes	West	Male	Below Poverty Level	15	0.1828	0.0233	0.1371	0.2396
Yes	West	Male	Below Poverty Level	16	0.1881	0.0242	0.1405	0.2471
Yes	West	Male	Below Poverty Level	17	0.1964	0.0396	0.1234	0.2978

Appendix 5C, Attachment C, Table 5CC-2. Smoothed prevalence for children "STILL" having asthma.								
Smoothed	Region	Gender	Poverty Status	Age	Prevalence	SE	LowerCI	UpperCI
Yes	Midwest	Female	Above Poverty Level	0	0.0082	0.0051	0.0021	0.0319
Yes	Midwest	Female	Above Poverty Level	1	0.0168	0.0064	0.0073	0.0382
Yes	Midwest	Female	Above Poverty Level	2	0.0289	0.0070	0.0169	0.0490
Yes	Midwest	Female	Above Poverty Level	3	0.0420	0.0086	0.0267	0.0655
Yes	Midwest	Female	Above Poverty Level	4	0.0509	0.0103	0.0326	0.0788
Yes	Midwest	Female	Above Poverty Level	5	0.0573	0.0108	0.0378	0.0859
Yes	Midwest	Female	Above Poverty Level	6	0.0611	0.0109	0.0412	0.0897
Yes	Midwest	Female	Above Poverty Level	7	0.0624	0.0107	0.0427	0.0902
Yes	Midwest	Female	Above Poverty Level	8	0.0629	0.0100	0.0443	0.0886
Yes	Midwest	Female	Above Poverty Level	9	0.0663	0.0096	0.0481	0.0907
Yes	Midwest	Female	Above Poverty Level	10	0.0737	0.0108	0.0533	0.1012
Yes	Midwest	Female	Above Poverty Level	11	0.0889	0.0126	0.0649	0.1206
Yes	Midwest	Female	Above Poverty Level	12	0.1056	0.0151	0.0768	0.1435
Yes	Midwest	Female	Above Poverty Level	13	0.1157	0.0163	0.0845	0.1565
Yes	Midwest	Female	Above Poverty Level	14	0.1191	0.0160	0.0882	0.1588
Yes	Midwest	Female	Above Poverty Level	15	0.1177	0.0144	0.0896	0.1530
Yes	Midwest	Female	Above Poverty Level	16	0.1107	0.0143	0.0831	0.1461
Yes	Midwest	Female	Above Poverty Level	17	0.0999	0.0205	0.0632	0.1544
Yes	Midwest	Female	Below Poverty Level	0	0.0381	0.0164	0.0146	0.0956
Yes	Midwest	Female	Below Poverty Level	1	0.0620	0.0160	0.0349	0.1076
Yes	Midwest	Female	Below Poverty Level	2	0.0875	0.0160	0.0581	0.1295
Yes	Midwest	Female	Below Poverty Level	3	0.1079	0.0183	0.0738	0.1550
Yes	Midwest	Female	Below Poverty Level	4	0.1187	0.0202	0.0811	0.1704
Yes	Midwest	Female	Below Poverty Level	5	0.1117	0.0194	0.0758	0.1616
Yes	Midwest	Female	Below Poverty Level	6	0.0940	0.0188	0.0602	0.1439
Yes	Midwest	Female	Below Poverty Level	7	0.0974	0.0187	0.0634	0.1469
Yes	Midwest	Female	Below Poverty Level	8	0.1144	0.0205	0.0765	0.1676
Yes	Midwest	Female	Below Poverty Level	9	0.1237	0.0220	0.0830	0.1805
Yes	Midwest	Female	Below Poverty Level	10	0.1196	0.0237	0.0766	0.1821
Yes	Midwest	Female	Below Poverty Level	11	0.1074	0.0225	0.0672	0.1673
Yes	Midwest	Female	Below Poverty Level	12	0.1025	0.0199	0.0664	0.1551
Yes	Midwest	Female	Below Poverty Level	13	0.1096	0.0211	0.0712	0.1649
Yes	Midwest	Female	Below Poverty Level	14	0.1236	0.0229	0.0815	0.1830
Yes	Midwest	Female	Below Poverty Level	15	0.1412	0.0266	0.0924	0.2099
Yes	Midwest	Female	Below Poverty Level	16	0.1633	0.0413	0.0914	0.2746
Yes	Midwest	Female	Below Poverty Level	17	0.1906	0.0779	0.0722	0.4158
Yes	Midwest	Male	Above Poverty Level	0	0.0122	0.0064	0.0038	0.0384
Yes	Midwest	Male	Above Poverty Level	1	0.0268	0.0083	0.0135	0.0525
Yes	Midwest	Male	Above Poverty Level	2	0.0480	0.0091	0.0315	0.0725
Yes	Midwest	Male	Above Poverty Level	3	0.0710	0.0113	0.0500	0.1001
Yes	Midwest	Male	Above Poverty Level	4	0.0842	0.0134	0.0591	0.1187
Yes	Midwest	Male	Above Poverty Level	5	0.0934	0.0138	0.0673	0.1282
Yes	Midwest	Male	Above Poverty Level	6	0.1056	0.0144	0.0779	0.1416
Yes	Midwest	Male	Above Poverty Level	7	0.1117	0.0149	0.0829	0.1489
Yes	Midwest	Male	Above Poverty Level	8	0.1111	0.0152	0.0820	0.1489
Yes	Midwest	Male	Above Poverty Level	9	0.1138	0.0155	0.0840	0.1525
Yes	Midwest	Male	Above Poverty Level	10	0.1126	0.0153	0.0831	0.1507
Yes	Midwest	Male	Above Poverty Level	11	0.1108	0.0146	0.0826	0.1472
Yes	Midwest	Male	Above Poverty Level	12	0.1129	0.0137	0.0861	0.1466
Yes	Midwest	Male	Above Poverty Level	13	0.1139	0.0132	0.0880	0.1462
Yes	Midwest	Male	Above Poverty Level	14	0.1128	0.0127	0.0878	0.1438
Yes	Midwest	Male	Above Poverty Level	15	0.1054	0.0118	0.0822	0.1343
Yes	Midwest	Male	Above Poverty Level	16	0.0935	0.0133	0.0682	0.1269
Yes	Midwest	Male	Above Poverty Level	17	0.0782	0.0184	0.0462	0.1292
Yes	Midwest	Male	Below Poverty Level	0	0.0402	0.0177	0.0151	0.1028
Yes	Midwest	Male	Below Poverty Level	1	0.0824	0.0213	0.0463	0.1425
Yes	Midwest	Male	Below Poverty Level	2	0.1338	0.0225	0.0917	0.1911
Yes	Midwest	Male	Below Poverty Level	3	0.1774	0.0255	0.1282	0.2401
Yes	Midwest	Male	Below Poverty Level	4	0.1949	0.0267	0.1429	0.2601
Yes	Midwest	Male	Below Poverty Level	5	0.1867	0.0237	0.1402	0.2443
Yes	Midwest	Male	Below Poverty Level	6	0.1807	0.0222	0.1371	0.2344
Yes	Midwest	Male	Below Poverty Level	7	0.1734	0.0221	0.1301	0.2273

Appendix 5C, Attachment C, Table 5CC-2. Smoothed prevalence for children "STILL" having asthma.								
Smoothed	**Region**	**Gender**	**Poverty Status**	**Age**	**Prevalence**	**SE**	**LowerCl**	**UpperCl**
Yes	Midwest	Male	Below Poverty Level	8	0.1739	0.0248	0.1260	0.2350
Yes	Midwest	Male	Below Poverty Level	9	0.1814	0.0269	0.1297	0.2478
Yes	Midwest	Male	Below Poverty Level	10	0.1813	0.0282	0.1275	0.2514
Yes	Midwest	Male	Below Poverty Level	11	0.1749	0.0282	0.1214	0.2454
Yes	Midwest	Male	Below Poverty Level	12	0.1702	0.0298	0.1143	0.2457
Yes	Midwest	Male	Below Poverty Level	13	0.1499	0.0296	0.0959	0.2268
Yes	Midwest	Male	Below Poverty Level	14	0.1366	0.0269	0.0876	0.2066
Yes	Midwest	Male	Below Poverty Level	15	0.1484	0.0268	0.0987	0.2169
Yes	Midwest	Male	Below Poverty Level	16	0.1846	0.0359	0.1185	0.2761
Yes	Midwest	Male	Below Poverty Level	17	0.2590	0.0740	0.1306	0.4484
Yes	Northeast	Female	Above Poverty Level	0	0.0153	0.0089	0.0042	0.0537
Yes	Northeast	Female	Above Poverty Level	1	0.0281	0.0096	0.0132	0.0589
Yes	Northeast	Female	Above Poverty Level	2	0.0437	0.0090	0.0276	0.0683
Yes	Northeast	Female	Above Poverty Level	3	0.0584	0.0098	0.0402	0.0840
Yes	Northeast	Female	Above Poverty Level	4	0.0657	0.0112	0.0449	0.0950
Yes	Northeast	Female	Above Poverty Level	5	0.0668	0.0111	0.0461	0.0958
Yes	Northeast	Female	Above Poverty Level	6	0.0678	0.0111	0.0471	0.0967
Yes	Northeast	Female	Above Poverty Level	7	0.0696	0.0114	0.0482	0.0993
Yes	Northeast	Female	Above Poverty Level	8	0.0737	0.0124	0.0506	0.1062
Yes	Northeast	Female	Above Poverty Level	9	0.0840	0.0147	0.0569	0.1224
Yes	Northeast	Female	Above Poverty Level	10	0.0807	0.0144	0.0541	0.1187
Yes	Northeast	Female	Above Poverty Level	11	0.0710	0.0134	0.0466	0.1068
Yes	Northeast	Female	Above Poverty Level	12	0.0629	0.0116	0.0416	0.0938
Yes	Northeast	Female	Above Poverty Level	13	0.0680	0.0113	0.0469	0.0976
Yes	Northeast	Female	Above Poverty Level	14	0.0786	0.0117	0.0564	0.1085
Yes	Northeast	Female	Above Poverty Level	15	0.0913	0.0120	0.0681	0.1214
Yes	Northeast	Female	Above Poverty Level	16	0.1095	0.0165	0.0781	0.1513
Yes	Northeast	Female	Above Poverty Level	17	0.1328	0.0330	0.0753	0.2234
Yes	Northeast	Female	Below Poverty Level	0	0.0234	0.0142	0.0061	0.0856
Yes	Northeast	Female	Below Poverty Level	1	0.0564	0.0190	0.0266	0.1157
Yes	Northeast	Female	Below Poverty Level	2	0.1040	0.0219	0.0648	0.1627
Yes	Northeast	Female	Below Poverty Level	3	0.1466	0.0272	0.0964	0.2167
Yes	Northeast	Female	Below Poverty Level	4	0.1618	0.0304	0.1056	0.2400
Yes	Northeast	Female	Below Poverty Level	5	0.1441	0.0280	0.0928	0.2168
Yes	Northeast	Female	Below Poverty Level	6	0.1124	0.0238	0.0698	0.1761
Yes	Northeast	Female	Below Poverty Level	7	0.0751	0.0174	0.0447	0.1234
Yes	Northeast	Female	Below Poverty Level	8	0.0633	0.0157	0.0364	0.1078
Yes	Northeast	Female	Below Poverty Level	9	0.0838	0.0188	0.0507	0.1355
Yes	Northeast	Female	Below Poverty Level	10	0.1288	0.0270	0.0802	0.2004
Yes	Northeast	Female	Below Poverty Level	11	0.1778	0.0336	0.1154	0.2638
Yes	Northeast	Female	Below Poverty Level	12	0.2073	0.0349	0.1410	0.2941
Yes	Northeast	Female	Below Poverty Level	13	0.2063	0.0328	0.1435	0.2873
Yes	Northeast	Female	Below Poverty Level	14	0.1929	0.0287	0.1375	0.2637
Yes	Northeast	Female	Below Poverty Level	15	0.1703	0.0235	0.1248	0.2281
Yes	Northeast	Female	Below Poverty Level	16	0.1414	0.0234	0.0974	0.2009
Yes	Northeast	Female	Below Poverty Level	17	0.1108	0.0327	0.0567	0.2051
Yes	Northeast	Male	Above Poverty Level	0	0.0225	0.0108	0.0078	0.0633
Yes	Northeast	Male	Above Poverty Level	1	0.0368	0.0105	0.0195	0.0682
Yes	Northeast	Male	Above Poverty Level	2	0.0562	0.0104	0.0373	0.0838
Yes	Northeast	Male	Above Poverty Level	3	0.0797	0.0127	0.0559	0.1123
Yes	Northeast	Male	Above Poverty Level	4	0.1035	0.0162	0.0730	0.1449
Yes	Northeast	Male	Above Poverty Level	5	0.1289	0.0187	0.0931	0.1757
Yes	Northeast	Male	Above Poverty Level	6	0.1472	0.0190	0.1102	0.1938
Yes	Northeast	Male	Above Poverty Level	7	0.1423	0.0181	0.1070	0.1868
Yes	Northeast	Male	Above Poverty Level	8	0.1290	0.0163	0.0973	0.1690
Yes	Northeast	Male	Above Poverty Level	9	0.1251	0.0159	0.0943	0.1641
Yes	Northeast	Male	Above Poverty Level	10	0.1288	0.0155	0.0985	0.1668
Yes	Northeast	Male	Above Poverty Level	11	0.1262	0.0139	0.0989	0.1598
Yes	Northeast	Male	Above Poverty Level	12	0.1246	0.0139	0.0971	0.1584
Yes	Northeast	Male	Above Poverty Level	13	0.1230	0.0149	0.0939	0.1594
Yes	Northeast	Male	Above Poverty Level	14	0.1207	0.0144	0.0925	0.1560
Yes	Northeast	Male	Above Poverty Level	15	0.1114	0.0126	0.0868	0.1420
Yes	Northeast	Male	Above Poverty Level	16	0.0983	0.0124	0.0743	0.1291
Yes	Northeast	Male	Above Poverty Level	17	0.0823	0.0171	0.0518	0.1285

Appendix 5C, Attachment C, Table 5CC-2. Smoothed prevalence for children "STILL" having asthma.								
Smoothed	**Region**	**Gender**	**Poverty Status**	**Age**	**Prevalence**	**SE**	**LowerCI**	**UpperCI**
Yes	Northeast	Male	Below Poverty Level	0	0.0930	0.0402	0.0347	0.2262
Yes	Northeast	Male	Below Poverty Level	1	0.1202	0.0280	0.0710	0.1964
Yes	Northeast	Male	Below Poverty Level	2	0.1475	0.0256	0.0997	0.2130
Yes	Northeast	Male	Below Poverty Level	3	0.1714	0.0311	0.1134	0.2508
Yes	Northeast	Male	Below Poverty Level	4	0.1860	0.0335	0.1232	0.2708
Yes	Northeast	Male	Below Poverty Level	5	0.2060	0.0276	0.1519	0.2732
Yes	Northeast	Male	Below Poverty Level	6	0.2256	0.0276	0.1708	0.2919
Yes	Northeast	Male	Below Poverty Level	7	0.2496	0.0317	0.1866	0.3255
Yes	Northeast	Male	Below Poverty Level	8	0.2727	0.0387	0.1964	0.3653
Yes	Northeast	Male	Below Poverty Level	9	0.2579	0.0395	0.1810	0.3535
Yes	Northeast	Male	Below Poverty Level	10	0.2318	0.0366	0.1611	0.3216
Yes	Northeast	Male	Below Poverty Level	11	0.1902	0.0310	0.1311	0.2678
Yes	Northeast	Male	Below Poverty Level	12	0.1624	0.0268	0.1116	0.2302
Yes	Northeast	Male	Below Poverty Level	13	0.1641	0.0254	0.1155	0.2278
Yes	Northeast	Male	Below Poverty Level	14	0.1699	0.0251	0.1216	0.2323
Yes	Northeast	Male	Below Poverty Level	15	0.1797	0.0244	0.1321	0.2396
Yes	Northeast	Male	Below Poverty Level	16	0.1933	0.0276	0.1397	0.2612
Yes	Northeast	Male	Below Poverty Level	17	0.2097	0.0451	0.1274	0.3253
Yes	South	Female	Above Poverty Level	0	0.0131	0.0059	0.0048	0.0349
Yes	South	Female	Above Poverty Level	1	0.0228	0.0063	0.0124	0.0415
Yes	South	Female	Above Poverty Level	2	0.0352	0.0064	0.0236	0.0522
Yes	South	Female	Above Poverty Level	3	0.0495	0.0074	0.0355	0.0685
Yes	South	Female	Above Poverty Level	4	0.0633	0.0089	0.0464	0.0857
Yes	South	Female	Above Poverty Level	5	0.0740	0.0092	0.0561	0.0969
Yes	South	Female	Above Poverty Level	6	0.0826	0.0096	0.0638	0.1063
Yes	South	Female	Above Poverty Level	7	0.0888	0.0099	0.0695	0.1129
Yes	South	Female	Above Poverty Level	8	0.0860	0.0100	0.0666	0.1105
Yes	South	Female	Above Poverty Level	9	0.0791	0.0095	0.0606	0.1025
Yes	South	Female	Above Poverty Level	10	0.0747	0.0088	0.0576	0.0963
Yes	South	Female	Above Poverty Level	11	0.0736	0.0085	0.0570	0.0944
Yes	South	Female	Above Poverty Level	12	0.0776	0.0087	0.0606	0.0989
Yes	South	Female	Above Poverty Level	13	0.0851	0.0093	0.0669	0.1078
Yes	South	Female	Above Poverty Level	14	0.0871	0.0093	0.0688	0.1099
Yes	South	Female	Above Poverty Level	15	0.0876	0.0087	0.0702	0.1087
Yes	South	Female	Above Poverty Level	16	0.0859	0.0091	0.0681	0.1080
Yes	South	Female	Above Poverty Level	17	0.0819	0.0136	0.0567	0.1169
Yes	South	Female	Below Poverty Level	0	0.0396	0.0135	0.0186	0.0823
Yes	South	Female	Below Poverty Level	1	0.0573	0.0113	0.0371	0.0876
Yes	South	Female	Below Poverty Level	2	0.0772	0.0109	0.0564	0.1048
Yes	South	Female	Below Poverty Level	3	0.0963	0.0136	0.0704	0.1306
Yes	South	Female	Below Poverty Level	4	0.1120	0.0165	0.0805	0.1536
Yes	South	Female	Below Poverty Level	5	0.1206	0.0174	0.0874	0.1641
Yes	South	Female	Below Poverty Level	6	0.1219	0.0173	0.0888	0.1652
Yes	South	Female	Below Poverty Level	7	0.1152	0.0162	0.0842	0.1556
Yes	South	Female	Below Poverty Level	8	0.1131	0.0157	0.0829	0.1524
Yes	South	Female	Below Poverty Level	9	0.1190	0.0161	0.0880	0.1591
Yes	South	Female	Below Poverty Level	10	0.1208	0.0175	0.0874	0.1646
Yes	South	Female	Below Poverty Level	11	0.1195	0.0178	0.0857	0.1642
Yes	South	Female	Below Poverty Level	12	0.1275	0.0192	0.0910	0.1757
Yes	South	Female	Below Poverty Level	13	0.1405	0.0197	0.1026	0.1893
Yes	South	Female	Below Poverty Level	14	0.1394	0.0184	0.1037	0.1848
Yes	South	Female	Below Poverty Level	15	0.1296	0.0166	0.0973	0.1706
Yes	South	Female	Below Poverty Level	16	0.1136	0.0184	0.0791	0.1605
Yes	South	Female	Below Poverty Level	17	0.0923	0.0249	0.0503	0.1634
Yes	South	Male	Above Poverty Level	0	0.0228	0.0070	0.0116	0.0443
Yes	South	Male	Above Poverty Level	1	0.0476	0.0082	0.0325	0.0693
Yes	South	Male	Above Poverty Level	2	0.0793	0.0089	0.0619	0.1011
Yes	South	Male	Above Poverty Level	3	0.1076	0.0109	0.0859	0.1341
Yes	South	Male	Above Poverty Level	4	0.1193	0.0123	0.0949	0.1490
Yes	South	Male	Above Poverty Level	5	0.1194	0.0117	0.0960	0.1475
Yes	South	Male	Above Poverty Level	6	0.1145	0.0111	0.0924	0.1411
Yes	South	Male	Above Poverty Level	7	0.1071	0.0105	0.0861	0.1323
Yes	South	Male	Above Poverty Level	8	0.1011	0.0099	0.0813	0.1251
Yes	South	Male	Above Poverty Level	9	0.1000	0.0098	0.0806	0.1236

Appendix 5C, Attachment C, Table 5CC-2. Smoothed prevalence for children "STILL" having asthma.								
Smoothed	Region	Gender	Poverty Status	Age	Prevalence	SE	LowerCl	UpperCl
Yes	South	Male	Above Poverty Level	10	0.1059	0.0102	0.0855	0.1305
Yes	South	Male	Above Poverty Level	11	0.1122	0.0106	0.0910	0.1376
Yes	South	Male	Above Poverty Level	12	0.1103	0.0105	0.0893	0.1356
Yes	South	Male	Above Poverty Level	13	0.1052	0.0105	0.0843	0.1305
Yes	South	Male	Above Poverty Level	14	0.0983	0.0094	0.0795	0.1210
Yes	South	Male	Above Poverty Level	15	0.0899	0.0081	0.0737	0.1093
Yes	South	Male	Above Poverty Level	16	0.0811	0.0089	0.0636	0.1028
Yes	South	Male	Above Poverty Level	17	0.0727	0.0136	0.0479	0.1089
Yes	South	Male	Below Poverty Level	0	0.0499	0.0126	0.0285	0.0860
Yes	South	Male	Below Poverty Level	1	0.0749	0.0110	0.0542	0.1027
Yes	South	Male	Below Poverty Level	2	0.1033	0.0116	0.0805	0.1316
Yes	South	Male	Below Poverty Level	3	0.1305	0.0149	0.1012	0.1666
Yes	South	Male	Below Poverty Level	4	0.1519	0.0177	0.1171	0.1948
Yes	South	Male	Below Poverty Level	5	0.1595	0.0180	0.1240	0.2029
Yes	South	Male	Below Poverty Level	6	0.1598	0.0185	0.1234	0.2045
Yes	South	Male	Below Poverty Level	7	0.1540	0.0180	0.1186	0.1977
Yes	South	Male	Below Poverty Level	8	0.1466	0.0170	0.1130	0.1879
Yes	South	Male	Below Poverty Level	9	0.1457	0.0170	0.1122	0.1870
Yes	South	Male	Below Poverty Level	10	0.1504	0.0171	0.1167	0.1917
Yes	South	Male	Below Poverty Level	11	0.1508	0.0171	0.1171	0.1921
Yes	South	Male	Below Poverty Level	12	0.1506	0.0184	0.1146	0.1955
Yes	South	Male	Below Poverty Level	13	0.1470	0.0192	0.1097	0.1943
Yes	South	Male	Below Poverty Level	14	0.1345	0.0179	0.0999	0.1788
Yes	South	Male	Below Poverty Level	15	0.1215	0.0159	0.0907	0.1607
Yes	South	Male	Below Poverty Level	16	0.1080	0.0164	0.0770	0.1494
Yes	South	Male	Below Poverty Level	17	0.0948	0.0227	0.0555	0.1573
Yes	West	Female	Above Poverty Level	0	0.0077	0.0049	0.0019	0.0306
Yes	West	Female	Above Poverty Level	1	0.0122	0.0046	0.0053	0.0278
Yes	West	Female	Above Poverty Level	2	0.0181	0.0045	0.0105	0.0310
Yes	West	Female	Above Poverty Level	3	0.0248	0.0055	0.0153	0.0401
Yes	West	Female	Above Poverty Level	4	0.0305	0.0068	0.0186	0.0494
Yes	West	Female	Above Poverty Level	5	0.0382	0.0077	0.0245	0.0590
Yes	West	Female	Above Poverty Level	6	0.0482	0.0091	0.0318	0.0724
Yes	West	Female	Above Poverty Level	7	0.0573	0.0098	0.0393	0.0829
Yes	West	Female	Above Poverty Level	8	0.0628	0.0106	0.0432	0.0904
Yes	West	Female	Above Poverty Level	9	0.0697	0.0106	0.0497	0.0970
Yes	West	Female	Above Poverty Level	10	0.0768	0.0099	0.0577	0.1016
Yes	West	Female	Above Poverty Level	11	0.0786	0.0094	0.0603	0.1018
Yes	West	Female	Above Poverty Level	12	0.0808	0.0100	0.0615	0.1056
Yes	West	Female	Above Poverty Level	13	0.0829	0.0108	0.0621	0.1100
Yes	West	Female	Above Poverty Level	14	0.0845	0.0111	0.0632	0.1121
Yes	West	Female	Above Poverty Level	15	0.0908	0.0110	0.0694	0.1179
Yes	West	Female	Above Poverty Level	16	0.1016	0.0129	0.0766	0.1337
Yes	West	Female	Above Poverty Level	17	0.1180	0.0236	0.0753	0.1803
Yes	West	Female	Below Poverty Level	0	0.0244	0.0144	0.0066	0.0862
Yes	West	Female	Below Poverty Level	1	0.0270	0.0091	0.0128	0.0561
Yes	West	Female	Below Poverty Level	2	0.0306	0.0074	0.0179	0.0518
Yes	West	Female	Below Poverty Level	3	0.0354	0.0090	0.0201	0.0615
Yes	West	Female	Below Poverty Level	4	0.0407	0.0112	0.0221	0.0738
Yes	West	Female	Below Poverty Level	5	0.0577	0.0146	0.0328	0.0996
Yes	West	Female	Below Poverty Level	6	0.0807	0.0185	0.0483	0.1319
Yes	West	Female	Below Poverty Level	7	0.0954	0.0181	0.0624	0.1434
Yes	West	Female	Below Poverty Level	8	0.0876	0.0159	0.0583	0.1296
Yes	West	Female	Below Poverty Level	9	0.0648	0.0127	0.0419	0.0989
Yes	West	Female	Below Poverty Level	10	0.0495	0.0107	0.0306	0.0792
Yes	West	Female	Below Poverty Level	11	0.0473	0.0110	0.0282	0.0781
Yes	West	Female	Below Poverty Level	12	0.0606	0.0137	0.0366	0.0988
Yes	West	Female	Below Poverty Level	13	0.0845	0.0179	0.0526	0.1329
Yes	West	Female	Below Poverty Level	14	0.0931	0.0180	0.0603	0.1411
Yes	West	Female	Below Poverty Level	15	0.0846	0.0154	0.0562	0.1253
Yes	West	Female	Below Poverty Level	16	0.0629	0.0143	0.0379	0.1026
Yes	West	Female	Below Poverty Level	17	0.0376	0.0146	0.0158	0.0868
Yes	West	Male	Above Poverty Level	0	0.0007	0.0007	0.0001	0.0067
Yes	West	Male	Above Poverty Level	1	0.0052	0.0027	0.0014	0.0192

Appendix 5C, Attachment C, Table 5CC-2. Smoothed prevalence for children "STILL" having asthma.								
Smoothed	Region	Gender	Poverty Status	Age	Prevalence	SE	LowerCI	UpperCI
Yes	West	Male	Above Poverty Level	2	0.0225	0.0063	0.0112	0.0447
Yes	West	Male	Above Poverty Level	3	0.0596	0.0095	0.0398	0.0884
Yes	West	Male	Above Poverty Level	4	0.0989	0.0140	0.0691	0.1397
Yes	West	Male	Above Poverty Level	5	0.1070	0.0147	0.0754	0.1496
Yes	West	Male	Above Poverty Level	6	0.0959	0.0141	0.0660	0.1372
Yes	West	Male	Above Poverty Level	7	0.0830	0.0126	0.0565	0.1203
Yes	West	Male	Above Poverty Level	8	0.0877	0.0124	0.0613	0.1239
Yes	West	Male	Above Poverty Level	9	0.1029	0.0135	0.0737	0.1419
Yes	West	Male	Above Poverty Level	10	0.1189	0.0140	0.0883	0.1584
Yes	West	Male	Above Poverty Level	11	0.1292	0.0153	0.0955	0.1724
Yes	West	Male	Above Poverty Level	12	0.1214	0.0154	0.0879	0.1653
Yes	West	Male	Above Poverty Level	13	0.1050	0.0139	0.0749	0.1452
Yes	West	Male	Above Poverty Level	14	0.0981	0.0127	0.0707	0.1346
Yes	West	Male	Above Poverty Level	15	0.0997	0.0116	0.0742	0.1327
Yes	West	Male	Above Poverty Level	16	0.1091	0.0128	0.0810	0.1454
Yes	West	Male	Above Poverty Level	17	0.1290	0.0231	0.0814	0.1984
Yes	West	Male	Below Poverty Level	0	0.0263	0.0130	0.0088	0.0761
Yes	West	Male	Below Poverty Level	1	0.0374	0.0101	0.0204	0.0673
Yes	West	Male	Below Poverty Level	2	0.0518	0.0086	0.0358	0.0742
Yes	West	Male	Below Poverty Level	3	0.0681	0.0105	0.0483	0.0952
Yes	West	Male	Below Poverty Level	4	0.0871	0.0143	0.0604	0.1240
Yes	West	Male	Below Poverty Level	5	0.1074	0.0173	0.0749	0.1517
Yes	West	Male	Below Poverty Level	6	0.1167	0.0183	0.0820	0.1635
Yes	West	Male	Below Poverty Level	7	0.1138	0.0186	0.0789	0.1615
Yes	West	Male	Below Poverty Level	8	0.1073	0.0177	0.0741	0.1529
Yes	West	Male	Below Poverty Level	9	0.0964	0.0164	0.0659	0.1389
Yes	West	Male	Below Poverty Level	10	0.0830	0.0149	0.0557	0.1221
Yes	West	Male	Below Poverty Level	11	0.0745	0.0151	0.0474	0.1152
Yes	West	Male	Below Poverty Level	12	0.0825	0.0165	0.0527	0.1268
Yes	West	Male	Below Poverty Level	13	0.1000	0.0197	0.0643	0.1524
Yes	West	Male	Below Poverty Level	14	0.1074	0.0200	0.0707	0.1600
Yes	West	Male	Below Poverty Level	15	0.1120	0.0193	0.0760	0.1620
Yes	West	Male	Below Poverty Level	16	0.1127	0.0222	0.0724	0.1714
Yes	West	Male	Below Poverty Level	17	0.1084	0.0340	0.0531	0.2088

Appendix 5C, Attachment C, Table 5CC-3. Smoothed prevalence for adults "EVER" having asthma								
Smoothed	**Region**	**Gender**	**Poverty Status**	**Age_group**	**Prevalence**	**SE**	**LowerCI**	**UpperCI**
Yes	Midwest	Female	Above Poverty Level	18-24	0.1642	0.0141	0.1219	0.2176
Yes	Midwest	Female	Above Poverty Level	25-34	0.1341	0.0063	0.1142	0.1568
Yes	Midwest	Female	Above Poverty Level	35-44	0.1193	0.0058	0.1012	0.1402
Yes	Midwest	Female	Above Poverty Level	45-54	0.1204	0.0057	0.1025	0.1409
Yes	Midwest	Female	Above Poverty Level	55-64	0.1246	0.0066	0.1040	0.1486
Yes	Midwest	Female	Above Poverty Level	65-74	0.1165	0.0062	0.0971	0.1392
Yes	Midwest	Female	Above Poverty Level	75+	0.0980	0.0089	0.0719	0.1322
Yes	Midwest	Female	Below Poverty Level	18-24	0.2014	0.0153	0.1531	0.2603
Yes	Midwest	Female	Below Poverty Level	25-34	0.1812	0.0114	0.1445	0.2248
Yes	Midwest	Female	Below Poverty Level	35-44	0.1782	0.0130	0.1370	0.2284
Yes	Midwest	Female	Below Poverty Level	45-54	0.2104	0.0146	0.1638	0.2662
Yes	Midwest	Female	Below Poverty Level	55-64	0.2295	0.0164	0.1770	0.2920
Yes	Midwest	Female	Below Poverty Level	65-74	0.1892	0.0145	0.1435	0.2453
Yes	Midwest	Female	Below Poverty Level	75+	0.1176	0.0173	0.0690	0.1933
Yes	Midwest	Male	Above Poverty Level	18-24	0.1705	0.0149	0.1249	0.2284
Yes	Midwest	Male	Above Poverty Level	25-34	0.1209	0.0063	0.1008	0.1444
Yes	Midwest	Male	Above Poverty Level	35-44	0.0886	0.0053	0.0719	0.1087
Yes	Midwest	Male	Above Poverty Level	45-54	0.0727	0.0046	0.0583	0.0904
Yes	Midwest	Male	Above Poverty Level	55-64	0.0770	0.0054	0.0602	0.0980
Yes	Midwest	Male	Above Poverty Level	65-74	0.0828	0.0058	0.0647	0.1053
Yes	Midwest	Male	Above Poverty Level	75+	0.0847	0.0106	0.0545	0.1292
Yes	Midwest	Male	Below Poverty Level	18-24	0.1654	0.0175	0.1122	0.2370
Yes	Midwest	Male	Below Poverty Level	25-34	0.1143	0.0109	0.0808	0.1593
Yes	Midwest	Male	Below Poverty Level	35-44	0.1066	0.0122	0.0703	0.1585
Yes	Midwest	Male	Below Poverty Level	45-54	0.1376	0.0146	0.0936	0.1979
Yes	Midwest	Male	Below Poverty Level	55-64	0.1643	0.0164	0.1141	0.2309
Yes	Midwest	Male	Below Poverty Level	65-74	0.1396	0.0160	0.0918	0.2068
Yes	Midwest	Male	Below Poverty Level	75+	0.0853	0.0205	0.0353	0.1920
Yes	Northeast	Female	Above Poverty Level	18-24	0.1791	0.0176	0.1265	0.2474
Yes	Northeast	Female	Above Poverty Level	25-34	0.1423	0.0076	0.1183	0.1701
Yes	Northeast	Female	Above Poverty Level	35-44	0.1256	0.0072	0.1029	0.1525
Yes	Northeast	Female	Above Poverty Level	45-54	0.1246	0.0071	0.1024	0.1509
Yes	Northeast	Female	Above Poverty Level	55-64	0.1281	0.0076	0.1043	0.1565
Yes	Northeast	Female	Above Poverty Level	65-74	0.1151	0.0070	0.0934	0.1412
Yes	Northeast	Female	Above Poverty Level	75+	0.0879	0.0098	0.0598	0.1273
Yes	Northeast	Female	Below Poverty Level	18-24	0.1646	0.0182	0.1104	0.2383
Yes	Northeast	Female	Below Poverty Level	25-34	0.1705	0.0110	0.1356	0.2123
Yes	Northeast	Female	Below Poverty Level	35-44	0.1842	0.0126	0.1442	0.2323
Yes	Northeast	Female	Below Poverty Level	45-54	0.2084	0.0143	0.1629	0.2627
Yes	Northeast	Female	Below Poverty Level	55-64	0.2180	0.0156	0.1684	0.2773
Yes	Northeast	Female	Below Poverty Level	65-74	0.1695	0.0118	0.1321	0.2149
Yes	Northeast	Female	Below Poverty Level	75+	0.0960	0.0125	0.0603	0.1495
Yes	Northeast	Male	Above Poverty Level	18-24	0.1728	0.0210	0.1126	0.2560
Yes	Northeast	Male	Above Poverty Level	25-34	0.1163	0.0081	0.0914	0.1469
Yes	Northeast	Male	Above Poverty Level	35-44	0.0932	0.0070	0.0721	0.1197
Yes	Northeast	Male	Above Poverty Level	45-54	0.0901	0.0063	0.0710	0.1139
Yes	Northeast	Male	Above Poverty Level	55-64	0.0963	0.0072	0.0744	0.1237
Yes	Northeast	Male	Above Poverty Level	65-74	0.0874	0.0073	0.0656	0.1155
Yes	Northeast	Male	Above Poverty Level	75+	0.0708	0.0118	0.0398	0.1229
Yes	Northeast	Male	Below Poverty Level	18-24	0.1734	0.0193	0.1138	0.2552
Yes	Northeast	Male	Below Poverty Level	25-34	0.1323	0.0138	0.0896	0.1911
Yes	Northeast	Male	Below Poverty Level	35-44	0.1182	0.0135	0.0772	0.1768
Yes	Northeast	Male	Below Poverty Level	45-54	0.1254	0.0144	0.0816	0.1879
Yes	Northeast	Male	Below Poverty Level	55-64	0.1361	0.0198	0.0786	0.2253
Yes	Northeast	Male	Below Poverty Level	65-74	0.1305	0.0195	0.0743	0.2191
Yes	Northeast	Male	Below Poverty Level	75+	0.0988	0.0255	0.0373	0.2366
Yes	South	Female	Above Poverty Level	18-24	0.1533	0.0114	0.1185	0.1959
Yes	South	Female	Above Poverty Level	25-34	0.1235	0.0054	0.1065	0.1429
Yes	South	Female	Above Poverty Level	35-44	0.1114	0.0050	0.0956	0.1295
Yes	South	Female	Above Poverty Level	45-54	0.1149	0.0047	0.0998	0.1320
Yes	South	Female	Above Poverty Level	55-64	0.1261	0.0058	0.1077	0.1472
Yes	South	Female	Above Poverty Level	65-74	0.1188	0.0058	0.1004	0.1400

Appendix 5C, Attachment C, Table 5CC-3. Smoothed prevalence for adults "EVER" having asthma								
Smoothed	Region	Gender	Poverty Status	Age group	Prevalence	SE	LowerCI	UpperCI
Yes	South	Female	Above Poverty Level	75+	0.0959	0.0087	0.0701	0.1297
Yes	South	Female	Below Poverty Level	18-24	0.1491	0.0122	0.1107	0.1978
Yes	South	Female	Below Poverty Level	25-34	0.1365	0.0066	0.1149	0.1614
Yes	South	Female	Below Poverty Level	35-44	0.1414	0.0078	0.1159	0.1714
Yes	South	Female	Below Poverty Level	45-54	0.1686	0.0097	0.1369	0.2059
Yes	South	Female	Below Poverty Level	55-64	0.1881	0.0115	0.1505	0.2324
Yes	South	Female	Below Poverty Level	65-74	0.1651	0.0101	0.1325	0.2039
Yes	South	Female	Below Poverty Level	75+	0.1125	0.0124	0.0755	0.1644
Yes	South	Male	Above Poverty Level	18-24	0.1445	0.0095	0.1147	0.1805
Yes	South	Male	Above Poverty Level	25-34	0.1086	0.0050	0.0926	0.1269
Yes	South	Male	Above Poverty Level	35-44	0.0860	0.0044	0.0720	0.1025
Yes	South	Male	Above Poverty Level	45-54	0.0742	0.0040	0.0616	0.0891
Yes	South	Male	Above Poverty Level	55-64	0.0733	0.0045	0.0594	0.0902
Yes	South	Male	Above Poverty Level	65-74	0.0790	0.0048	0.0639	0.0974
Yes	South	Male	Above Poverty Level	75+	0.0900	0.0102	0.0606	0.1316
Yes	South	Male	Below Poverty Level	18-24	0.1433	0.0144	0.1000	0.2013
Yes	South	Male	Below Poverty Level	25-34	0.1031	0.0087	0.0766	0.1376
Yes	South	Male	Below Poverty Level	35-44	0.0934	0.0090	0.0664	0.1300
Yes	South	Male	Below Poverty Level	45-54	0.1055	0.0101	0.0751	0.1462
Yes	South	Male	Below Poverty Level	55-64	0.1072	0.0108	0.0750	0.1510
Yes	South	Male	Below Poverty Level	65-74	0.0942	0.0092	0.0666	0.1314
Yes	South	Male	Below Poverty Level	75+	0.0712	0.0123	0.0385	0.1279
Yes	West	Female	Above Poverty Level	18-24	0.1571	0.0135	0.1163	0.2089
Yes	West	Female	Above Poverty Level	25-34	0.1415	0.0067	0.1201	0.1660
Yes	West	Female	Above Poverty Level	35-44	0.1373	0.0070	0.1150	0.1631
Yes	West	Female	Above Poverty Level	45-54	0.1423	0.0067	0.1207	0.1670
Yes	West	Female	Above Poverty Level	55-64	0.1497	0.0071	0.1268	0.1758
Yes	West	Female	Above Poverty Level	65-74	0.1445	0.0070	0.1220	0.1704
Yes	West	Female	Above Poverty Level	75+	0.1266	0.0112	0.0929	0.1702
Yes	West	Female	Below Poverty Level	18-24	0.1434	0.0164	0.0945	0.2117
Yes	West	Female	Below Poverty Level	25-34	0.1318	0.0092	0.1026	0.1678
Yes	West	Female	Below Poverty Level	35-44	0.1440	0.0117	0.1074	0.1903
Yes	West	Female	Below Poverty Level	45-54	0.1806	0.0144	0.1350	0.2374
Yes	West	Female	Below Poverty Level	55-64	0.1713	0.0136	0.1284	0.2248
Yes	West	Female	Below Poverty Level	65-74	0.1511	0.0117	0.1141	0.1974
Yes	West	Female	Below Poverty Level	75+	0.1292	0.0177	0.0785	0.2054
Yes	West	Male	Above Poverty Level	18-24	0.1566	0.0173	0.1067	0.2240
Yes	West	Male	Above Poverty Level	25-34	0.1233	0.0069	0.1019	0.1485
Yes	West	Male	Above Poverty Level	35-44	0.1025	0.0060	0.0839	0.1247
Yes	West	Male	Above Poverty Level	45-54	0.0908	0.0054	0.0741	0.1107
Yes	West	Male	Above Poverty Level	55-64	0.0955	0.0059	0.0774	0.1174
Yes	West	Male	Above Poverty Level	65-74	0.1067	0.0068	0.0860	0.1318
Yes	West	Male	Above Poverty Level	75+	0.1265	0.0152	0.0834	0.1871
Yes	West	Male	Below Poverty Level	18-24	0.1521	0.0204	0.0938	0.2373
Yes	West	Male	Below Poverty Level	25-34	0.0942	0.0095	0.0660	0.1327
Yes	West	Male	Below Poverty Level	35-44	0.0885	0.0102	0.0590	0.1308
Yes	West	Male	Below Poverty Level	45-54	0.1133	0.0130	0.0753	0.1670
Yes	West	Male	Below Poverty Level	55-64	0.1237	0.0156	0.0789	0.1888
Yes	West	Male	Below Poverty Level	65-74	0.1134	0.0142	0.0726	0.1727
Yes	West	Male	Below Poverty Level	75+	0.0961	0.0190	0.0474	0.1849

Appendix 5C, Attachment C, Table 5CC-4. Smoothed prevalence for adults "STILL" having asthma								
Smoothed	Region	Gender	Poverty Status	Age_group	Prevalence	SE	LowerCI	UpperCI
Yes	Midwest	Female	Above Poverty Level	18-24	0.1046	0.0121	0.0703	0.1528
Yes	Midwest	Female	Above Poverty Level	25-34	0.0888	0.0057	0.0714	0.1100
Yes	Midwest	Female	Above Poverty Level	35-44	0.0835	0.0052	0.0675	0.1030
Yes	Midwest	Female	Above Poverty Level	45-44	0.0893	0.0050	0.0738	0.1077
Yes	Midwest	Female	Above Poverty Level	55-64	0.0909	0.0057	0.0736	0.1118
Yes	Midwest	Female	Above Poverty Level	65-74	0.0811	0.0051	0.0654	0.1002
Yes	Midwest	Female	Above Poverty Level	75+	0.0630	0.0067	0.0438	0.0898
Yes	Midwest	Female	Below Poverty Level	18-24	0.1327	0.0139	0.0907	0.1899
Yes	Midwest	Female	Below Poverty Level	25-34	0.1280	0.0095	0.0980	0.1656
Yes	Midwest	Female	Below Poverty Level	35-44	0.1315	0.0114	0.0961	0.1772
Yes	Midwest	Female	Below Poverty Level	45-44	0.1600	0.0134	0.1181	0.2132
Yes	Midwest	Female	Below Poverty Level	55-64	0.1777	0.0146	0.1318	0.2352
Yes	Midwest	Female	Below Poverty Level	65-74	0.1488	0.0128	0.1091	0.1998
Yes	Midwest	Female	Below Poverty Level	75+	0.0940	0.0157	0.0513	0.1659
Yes	Midwest	Male	Above Poverty Level	18-24	0.0807	0.0115	0.0491	0.1299
Yes	Midwest	Male	Above Poverty Level	25-34	0.0584	0.0045	0.0448	0.0758
Yes	Midwest	Male	Above Poverty Level	35-44	0.0479	0.0040	0.0359	0.0637
Yes	Midwest	Male	Above Poverty Level	45-44	0.0472	0.0038	0.0358	0.0620
Yes	Midwest	Male	Above Poverty Level	55-64	0.0522	0.0042	0.0395	0.0687
Yes	Midwest	Male	Above Poverty Level	65-74	0.0528	0.0045	0.0393	0.0706
Yes	Midwest	Male	Above Poverty Level	75+	0.0481	0.0081	0.0268	0.0847
Yes	Midwest	Male	Below Poverty Level	18-24	0.0912	0.0136	0.0542	0.1496
Yes	Midwest	Male	Below Poverty Level	25-34	0.0683	0.0091	0.0430	0.1067
Yes	Midwest	Male	Below Poverty Level	35-44	0.0694	0.0109	0.0402	0.1173
Yes	Midwest	Male	Below Poverty Level	45-44	0.1015	0.0141	0.0624	0.1610
Yes	Midwest	Male	Below Poverty Level	55-64	0.1338	0.0165	0.0866	0.2010
Yes	Midwest	Male	Below Poverty Level	65-74	0.1202	0.0161	0.0751	0.1869
Yes	Midwest	Male	Below Poverty Level	75+	0.0709	0.0210	0.0250	0.1850
Yes	Northeast	Female	Above Poverty Level	18-24	0.1098	0.0134	0.0721	0.1638
Yes	Northeast	Female	Above Poverty Level	25-34	0.0965	0.0065	0.0765	0.1210
Yes	Northeast	Female	Above Poverty Level	35-44	0.0899	0.0063	0.0708	0.1136
Yes	Northeast	Female	Above Poverty Level	45-44	0.0901	0.0060	0.0718	0.1124
Yes	Northeast	Female	Above Poverty Level	55-64	0.0917	0.0062	0.0727	0.1151
Yes	Northeast	Female	Above Poverty Level	65-74	0.0862	0.0059	0.0681	0.1085
Yes	Northeast	Female	Above Poverty Level	75+	0.0726	0.0093	0.0467	0.1110
Yes	Northeast	Female	Below Poverty Level	18-24	0.1212	0.0166	0.0744	0.1915
Yes	Northeast	Female	Below Poverty Level	25-34	0.1199	0.0093	0.0914	0.1559
Yes	Northeast	Female	Below Poverty Level	35-44	0.1338	0.0106	0.1013	0.1747
Yes	Northeast	Female	Below Poverty Level	45-44	0.1655	0.0127	0.1260	0.2143
Yes	Northeast	Female	Below Poverty Level	55-64	0.1824	0.0143	0.1381	0.2370
Yes	Northeast	Female	Below Poverty Level	65-74	0.1273	0.0098	0.0972	0.1650
Yes	Northeast	Female	Below Poverty Level	75+	0.0529	0.0086	0.0300	0.0917
Yes	Northeast	Male	Above Poverty Level	18-24	0.0922	0.0154	0.0509	0.1616
Yes	Northeast	Male	Above Poverty Level	25-34	0.0600	0.0058	0.0428	0.0836
Yes	Northeast	Male	Above Poverty Level	35-44	0.0488	0.0050	0.0340	0.0696
Yes	Northeast	Male	Above Poverty Level	45-44	0.0483	0.0051	0.0334	0.0693
Yes	Northeast	Male	Above Poverty Level	55-64	0.0563	0.0065	0.0376	0.0834
Yes	Northeast	Male	Above Poverty Level	65-74	0.0576	0.0063	0.0393	0.0837
Yes	Northeast	Male	Above Poverty Level	75+	0.0554	0.0106	0.0281	0.1062
Yes	Northeast	Male	Below Poverty Level	18-24	0.0791	0.0128	0.0430	0.1409
Yes	Northeast	Male	Below Poverty Level	25-34	0.0800	0.0119	0.0459	0.1360
Yes	Northeast	Male	Below Poverty Level	35-44	0.0805	0.0135	0.0427	0.1465
Yes	Northeast	Male	Below Poverty Level	45-44	0.0857	0.0162	0.0419	0.1672
Yes	Northeast	Male	Below Poverty Level	55-64	0.1064	0.0224	0.0475	0.2211
Yes	Northeast	Male	Below Poverty Level	65-74	0.1040	0.0200	0.0501	0.2035
Yes	Northeast	Male	Below Poverty Level	75+	0.0771	0.0236	0.0241	0.2203
Yes	South	Female	Above Poverty Level	18-24	0.0891	0.0083	0.0649	0.1212
Yes	South	Female	Above Poverty Level	25-34	0.0735	0.0039	0.0615	0.0876
Yes	South	Female	Above Poverty Level	35-44	0.0684	0.0036	0.0571	0.0817
Yes	South	Female	Above Poverty Level	45-44	0.0732	0.0037	0.0617	0.0866
Yes	South	Female	Above Poverty Level	55-64	0.0846	0.0046	0.0705	0.1012
Yes	South	Female	Above Poverty Level	65-74	0.0817	0.0047	0.0674	0.0987

Appendix 5C, Attachment C, Table 5CC-4. Smoothed prevalence for adults "STILL" having asthma								
Smoothed	Region	Gender	Poverty Status	Age group	Prevalence	SE	LowerCI	UpperCI
Yes	South	Female	Above Poverty Level	75+	0.0641	0.0070	0.0443	0.0920
Yes	South	Female	Below Poverty Level	18-24	0.0948	0.0105	0.0641	0.1380
Yes	South	Female	Below Poverty Level	25-34	0.0942	0.0059	0.0758	0.1166
Yes	South	Female	Below Poverty Level	35-44	0.1086	0.0073	0.0859	0.1365
Yes	South	Female	Below Poverty Level	45-44	0.1446	0.0095	0.1149	0.1806
Yes	South	Female	Below Poverty Level	55-64	0.1618	0.0112	0.1267	0.2043
Yes	South	Female	Below Poverty Level	65-74	0.1379	0.0095	0.1082	0.1742
Yes	South	Female	Below Poverty Level	75+	0.0881	0.0109	0.0570	0.1337
Yes	South	Male	Above Poverty Level	18-24	0.0600	0.0073	0.0392	0.0907
Yes	South	Male	Above Poverty Level	25-34	0.0490	0.0035	0.0381	0.0629
Yes	South	Male	Above Poverty Level	35-44	0.0421	0.0033	0.0322	0.0550
Yes	South	Male	Above Poverty Level	45-44	0.0386	0.0031	0.0292	0.0510
Yes	South	Male	Above Poverty Level	55-64	0.0384	0.0034	0.0282	0.0520
Yes	South	Male	Above Poverty Level	65-74	0.0457	0.0038	0.0343	0.0607
Yes	South	Male	Above Poverty Level	75+	0.0627	0.0089	0.0382	0.1013
Yes	South	Male	Below Poverty Level	18-24	0.0583	0.0080	0.0358	0.0937
Yes	South	Male	Below Poverty Level	25-34	0.0443	0.0053	0.0290	0.0672
Yes	South	Male	Below Poverty Level	35-44	0.0492	0.0067	0.0303	0.0790
Yes	South	Male	Below Poverty Level	45-44	0.0720	0.0090	0.0460	0.1112
Yes	South	Male	Below Poverty Level	55-64	0.0771	0.0096	0.0492	0.1188
Yes	South	Male	Below Poverty Level	65-74	0.0608	0.0075	0.0390	0.0937
Yes	South	Male	Below Poverty Level	75+	0.0353	0.0082	0.0154	0.0787
Yes	West	Female	Above Poverty Level	18-24	0.0842	0.0115	0.0522	0.1328
Yes	West	Female	Above Poverty Level	25-34	0.0876	0.0054	0.0708	0.1080
Yes	West	Female	Above Poverty Level	35-44	0.0931	0.0062	0.0742	0.1163
Yes	West	Female	Above Poverty Level	45-44	0.0981	0.0065	0.0781	0.1226
Yes	West	Female	Above Poverty Level	55-64	0.1028	0.0067	0.0820	0.1281
Yes	West	Female	Above Poverty Level	65-74	0.0984	0.0061	0.0795	0.1213
Yes	West	Female	Above Poverty Level	75+	0.0825	0.0090	0.0565	0.1189
Yes	West	Female	Below Poverty Level	18-24	0.0863	0.0121	0.0524	0.1387
Yes	West	Female	Below Poverty Level	25-34	0.0934	0.0078	0.0695	0.1243
Yes	West	Female	Below Poverty Level	35-44	0.1091	0.0100	0.0789	0.1489
Yes	West	Female	Below Poverty Level	45-44	0.1332	0.0120	0.0967	0.1806
Yes	West	Female	Below Poverty Level	55-64	0.1292	0.0120	0.0929	0.1770
Yes	West	Female	Below Poverty Level	65-74	0.1169	0.0104	0.0854	0.1580
Yes	West	Female	Below Poverty Level	75+	0.1021	0.0148	0.0609	0.1662
Yes	West	Male	Above Poverty Level	18-24	0.0597	0.0092	0.0351	0.0998
Yes	West	Male	Above Poverty Level	25-34	0.0569	0.0046	0.0432	0.0745
Yes	West	Male	Above Poverty Level	35-44	0.0549	0.0045	0.0414	0.0723
Yes	West	Male	Above Poverty Level	45-44	0.0525	0.0046	0.0389	0.0704
Yes	West	Male	Above Poverty Level	55-64	0.0562	0.0053	0.0407	0.0770
Yes	West	Male	Above Poverty Level	65-74	0.0660	0.0058	0.0487	0.0889
Yes	West	Male	Above Poverty Level	75+	0.0783	0.0131	0.0437	0.1364
Yes	West	Male	Below Poverty Level	18-24	0.0720	0.0125	0.0389	0.1295
Yes	West	Male	Below Poverty Level	25-34	0.0484	0.0068	0.0294	0.0787
Yes	West	Male	Below Poverty Level	35-44	0.0539	0.0084	0.0311	0.0919
Yes	West	Male	Below Poverty Level	45-44	0.0784	0.0115	0.0465	0.1293
Yes	West	Male	Below Poverty Level	55-64	0.0936	0.0155	0.0517	0.1635
Yes	West	Male	Below Poverty Level	65-74	0.0758	0.0129	0.0413	0.1350
Yes	West	Male	Below Poverty Level	75+	0.0489	0.0136	0.0182	0.1250

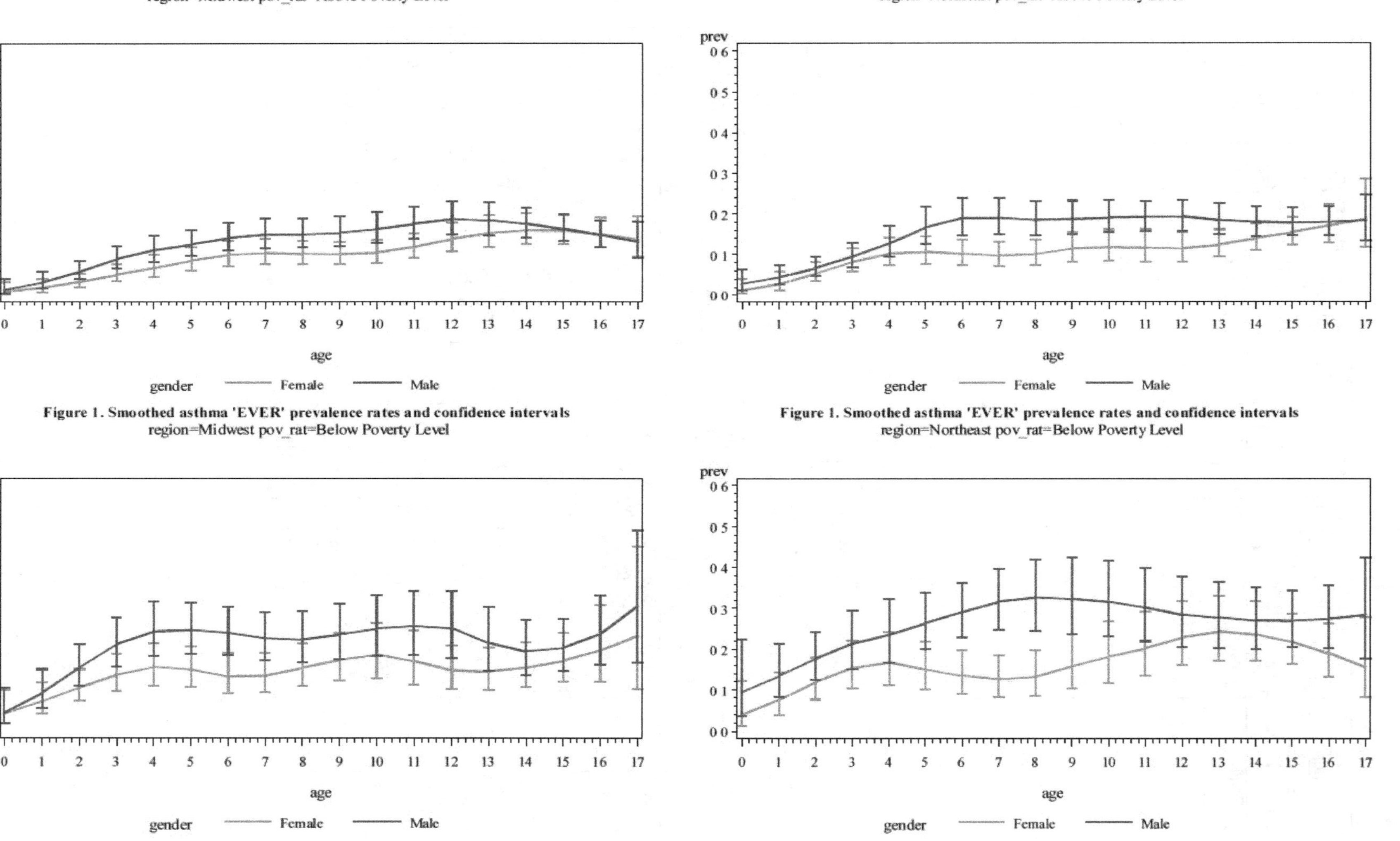

Appendix 5C, Attachment C, Figure 5CC-1. Smoothed prevalence and confidence intervals for children 'EVER' having asthma.

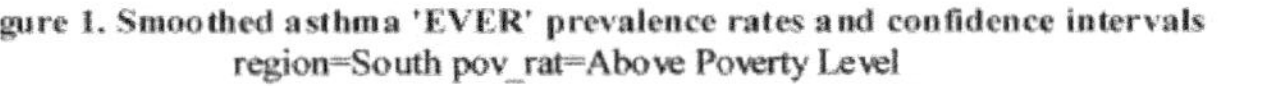

Figure 1. Smoothed asthma 'EVER' prevalence rates and confidence intervals
region=South pov_rat=Above Poverty Level

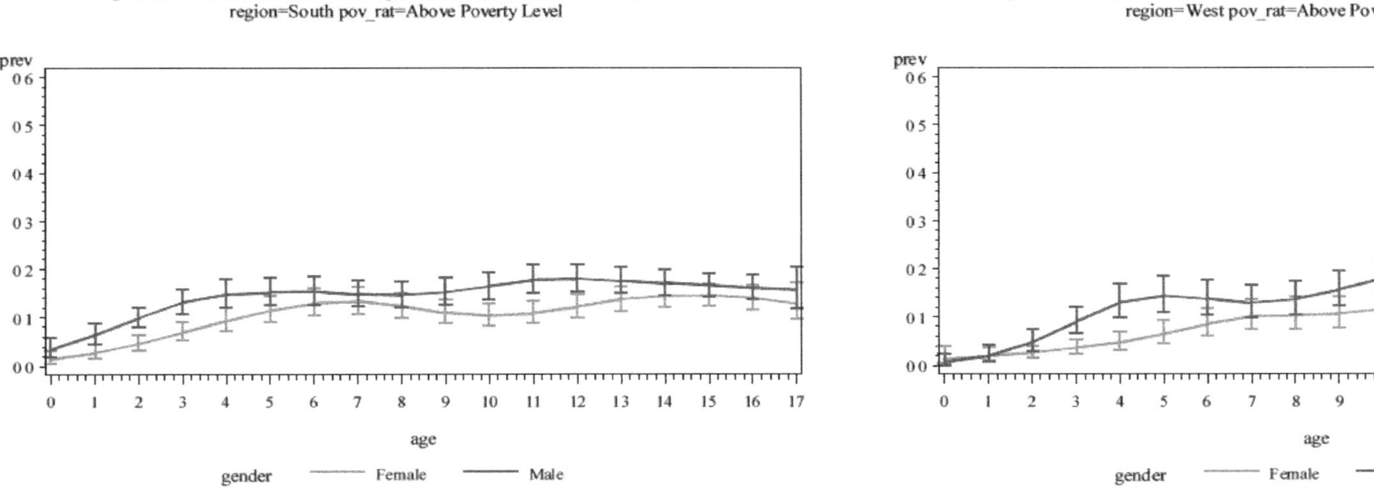

Figure 1. Smoothed asthma 'EVER' prevalence rates and confidence intervals
region=West pov_rat=Above Poverty Level

Figure 1. Smoothed asthma 'EVER' prevalence rates and confidence intervals
region=South pov_rat=Below Poverty Level

Figure 1. Smoothed asthma 'EVER' prevalence rates and confidence intervals
region=West pov_rat=Below Poverty Level

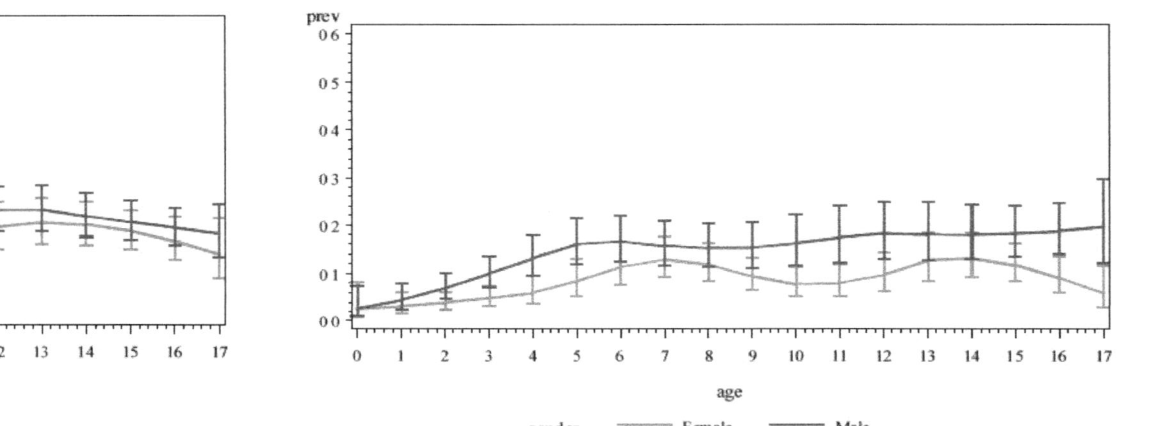

Appendix 5C, Attachment C, Figure 5CC-1, cont. Smoothed prevalence and confidence intervals for children 'EVER' having asthma.

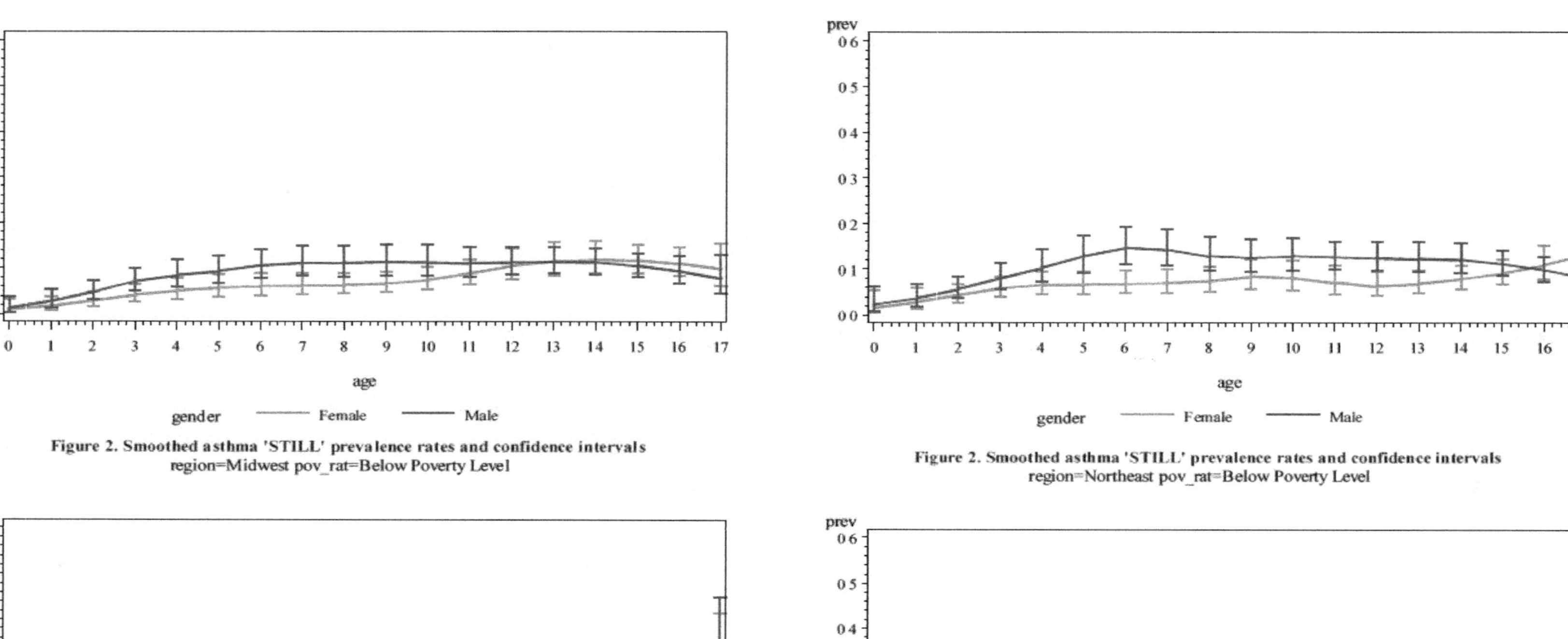

Appendix 5C, Attachment C, Figure 5CC-2. Smoothed prevalence and confidence intervals for children 'STILL' having asthma.

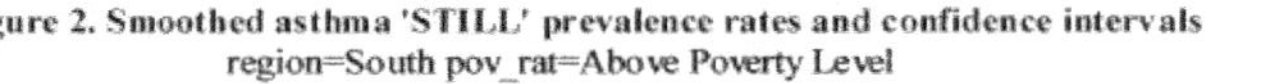

Figure 2. Smoothed asthma 'STILL' prevalence rates and confidence intervals
region=South pov_rat=Above Poverty Level

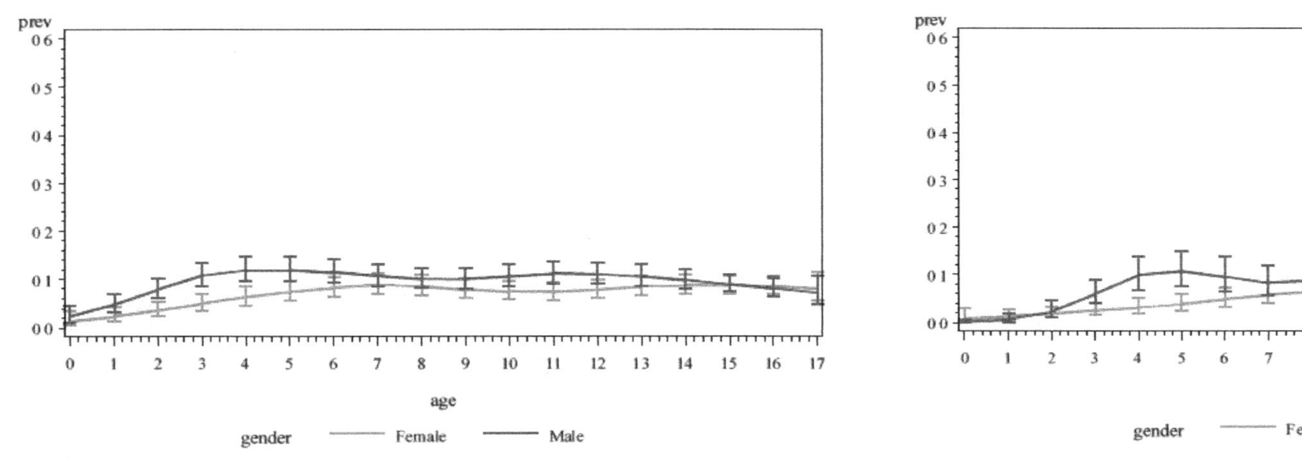

Figure 2. Smoothed asthma 'STILL' prevalence rates and confidence intervals
region=West pov_rat=Above Poverty Level

Figure 2. Smoothed asthma 'STILL' prevalence rates and confidence intervals
region=South pov_rat=Below Poverty Level

Figure 2. Smoothed asthma 'STILL' prevalence rates and confidence intervals
region=West pov_rat=Below Poverty Level

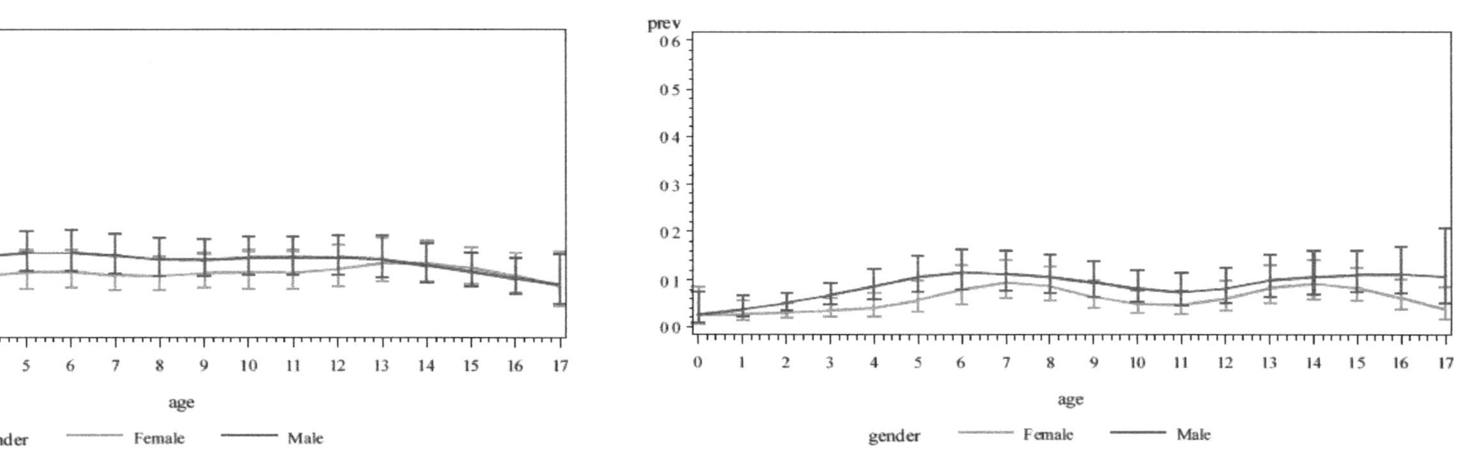

Appendix 5C, Attachment C, Figure 5CC-2, cont. Smoothed prevalence and confidence intervals for children 'STILL' having asthma.

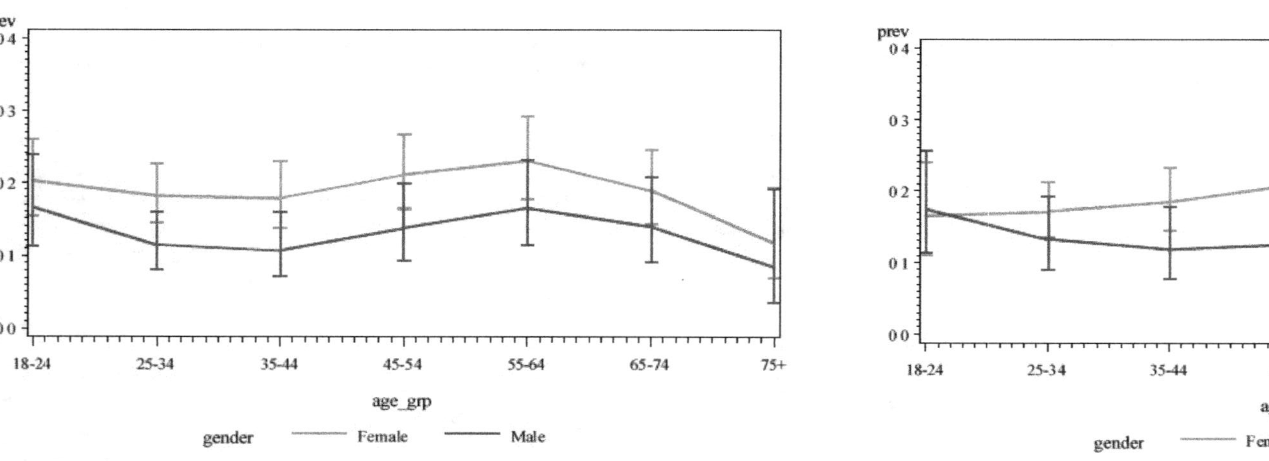

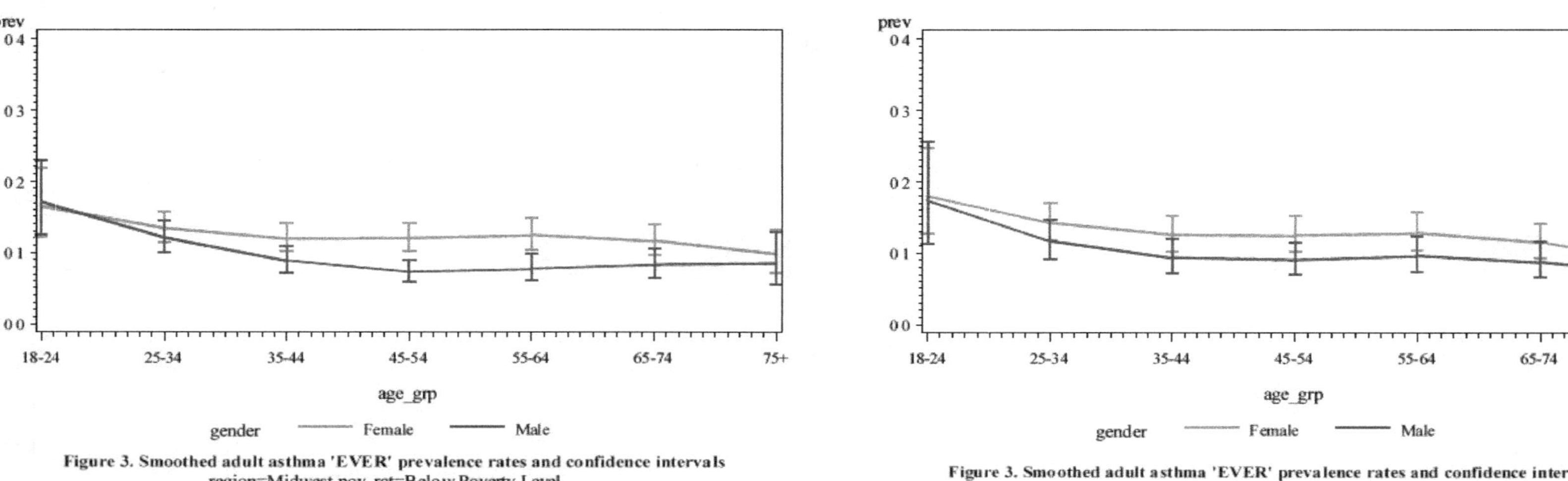

Appendix 5C, Attachment C, Figure 5CC-3. Smoothed prevalence and confidence intervals for Adults 'EVER' having asthma.

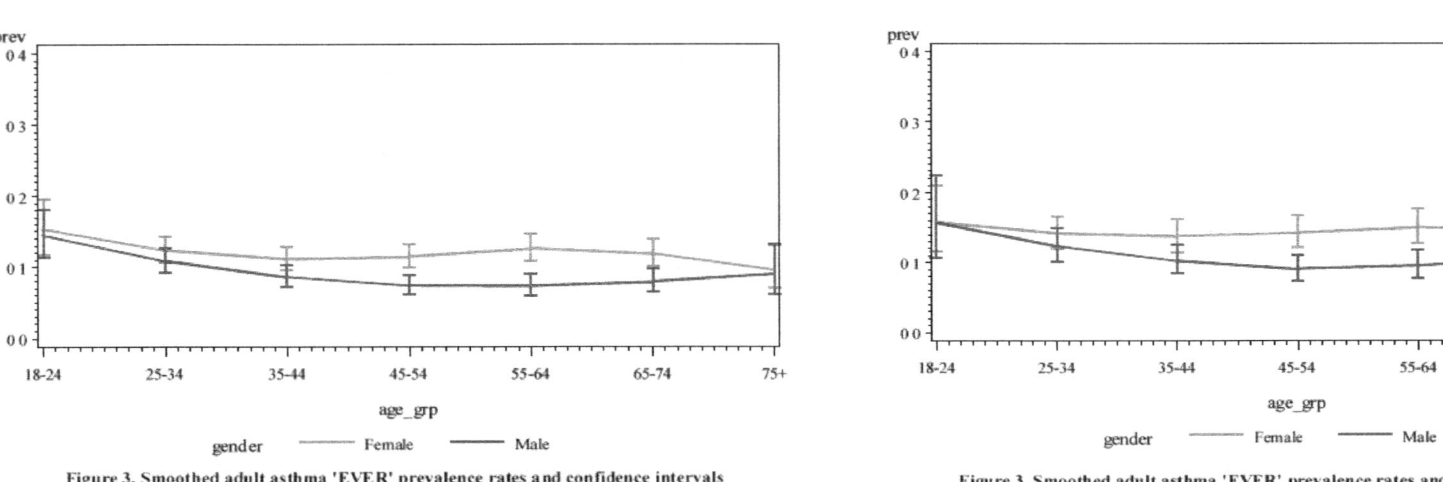

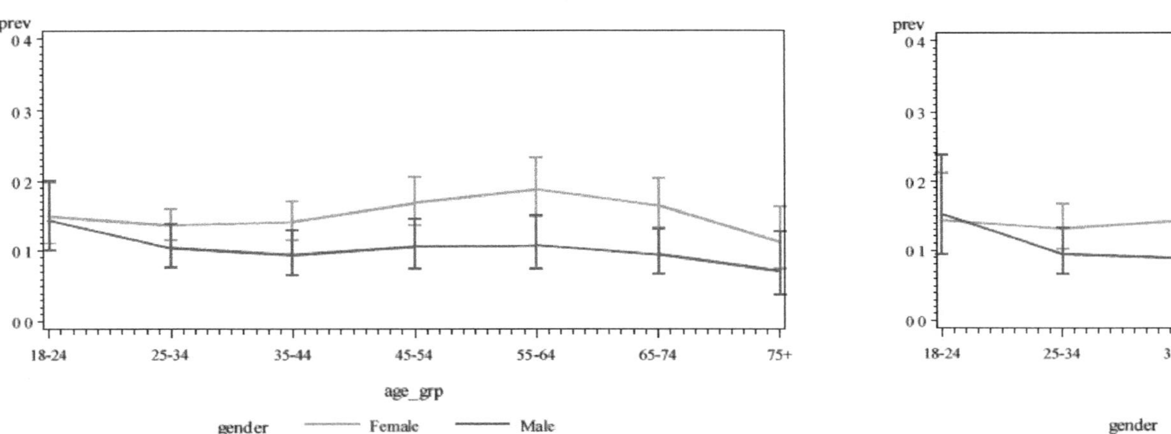

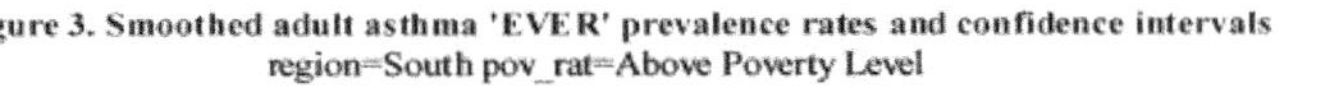

Appendix 5C, Attachment C, Figure 5CC-3, cont. Smoothed prevalence and confidence intervals for Adults 'EVER' having asthma.

Figure 4. Smoothed adult asthma 'STILL' prevalence rates and confidence intervals
region=Midwest pov_rat=Above Poverty Level

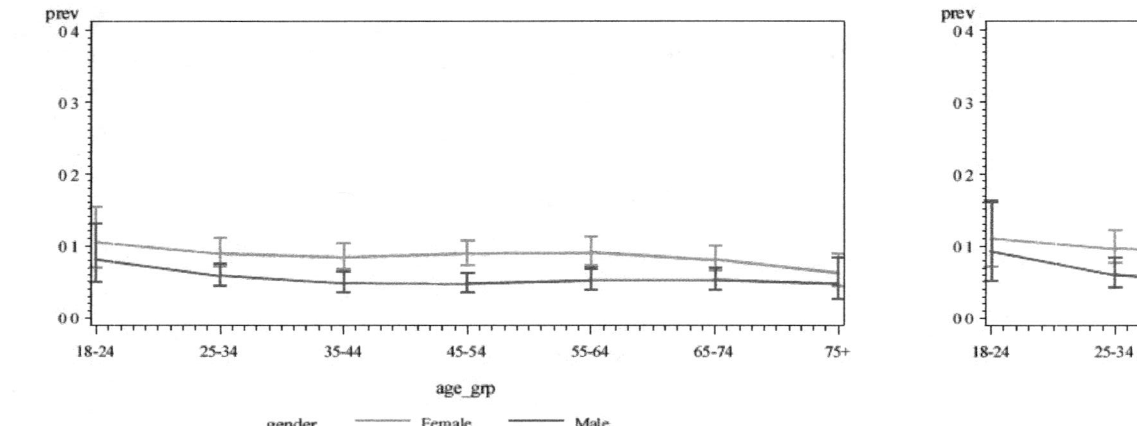

Figure 4. Smoothed adult asthma 'STILL' prevalence rates and confidence intervals
region=Northeast pov_rat=Above Poverty Level

Figure 4. Smoothed adult asthma 'STILL' prevalence rates and confidence intervals
region=Midwest pov_rat=Below Poverty Level

Figure 4. Smoothed adult asthma 'STILL' prevalence rates and confidence intervals
region=Northeast pov_rat=Below Poverty Level

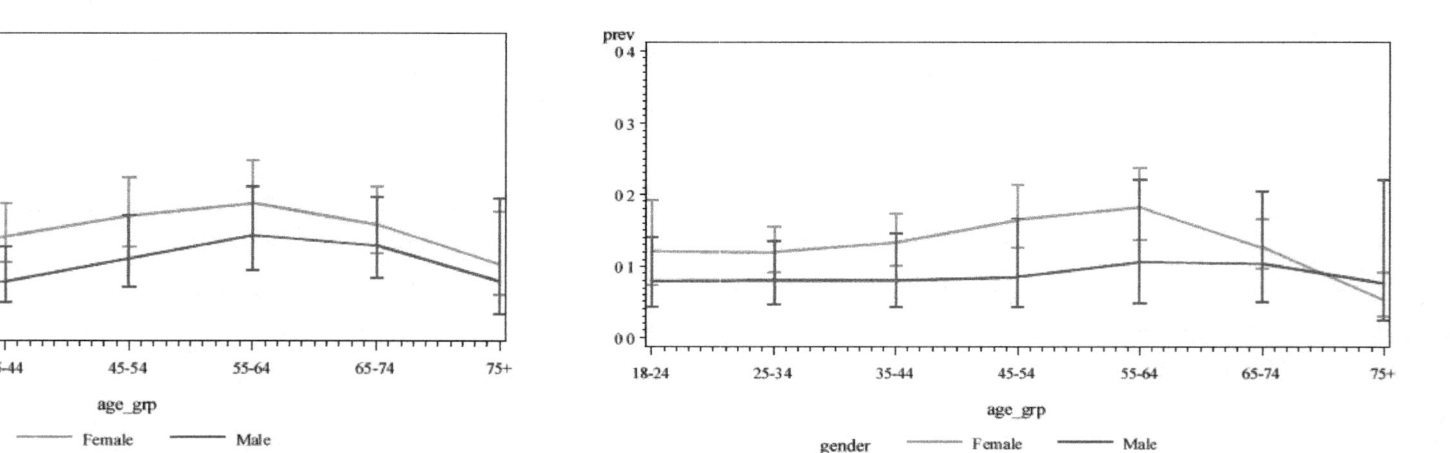

Appendix 5C, Attachment C, Figure 5CC-4. Smoothed prevalence and confidence intervals for Adults 'STILL' having asthma.

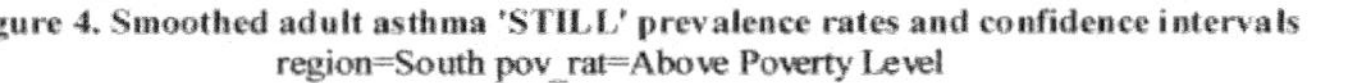

Figure 4. Smoothed adult asthma 'STILL' prevalence rates and confidence intervals
region=South pov_rat=Above Poverty Level

Figure 4. Smoothed adult asthma 'STILL' prevalence rates and confidence intervals
region=West pov_rat=Above Poverty Level

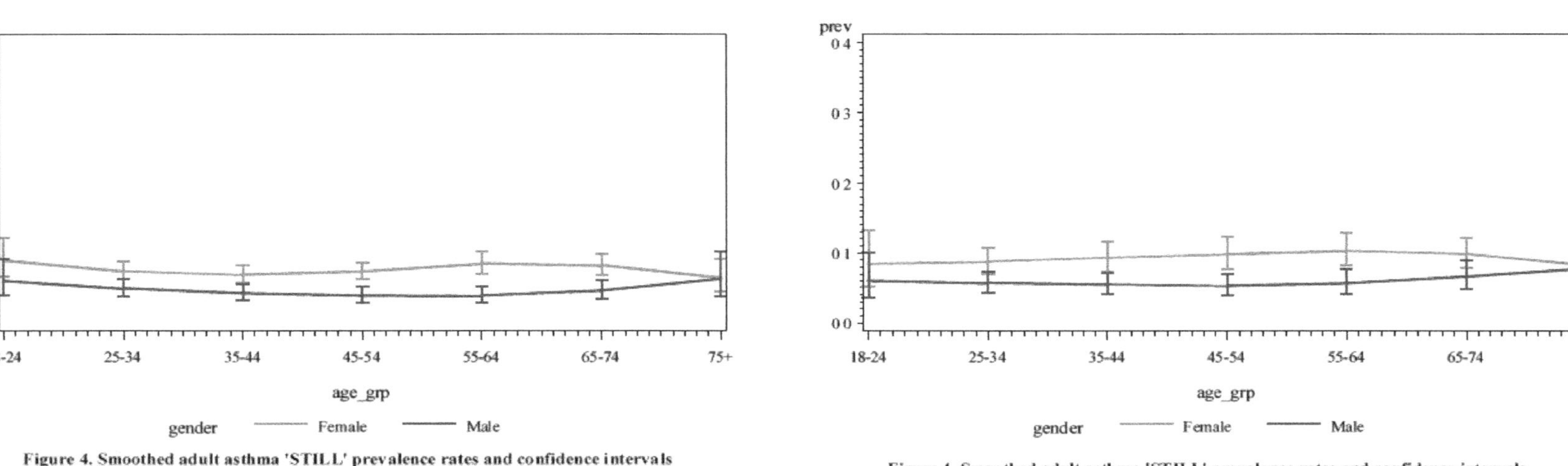

Figure 4. Smoothed adult asthma 'STILL' prevalence rates and confidence intervals
region=South pov_rat=Below Poverty Level

Figure 4. Smoothed adult asthma 'STILL' prevalence rates and confidence intervals
region=West pov_rat=Below Poverty Level

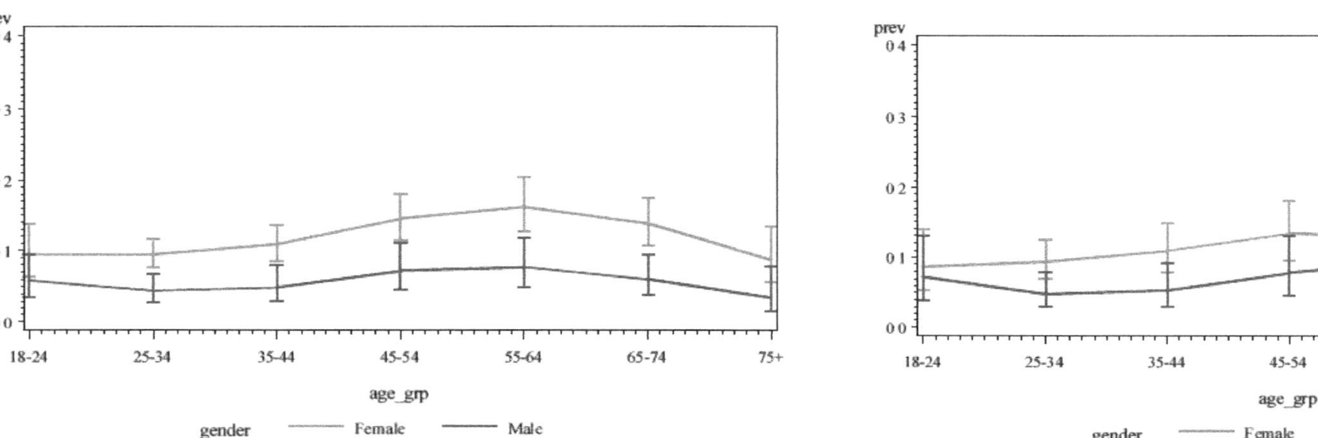

Appendix 5C, Attachment C, Figure 5CC-4, cont. Smoothed prevalence and confidence intervals for Adults 'STILL' having asthma.

Appendix 5-D

Variability Analysis and Uncertainty Characterization

Table of Contents

5D-1. OVERVIEW 2
5D-2. TREATMENT OF VARIABILITY AND CO-VARIABILITY 2
5D-3. CHARACTERIZATION OF UNCERTAINTY 8
5D-4. REFERENCES 11

List of Tables

Table 5D-1. Components of exposure variability modeled by APEX. 5
Table 5D-2. Important components of co-variability in exposure modeling. 7

5D-1. OVERVIEW

An important issue associated with any population exposure or risk assessment is the characterization of variability and uncertainty. *Variability* refers to the inherent heterogeneity in a population or variable of interest (e.g., residential air exchange rates). The degree of variability cannot be reduced through further research, only better characterized with additional measurement. *Uncertainty* refers to the lack of knowledge regarding the values of model input variables (i.e., *parameter uncertainty*), the physical systems or relationships used (i.e., use of input variables to estimate exposure or risk or *model uncertainty*), and in specifying the scenario that is consistent with purpose of the assessment (i.e., *scenario uncertainty*). Uncertainty is, ideally, reduced to the maximum extent possible through improved measurement of key parameters and iterative model refinement. The approaches used to assess variability and to characterize uncertainty in this REA are discussed in the following two sections. The primary purpose of this characterization is to provide a summary of variability and uncertainty evaluations conducted to date regarding our O_3 exposure assessments and APEX exposure modeling and to identify the most important elements of uncertainty in need of further characterization. Each section contains a concise tabular summary of the identified components and how, for elements of uncertainty, each source may affect the estimated exposures.

5D-2. TREATMENT OF VARIABILITY AND CO-VARIABILITY

The purpose for addressing variability in this REA is to ensure that the estimates of exposure and risk reflect the variability of ambient O_3 concentrations, population characteristics, associated O_3 exposure and intake dose, and potential health risk across the study area and for the simulated at-risk populations. In this REA, there are several algorithms that account for variability of input data when generating the number of estimated benchmark exceedances or health risk outputs. For example, variability may arise from differences in the population residing within census tracts (e.g., age distribution) and the activities that may affect population exposure to O_3 and the resulting intake dose estimate (e.g., time spent outdoors, performing moderate or greater exertion level activities outdoors). A complete range of potential exposure levels and associated risk estimates can be generated when appropriately addressing variability in exposure and risk assessments; note however that the range of values obtained would be within the constraints of the input parameters, algorithms, or modeling system used, not necessarily the complete range of the true exposure or risk values.

Where possible, staff identified and incorporated the observed variability in input data sets rather than employing standard default assumptions and/or using point estimates to describe model inputs. The details regarding variability distributions used in data inputs are described in Appendix 5B, while details regarding the variability addressed within its algorithms and processes are found in the APEX TSD (US EPA, 2012).

Briefly, APEX has been designed to account for variability in most of the input data, including the physiological variables that are important inputs to determining exertion levels and associated ventilation rates. APEX simulates individuals and then calculates O_3 exposures for each of these simulated individuals. The individuals are selected to represent a random sample from a defined population. The collection of individuals represents the variability of the target population, and accounts for several types of variability, including demographic, physiological, and human behavior. In this assessment, we simulated 200,000 individuals to reasonably capture the variability expected in the population exposure distribution for each study area. APEX incorporates stochastic processes representing the natural variability of personal profile characteristics, activity patterns, and microenvironment parameters. In this way, APEX is able to represent much of the variability in the exposure estimates resulting from the variability of the factors effecting human exposure.

We note also that correlations and non-linear relationships between variables input to the model can result in the model producing incorrect results if the inherent relationships between these variables are not preserved. That is why APEX is also designed to account for co-variability, or linear and nonlinear correlation among the model inputs, provided that enough is known about these relationships to specify them. This is accomplished by providing inputs that enable the correlation to be modeled explicitly within APEX. For example, there is a non-linear relationship between the outdoor temperature and air exchange rate in homes. One factor that contributes to this non-linear relationship is that windows tend to be closed more often when temperatures are at either low or high extremes than when temperatures are moderate. This relationship is explicitly modeled in APEX by specifying different probability distributions of air exchange rates for different ambient temperatures. In any event, APEX models variability and co-variability in two ways:

- **Stochastically**. The user provides APEX with probability distributions characterizing the variability of many input parameters. These are treated stochastically in the model and the estimated exposure distributions reflect this variability. For example, the rate of O_3 removal in houses can depend on a number of factors which we are not able to explicitly model at this time, due to a lack of data. However, we can specify a distribution of removal rates which reflects observed variations in O_3 decay. APEX randomly samples from this distribution to obtain values which are used in the mass balance model. Further, co-variability can be modeled stochastically through the use of conditional distributions. If two or more parameters are related, conditional distributions that depend on the values of the related parameters are input to APEX. For example, the distribution of air exchange rates (AERs) in a house depends on the outdoor temperature and whether or not air conditioning (A/C) is in use. In this case, a set of AER distributions is provided to APEX for different ranges of temperatures and A/C use, and the selection of the distribution in APEX is driven by the temperature and A/C status at that time. The spatial variability of A/C prevalence is modeled by supplying APEX with A/C prevalence for each Census tract in the modeled area.

- **Explicitly**. For some variables used in modeling exposure, APEX models variability and co-variability explicitly and not stochastically. For example, hourly-average ambient O_3 concentrations and temperatures are used in model calculations. These are input to the model for every hour in the time period modeled at different spatial locations, and in this way the variability and co-variability of hourly concentrations and temperatures are modeled explicitly.

Important sources of the variability and co-variability accounted for by APEX and used for this exposure analysis are summarized in **Table 5D-1** and **Table 5D-2** below, respectively.

Table 5D-1. Components of exposure variability modeled by APEX.

Component	Variability Source	Comment
Simulated Individuals	Population data	Individuals are randomly sampled from US census tracts used in each model study area, stratified by age (single years), gender, and employment status probability distributions (US Census Bureau, 2007a).
	Commuting data	Employed individuals are probabilistically assigned ambient concentrations originating from either their home or work tract based on US Census derived commuter data (US Census Bureau, 2007a).
	Activity patterns	Data diaries are randomly selected from CHAD master (>38,000 diaries) using six diary pools stratified by two day-types (weekday, weekend) and three temperature ranges (< 55.0 °F, between 55.0 and 83.9°F, and ≥84.0 °F). The CHAD diaries capture real locations that people visit and the activities they perform, ranging from 1 minute to 1 hour in duration (US EPA, 2002).
	Longitudinal profiles	A sequence of diaries is linked together for each individual that preserves both the inter- and intra-personal variability in human activities (Glen et al., 2008).
	Asthma prevalence	Asthma prevalence is stratified by two genders, single age years (0-17), seven age groups, (18-24, 25-34, 35-44, 45-54, 55-64, 65-74, and, ≥75), four regions (Midwest, Northeast, South, and West), and US census tract level poverty ratios (CDC, 2011; US Census Bureau, 2007b).
Ambient Input	Measured ambient O_3 concentrations	Temporal: 1-hour concentrations for an entire O_3 season or year predicted using ambient monitoring data. Spatial: Several monitors are used to represent ambient conditions within each study area; each monitor was assigned a 30 km zone of influence, though value from closest monitor is used for each tract. Four US study areas assess regional differences in ambient conditions.
	Meteorological data	Spatial: Values from closest available local surface National Weather Service (NWS) station were used. Temporal: 1-hour temperature data input for each year; daily values calculated by APEX.
Microenvironmental Approach	Microenvironments: General	Twenty-eight total microenvironments are represented, including those expected to be associated with high exposure concentrations (i.e., outdoors and outdoor near-road). Where this type of variability is incorporated within particular microenvironmental algorithm inputs, this results in differential exposure estimates for each individual (and event) as persons spend varying time frequency within each microenvironment and ambient concentrations vary spatially within and between study areas.
	Microenvironments: Spatial Variability	Ambient concentrations used in microenvironmental algorithms vary spatially within (where more than one site available) and among study areas. Concentrations near roadways are adjusted to account for titration by NO.

Component	Variability Source	Comment
	Microenvironments: Temporal Variability	All exposure calculations are performed at the event-level when using either factors or mass balance approach (durations can be as short as one minute). In addition, for the indoor microenvironments, using a mass balance model accounts for O_3 concentrations occurring during a previous hour (and of ambient origin) to calculate a current event's indoor O_3 concentrations.
	Air exchange rates	Several lognormal distributions are sampled based on five daily mean temperature ranges, study area, and study-area specific A/C prevalence rates.
	Proximity factors for on- and near roads	Three distributions are used, stratified by road-type (urban, interstate, and rural), selected based on VMT to address expected ozone titration by NO near roads.
Physiological Factors and Algorithms	Resting metabolic rate (RMR)	Regression equations for three age-group (18-29, 30-59, and 60+) and two genders were used with body mass as the independent variable (see Johnson et al. (2000) and section 5.3 of APEX TSD).
	Maximum normalized oxygen consumption rate (NVO_2)	Single year age- and gender-specific normal distributions are randomly sampled for each person (Isaacs and Smith, 2005 and section 7.2 of APEX TSD). This variable is used to calculate maximum metabolic equivalents (METS).
	Maximum oxygen debt (MOXD)	Normal distributions for maximum obtainable oxygen, stratified by 3 age groups (ages 0-11, 12-18, 19-100) and two genders (Isaacs and Smith, 2007 and section 7.2 of APEX TSD). Used when adjusting METS to address fatigue and EPOC.
	Recovery time	One uniform distribution randomly sampled to estimate the time required to recover a maximum oxygen deficit (Isaacs and Smith, 2007 and section 7.2 of APEX TSD).
	METS by activity	Values randomly sampled from distributions developed for specific activities (a few are age-group specific) (McCurdy, 2000; US EPA, 2002).
	Oxygen uptake per unit of energy expended (UCF)	Values randomly sampled from a uniform distribution to convert energy expenditure to oxygen consumption (Johnson et al., 2000 and section 5.3 of APEX TSD).
	Body mass	Randomly selected from population-weighted lognormal distributions with age- and gender-specific geometric mean (GM) and geometric standard deviation (GSD) derived from the National Health and Nutrition Examination Survey (NHANES) for the years 1999-2004 (Isaacs and Smith (2005) and section 5.3 of APEX TSD).
	Height	Values randomly sampled from distributions used are based on equations developed for each gender by Johnson (1998) using height and weight data from Brainard and Burmaster (1992) (also see Appendix B of 2010 CO REA).
	Body surface area	Point estimates of exponential parameters used for calculating body surface area as a function of body mass (Burmaster, 1998)

Component	Variability Source	Comment
	Ventilation rate	Event-level activity-specific regression equations stratified by four age groups, using age, gender, body mass normalized oxygen consumption rate as independent variables, and accounting for intra and interpersonal variability (Graham and McCurdy, 2005).
	Fatigue and EPOC	APEX approximates the onset of fatigue, controlling for unrealistic or excessive exercise events in each persons activity time-series while also estimating excess post-exercise oxygen consumption (EPOC) that may occur following vigorous exertion activities (Isaacs et al., 2007 and section 7.2 of APEX TSD).

Table 5D-2. Important components of co-variability in exposure modeling.

Type of Co-variability	Modeled by APEX?	Treatment in APEX / Comments
Within-person correlations [1]	Yes	Sequence of activities performed, microenvironments visited, and general physiological parameters (body mass, height, ventilation rates).
Between-person correlations	No	Judged as not important.
Correlations between profile variables and microenvironment parameters	Yes	Profiles are assigned microenvironment parameters.
Correlations between demographic variables (e.g., age, gender) and activities	Yes	Age and gender are used in activity diary selection.
Correlations between activities and microenvironment parameters	No	Perhaps important, but do not have data. For example, frequency of opening windows when cooking or smoking tobacco products.
Correlations among microenvironment parameters in the same microenvironment	Yes	Modeled with joint conditional variables.
Correlations between demographic variables and air quality	Yes	Modeled with the spatially varying demographic variables and air quality input to APEX.
Correlations between meteorological variables and activities	Yes	Temperature is used in activity diary selection.
Correlations between meteorological variables and microenvironment parameters	Yes	The distributions of microenvironment parameters can be functions of temperature.
Correlations between drive times in CHAD and commute distances traveled	Yes	CHAD diary selection is weighted by commute times for employed persons during weekdays.
Consistency of occupation/school microenvironmental time and time spent commuting/busing for individuals from one working/school day to the next.	No	Simulated individuals are assigned activity diaries longitudinally without regard to occupation or school schedule (note though, longitudinal variable used to develop annual profile is time spent outdoors).

[1] The term correlation is used to represent linear and nonlinear relationships.

5D-3. CHARACTERIZATION OF UNCERTAINTY

While it may be possible to capture a range of exposure or risk values by accounting for variability inherent to influential factors, the true exposure or risk for any given individual within a study area is unknown, though can be estimated. To characterize health risks, exposure and risk assessors commonly use an iterative process of gathering data, developing models, and estimating exposures and risks, given the goals of the assessment, scale of the assessment performed, and limitations of the input data available. However, significant uncertainty often remains and emphasis is then placed on characterizing the nature of that uncertainty and its impact on exposure and risk estimates.

In the final 2008 O_3 NAAQS rule,[1] EPA staff performed such a characterization and at that time, identified the most important uncertainties affecting the exposure estimates. The key elements of uncertainty were 1) the modeling of human activity patterns over an O_3 season, 2) the modeling of variations in ambient O_3 concentrations near roadways, 3) the modeling of air exchange rates that affect the amount of O_3 that penetrates indoors, and 4) the characterization of energy expenditure (and related ventilation rate estimates) for children engaged in various activities. Further, the primary findings of a quantitative Monte Carlo analysis also performed at that time indicated that the overall uncertainty of the APEX estimated exposure distributions was relatively small: the percent of children or asthmatic children with exposures above 0.06, 0.07, or 0.08 ppm-8hr under moderate exertion have 95% were estimated by APEX to have uncertainty intervals of at most ±6 percentage points. Details for these previously identified uncertainties are discussed in the 2007 O_3 Staff Paper (section 4.6) and in a technical memorandum describing the 2007 O_3 exposure modeling uncertainty analysis (Langstaff, 2007).

The REA's conducted for the most recent NO_2 (US EPA, 2008), SO_2 (US EPA, 2009), and CO (US EPA, 2010) NAAQS reviews also presented characterizations of the uncertainties associated with APEX exposure modeling (among other pollutant specific issues), albeit mainly qualitative evaluations. Conclusions drawn from all of these assessments regarding exposure modeling uncertainty have been integrated here, following the standard approach used by EPA staff since 2008 and outlined by WHO (2008) to identify, evaluate, and prioritize the most important uncertainties relevant to the estimated potential health effect endpoints used in this O_3 REA. Staff selected the qualitative approach used for this first draft O_3 REA as a step towards developing an appropriate probabilistic uncertainty analysis, perhaps similar to that performed at the time of the 2007 O_3 REA by Langstaff (2007).

---

[1] Federal Register Vol. 73, No. 60. Available at: http://www.epa.gov/ttn/naaqs/standards/ozone/fr/20080327.pdf

The qualitative approach used in this first draft O_3 REA varies from that described by WHO (2008) in that a greater focus was placed on evaluating the direction and the magnitude[2] of the uncertainty; that is, qualitatively rating how the source of uncertainty, in the presence of alternative information, may affect the estimated exposures and health risk results. In addition and consistent with the WHO (2008) guidance, staff discuss the uncertainty in the knowledge base (e.g., the accuracy of the data used, acknowledgement of data gaps) and decisions made where possible (e.g., selection of particular model forms), although qualitative ratings were assigned only to uncertainty regarding the knowledge base.

First, staff identified the key aspects of the assessment approach that may contribute to uncertainty in the exposure and risk estimates and provided the rationale for their inclusion. Then, staff characterized the *magnitude* and *direction* of the influence on the assessment results for each of these identified sources of uncertainty. Consistent with the WHO (2008) guidance, staff subjectively scaled the overall impact of the uncertainty by considering the degree of uncertainty as implied by the relationship between the source of uncertainty and the exposure concentrations.

Where the magnitude of uncertainty was rated *low*, it was judged that changes within the source of uncertainty would have only a small effect on the exposure results. For example, we have commonly employed statistical procedure to substitute missing concentration values to complete the meteorological data sets. Staff has consistently compared the air quality distributions and found negligible differences between the substituted data set and the one with missing values (e.g., Tables 5-13 through 5-16 of US EPA, 2010), primarily because of the infrequency of missing value substitutions needed to complete a data set. There is still uncertainty in the approach used, and there may be alternative, and possibly better, methods available to perform such a task. However, in this instance, staff judged that the quantitative comparison of the ambient concentration data sets indicates that there would likely be little influence on exposure estimates by the data substitution procedure used.

A magnitude designation of *moderate* implies that a change within the source of uncertainty would likely have a moderate (or proportional) effect on the results. For example, the magnitude of uncertainty associated with using the quadratic approach to represent a hypothetical future air quality scenario was rated as *low-moderate*. While we do not have information regarding how the ambient O_3 concentration distribution might look in the future, we do know however what the distribution might look like based on historical trends and the emission sources. These historical data and trends serve to generate algorithms used to adjust air quality. If these trends in observed concentrations and emissions were to remain constant in the

---

[2] This is synonymous with the "level of uncertainty" discussed in WHO (2008), section 5.1.2.2.

future, then the magnitude of the impact to estimated exposures in this assessment would be judged as likely *low* or having negligible impact on the estimated exposures. However, if there are entirely new emission sources in the future or if the approach developed is not equally appropriate across the range of assessed study areas, the magnitude of influence might be judged as greater. For example, when comparing exposure estimates for one year that used three different 3-year periods to adjust that year's air quality levels to just meet the current standard, staff observed mainly proportional differences (e.g., a factor of two or three) in the estimated number of persons exposed in more than half of the twelve study areas (Langstaff, 2007). Assuming that these types of ambient concentration adjustments could reflect the addition of a new or unaccounted for emission source in a particular study area, staff also judged the magnitude of influence in using the quadratic approach to adjust air quality data to represent a hypothetical future scenario as *moderate*. A characterization of *high* implies that a small change in the source would have a large affect on results, potentially an order of magnitude or more. This rating would be used where the model estimates were extremely sensitive to the identified source of uncertainty.

In addition to characterizing the magnitude of uncertainty, staff also included the direction of influence, indicating how the source of uncertainty was judged to affect estimated exposures or risk estimates; either the estimated values were possibly *over-* or *under-estimated*. In the instance where the component of uncertainty can affect the assessment endpoint in either direction, the influence was judged as *both*. Staff characterized the direction of influence as *unknown* when there was no evidence available to judge the directional nature of uncertainty associated with the particular source. Staff also subjectively scaled the knowledge-base uncertainty associated with each identified source using a three-level scale: *low* indicated significant confidence in the data used and its applicability to the assessment endpoints, *moderate* implied that there were some limitations regarding consistency and completeness of the data used or scientific evidence presented, and *high* indicated the extent of the knowledge-base was extremely limited.

The output of the uncertainty characterization is a summary describing, for each identified source of uncertainty, the magnitude of the impact and the direction of influence the uncertainty may have on the exposure and risk characterization results.

5D-4. REFERENCES

Brainard J and Burmaster D. (1992). Bivariate distributions for height and weight of men and women in the United States. *Risk Analysis*. 12(2):267-275.

Burmaster DE. (1998). Lognormal distributions for skin area as a function of body weight. *Risk Analysis*. 18(1):27-32.

CDC. (2011). Summary Health Statistics for U.S. Adults: National Health Interview Survey, years 2006-10. U.S. Department of Health and Human Services, Hyattsville, MD. Data and documentation available at: http://www.cdc.gov/nchs/nhis.htm (accessed October 4, 2011).

Glen G, Smith L, Isaacs K, McCurdy T, Langstaff J. (2008). A new method of longitudinal diary assembly for human exposure modeling. *J Expos Sci Environ Epidem*. 18:299-311.

Graham SE and T McCurdy. (2005). Revised ventilation rate (VE) equations for use in inhalation-oriented exposure models. Report no. EPA/600/X-05/008. Report is found within Appendix A of US EPA (2009). Metabolically Derived Human Ventilation Rates: A Revised Approach Based Upon Oxygen Consumption Rates (Final Report). Report no. EPA/600/R-06/129F. Appendix D contains "Response to peer-review comments on Appendix A", prepared by S. Graham (US EPA). Available at: http://cfpub.epa.gov/ncea/cfm/recordisplay.cfm?deid=202543

Issacs K and Smith L. (2005). New Values for Physiological Parameters for the Exposure Model Input File Physiology.txt. Technical memorandum to Tom McCurdy, NERL WA10. December 20, 2005. Provided in Appendix A of the CO REA (US EPA, 2010).

Isaacs K, Glen G, McCurdy T., and Smith L. (2007). Modeling energy expenditure and oxygen consumption in human exposure models: Accounting for fatigue and EPOC. *J Expos Sci Environ Epidemiol*. 18(3):289-98.

Johnson T. (1998). Memo No. 5: Equations for Converting Weight to Height Proposed for the 1998 Version of pNEM/CO. Memorandum Submitted to U.S. Environmental Protection Agency. TRJ Environmental, Inc., 713 Shadylawn Road, Chapel Hill, North Carolina 27514.

Johnson T, Mihlan G, LaPointe J, Fletcher K, Capel J, Rosenbaum A, Cohen J, Stiefer P. (2000). Estimation of carbon monoxide exposures and associated carboxyhemoglobin levels for residents of Denver and Los Angeles using pNEM/CO. Appendices. EPA constract 68-D6-0064.

Langstaff JE. (2007). OAQPS Staff Memorandum to Ozone NAAQS Review Docket (OAR-2005-0172). Subject: Analysis of Uncertainty in Ozone Population Exposure Modeling. [January 31, 2007]. Available at: http://www.epa.gov/ttn/naaqs/standards/ozone/s_o3_cr_td.html

McCurdy T. (2000). Conceptual basis for multi-route intake dose modeling using an energy expenditure approach. *J Expos Anal Environ Epidemiol.* 10:1-12.

Schofield WN. (1985). Predicting basal metabolic rate, new standards, and review of previous work. *Hum Nutr Clin Nutr.* 39C(S1):5-41.

US Census Bureau. (2007a). Employment Status: 2000- Supplemental Tables. Available at: http://www.census.gov/population/www/cen2000/phc-t28.html.

US Census Bureau. (2007b). 2000 Census of Population and Housing. Summary File 3 (SF3) Technical Documentation, available at: http://www.census.gov/prod/cen2000/doc/sf3.pdf. Individual SF3 files '30' (for income/poverty variables pct49-pct51) for each state were downloaded from: http://www2.census.gov/census_2000/datasets/Summary_File_3/.

US EPA. (2002). EPA's Consolidated Human Activities Database. Available at: http://www.epa.gov/chad/.

US EPA. (2007). Ozone Population Exposure Analysis for Selected Urban Areas. Office of Air Quality Planning and Standards, U.S. Environmental Protection Agency, Research Triangle Park, NC. Available at: http://www.epa.gov/ttn/naaqs/standards/ozone/s_o3_cr_td.html

US EPA. (2008). Risk and Exposure Assessment to Support the Review of the NO_2 Primary National Ambient Air Quality Standard. Report no. EPA-452/R-08-008a. November 2008. Available at: http://www.epa.gov/ttn/naaqs/standards/nox/data/20081121_NO2_REA_final.pdf.

US EPA. (2009). Risk and Exposure Assessment to Support the Review of the SO_2 Primary National Ambient Air Quality Standard. Report no. EPA-452/R-09-007. August 2009. Available athttp://www.epa.gov/ttn/naaqs/standards/so2/data/200908SO2REAFinalReport.pdf.

US EPA. (2010). Quantitative Risk and Exposure Assessment for Carbon Monoxide – Amended. EPA Office of Air Quality Planning and Standards. EPA-452/R-10-009. July 2010. Available at: http://www.epa.gov/ttn/naaqs/standards/co/data/CO-REA-Amended-July2010.pdf

US EPA. (2012). Total Risk Integrated Methodology (TRIM) - Air Pollutants Exposure Model Documentation (TRIM.Expo / APEX, Version 4.4) Volume I: User's Guide. Office of Air Quality Planning and Standards, U.S. Environmental Protection Agency, Research Triangle Park, NC. EPA-452/B-12-001a. Available at: http://www.epa.gov/ttn/fera/human_apex.html

WHO. (2008). Harmonization Project Document No. 6. Part 1: Guidance document on characterizing and communicating uncertainty in exposure assessment. Available at: http://www.who.int/ipcs/methods/harmonization/areas/exposure/en/.

Appendix 5-E

Updated Analysis of Air Exchange Rate Data

MEMORANDUM

To: John Langstaff

From: Jonathan Cohen, Hemant Mallya, Arlene Rosenbaum

Date: 28 December, 2012

Re: Updated Analysis of Air Exchange Rate Data

EPA is planning to use the APEX exposure model to estimate ozone exposure in 16 cities / metropolitan areas: Atlanta, GA; Baltimore, MD; Boston, MA; Chicago, IL; Cleveland, OH; Dallas, TX: Detroit, MI; Denver, CO: Houston, TX; Los Angeles, CA; New York, NY; Philadelphia, PA; Sacramento, CA; Seattle, WA; St. Louis, MO-IL; Washington, DC. As part of this effort, ICF International has developed distributions of residential and non-residential air exchange rates (AER) for use as APEX inputs for the cities to be modeled. This memorandum describes the analysis of the AER data and the proposed APEX input distributions. Also included in this memorandum are proposed APEX inputs for penetration and proximity factors for selected microenvironments.

Residential Air Exchange Rates

Studies. Residential air exchange rate (AER) data were obtained from the following seven studies and summarized in Table 1:

> **Avol:** Avol et al., 1998. In this study, ozone concentrations and AERs were measured at 126 residences in the greater Los Angeles metropolitan area between February and December, 1994. Measurements were taken in four communities: Lancaster, Lake Gregory, Riverside, and San Dimas. Data included the daily average outdoor temperature, the presence or absence of an air conditioner (either central or room), and the presence or absence of a swamp (evaporative) cooler. Air exchange rates were computed based on the total house volume and based on the total house volume corrected for the furniture. These data analyses used the study corrected AERs.
>
> **RTP Panel:** Williams et al., 2003a, 2003b. In this study particulate matter concentrations and daily average AERs were measured at 37 residences in central North Carolina during 2000 and 2001 (averaging about 23 AER measurements per residence). The residences belong to two specific cohorts: a mostly Caucasian, non-smoking group aged at least 50 years having cardiac defibrillators living in Chapel Hill; a group of non-smoking, African Americans aged at least 50 years with controlled hypertension living in a low-to-moderate SES neighborhood in Raleigh. Data included the daily average outdoor temperature, and the number of air conditioner units (either central or room). Every residence had at least one air conditioner unit.

RIOPA: Meng et al., 2004, Weisel et al., 2004. The Relationship of Indoor, Outdoor, and Personal Air (RIOPA) study was undertaken to estimate the impact of outdoor sources of air toxics to indoor concentrations and personal exposures. Volatile organic compounds, carbonyls, fine particles and AERs were measured once or twice at 310 non-smoking residences from summer 1999 to spring 2001. Measurements were made at residences in Elizabeth, NJ, Houst on TX, and Los Angeles CA. Residences in California were randomly selected. Residences in New Jersey and Texas were preferentially selected to be close (< 0.5 km) to sources of air toxics. The AER measurements (generally over 24 hours) used a PMCH tracer. Data included the daily average outdoor temperature, and the presence or absence of central air conditioning, room air conditioning, or a swamp (evaporative) cooler.

TEACH: Chillrud at al., 2004, Kinney et al., 2002, Sax et al., 2004. The Toxic Exposure Assessment, a Columbia/Harvard (TEACH) study was designed to characterize levels of and factors influencing exposures to air toxics among high school students living in inner-city neighborhoods of New York City and Los Angeles, CA. Volatile organic compounds, aldehydes, fine particles, selected trace elements, and AER were measured at 87 high school student's residences in New York City and Los Angeles in 1999 and 2000. Data included the presence or absence of an air conditioner (central or room) and hourly outdoor temperatures (which were converted to daily averages for these analyses).

Wilson 1984: Wilson et al., 1986, 1996. In this 1984 study, AER and other data were collected at about 600 southern California homes with three seven-day tests (in March and July 1984, and January 1985) for each home. We obtained the data directly from Mr. Wilson. The available data consisted of the three seven-day averages, the month, the residence zip code, the presence or absence of a central air conditioner, and the presence or absence of a room air conditioner. We matched these data by month and zip code to the corresponding monthly average temperatures obtained from EPA's SCRAM website as well as from the archives in www.wunderground.com (personal and airport meteorological stations). Residences more than 25 miles away from the nearest available meteorological station were excluded from the analysis. For our analyses, the city/location was defined by the meteorological station, since grouping the data by zip code would not have produced sufficient data for most of the zip codes.

Wilson 1991: Wilson et al., 1996, Colome et al., 1993, 1994. In this 1991 study, AER and other data were collected at about 300 California homes with one two-day test in the winter for each home. We obtained the data directly from Mr. Wilson. The available data consisted of the two-day averages, the date, city name, the residence zip code, the presence or absence of a central air conditioner, the presence or absence of a swamp (evaporative) cooler, and the presence or absence of a room air conditioner. We matched these data by date, city, and zip code to the corresponding daily average temperatures obtained from EPA's SCRAM website as well as from the archives in www.wunderground.com (personal and airport meteorological stations). Residences more than 25 miles away from the nearest available meteorological station were excluded from the analysis. For our analyses, the city/location was defined by the meteorological station, since grouping the data by zip code would not have produced sufficient data for most of the zip codes.

Murray and Burmaster: Murray and Burmaster (1995). For this article, Murray and Burmaster corrected and compiled nationwide residential AER data from several studies conducted between 1982 and 1987. These data were originally compiled by the Lawrence Berkeley National Laboratory. We acknowledge Mr. Murray's assistance in obtaining these data for us. The available data consisted of AER measurements, dates, cities, and degree-days. Information on air conditioner presence or absence was not available.

DEARS: Sheldon (2007). The Detroit Exposure and Aerosol Research Study (DEARS) collected air exchange rate data as well as $PM_{2.5}$ and other air pollutant data at about 120 homes in Detroit, Michigan, for 3 years starting in the summer of 2004. Each home was sampled for 5 days in the winter and/or 5 days in the summer. The available data included AER measurements, dates, average, minimum, and maximum temperatures, the use or non-use of air/conditioners during each measurement day, the use or non-use of window fans during each measurement day, and the number of minutes per day that the windows were open.

For each of the studies, air conditioner usage, window status (open or closed), and fan status (on or off) was not part of the experimental design, although some of these studies included information on whether air conditioners or fans were used (and for how long) and whether windows were closed during the AER measurements (and for how long).

As indicated in the above summaries, random selection was not used to identify homes to include in some of the studies: The RTP Panel study selected two specific cohorts of older subjects with specific diseases. The RIOPA study was biased towards residences near air toxics sources. The TEACH study focused on inner-city neighborhoods. Nevertheless, we included all these studies because we determined that any potential selection bias would be likely to be small and we preferred to keep as much data as possible. The DEARS study selected homes from certain neighborhoods in Detroit. The proportion of the DEARS study homes that used A/C on one or more survey days was 57%, and the proportion of DEARS study homes with some daily temperatures above 25 °C that used A/C on one or more of those hot days was 73%. The American Housing Survey shows that for the Detroit metropolitan area, the proportion of homes with A/C was 90% (see Table 2 below), but 71% for the city of Detroit. This suggests that the DEARS study sample may be representative of an older housing stock than the overall Detroit metropolitan area.

All data and statistical results are compiled into the attached Excel spreadsheet Summary_Statistic.Dec 2012.xls.

Table 1. Summary of Studies of Residential Air Exchange Rates.

Study/ Attribute	Avol	RTP Panel	RIOPA	TEACH	Wilson 1984	Wilson 1991	Murray and Burmaster	DEARS
Locations	Lancaster, Lake Gregory, Riverside, San Dimas. All in Southern CA	Research Triangle Park, NC	CA; NJ; TX	Los Angeles, CA; New York City, NY	Southern CA	Southern CA	AZ, CA, CO, CT, FL, ID, MD, MN, MT, NJ	Detroit, MI
Years	1994	2000; 2001	1999; 2000; 2001	1999; 2000	1984, 1985	1984	1982 – 1987	2004 – 2006
Months/ Seasons	Feb; Mar; Apr; May; Jun; Jul; Aug; Sep; Oct; Nov	2000 (Jun; Jul; Aug; Sep; Oct; Nov), 2001 (Jan; Feb; Apr; May)	1999 (July to Dec); 2000 (all months); 2001 (Jan and Feb)	1999 (Feb; Mar; Apr; Jul; Aug); 2000 (Jan; Feb; Mar; Sep; Oct)	Mar 1984, Jul 1984, Jan 1985	Jan, Mar, Jul	Various	Jan. Feb, Mar, Jul. Aug
Homes with AER Measurements	86	37	284	85	581	288	1,884	127
Total AER Measurements	161	854	524	151	1,362	316	2,844	868
Average AER Measurements per Home	1.87	23.08	1.85	1.78	2.34	1.10	1.51	6.83
AER Measurement Duration	Not Available	24 hour	24 to 96 hours	Sample time (hours) reported. Ranges from about 1 to 7 days	7 days	7 days	Not available	24 hour
AER Measurement Technique	Not Available	Perflourocarbon tracer	PMCH tracer	Perflourocarbon tracer	Perflourocarbon tracer	Perflourocarbon tracer	Not available	Perflourocarbon tracer
Min AER (h^{-1})	0.01	0.02	0.08	0.12	0.03	0.01	0.01	0.08
Max AER (h^{-1})	2.70	21.44	87.50	8.87	11.77	2.91	11.77	13.56
Mean AER (h^{-1})	0.80	0.72	1.41	1.71	1.05	0.57	0.76	0.80
Min Temp. (°C)	-0.04	-2.18	-6.82	-1.36	11.00	3.00	Not available	-12.92
Max Temp. (°C)	36.25	30.81	32.50	32.00	28.00	25.00	Not available	30.79
Air Conditioner Categories	No A/C; Central or Room A/C; Swamp Cooler only; Swamp + [Central or Room]	Central or Room A/C (Y/N)	Window A/C (Y/N); Evap Coolers (Y/N)	Central or Room A/C (Y/N)	Central A/C (Y/N); Room A/C (Y/N);	Central A/C (Y/N); Room A/C (Y/N); Swamp Cooler(Y/N)	Not available	Central or Room A/C (Y/N)

Study/ Attribute	Avol	RTP Panel	RIOPA	TEACH	Wilson 1984	Wilson 1991	Murray and Burmaster	DEARS
Air Conditioner Measurements Made	A/C use in minutes	Not Available	Duration measurements in Hrs and Mins	Not Available	Not Available	Not Available	Not available	Not available
Fan Categories	Not available	Fan (Y/N)	Fan (Y/N)	Not Available	Not Available	Not Available	Not available	Fan (Y/N)
Fan Measurements Made	Time on or off for various fan types during sampling was recorded, but not included in database provided.	Not Available	Duration measurements in Hours and Minutes	Not Available	Not Available	Not Available	Not available	Daily duration in minutes
Window Open/ Closed Data	Duration open between times 6am-12 pm; 12pm - 6 pm; and 6pm - 6am	Windows (open / closed along with duration open in inch-hours units	Windows (Open / Closed) along with window open duration measurements	Not Available	Not Available	Not Available	Not available	Windows (Open / Closed) along with window open duration measurements
Comments			CA sample was a random sample of homes. NJ and TX homes were del berately chosen to be near to ambient sources.	Restricted to inner-city homes with high school students.	Contemporaneous temperature data obtained for these analyses from SCRAM and www.wunderground.com meteorological data.	Contemporaneous temperature data obtained for these analyses from SCRAM and www.wunderground.com meteorological data.		

We compiled the data from these eight studies to create the following variables, of which some had missing values:

- Study
- Date
- Time – Time of the day that the AER measurement was made
- House_ID – Residence identifier
- Measurement_ID – Uniquely identifies each AER measurement for a given study
- AER – Air Exchange Rate (per hour)
- AER_Duration – Length of AER measurement period (hours)
- Have_AC – Indicates if the residence has any type of air conditioner (A/C), either a room A/C or central A/C or swamp cooler or any of them in combination. “Y” = “Yes.” “N” = “No”
- Type_of_AC1 – Indicates the types of A/C or swamp cooler available in each house measured. Possible values: “Central A/C” “Central and Room A/C” “Central or Room A/C” “No A/C” “Swamp + (Central or Room)” “Swamp Cooler only” “Window A/C” “Window and Evap” “ ” (missing)
- Type_of_AC2 – Indicates if a house measured has either no A/C or some A/C. Possible values: “No A/C” “Central or Room A/C” “ ” (missing)
- Type_of_AC3 – Indicates if a house measured has either no A/C, central A/C, or room A/C. If separate A/C information is available for central A/C and for room A/C, then homes with both central and room A/C are coded as having “Central A/C.” Possible values: “No A/C” “Central A/C” “Room A/C“ “” (missing)
- Have_Fan – Indicates if the house studied has any fans
- Mean_Temp – Daily average outside temperature (°C)
- Min_Temp – Minimum hourly outside temperature (°C)
- Max_Temp – Maximum hourly outside temperature (°C)
- State
- City
- Location – Two character abbreviation
- Flag – Data status. Murray and Burmaster study: “Used” or “Not Used.” Other studies: “Used”; “Missing” (missing values for AER, Type_of_AC2, and/or Mean_Temp); “Outlier”.

Note that in the Wilson 1991 study, one of the Los Angeles values was recorded as having an “Unknown” air conditioning type. In this analysis, this measurement is treated as having missing A/C information and is not included in the main data analyses.

Note that for the DEARS study, the available data did not include the presence or absence of an air conditioner or window fan. Instead, the data set included a flag indicating whether or not an air conditioner or window fan was used on that day. For the data analysis, we assumed that if the air conditioner or window fan was used on at least one day, then the air conditioner or window fan was present; otherwise the air conditioner or window fan was absent. Since the maximum observed temperature during the study days was 31°C and some of the homes did not have measurements on hot summer days, some of the homes designated as having no A/C may have been mistakenly designated. For the DEARS study the type of air conditioner (central or room) was not provided.

The compiled data base is the attached PC SAS dataset complete_database2.sas7bdat.

The main data analysis was based on the seven studies other than the Murray and Burmaster study. The Murray and Burmaster data were excluded because of the absence of information on air conditioner presence. (However, a subset of these data was used for a supplementary analysis described below.)

Based on our review of the AER data we excluded nine values with unrealistically high AER values – above 10 per hour (see the worksheet "OUTLIERS"). The main data analysis used all the remaining data that had non-missing values for AER, Type_of_AC2, and Mean_Temp. For the main data analysis, we decided to base the A/C type variable on the broad characterization "No A/C" versus "Central or Room A/C" since this variable could be calculated from all of the studies (excluding Murray and Burmaster). Information on the presence or absence of swamp coolers was not available from all the studies, and, also importantly, the corresponding information on swamp cooler prevalence for the subsequent ozone modeling cities was not available from the American Housing Survey. It is plausible that AER distributions depend upon the presence or absence of a swamp cooler. It is also plausible that AER distributions also depend upon whether the residence specifically has a central A/C, room or window A/C, or both. However we determined to use the broader A/C type definition, which in effect assumes that the exact A/C type and the presence of a swamp cooler are approximately proportionately represented in the surveyed residences. The detailed A/C type variable Type_of_AC3 was used for the Los Angeles AER distributions.

Most of the studies had more than one AER measurement for the same house. It is reasonable to assume that the AER varies with the house as well as other factors such as the temperature. (The A/C type can be assumed to be the same for each measurement of the same house). We expected the temperature to be an important factor since the AER will be affected by the use of the available ventilation (air conditioners, windows, fans), which in turn will depend upon the outside meteorology. Therefore it is not appropriate to average data for the same house under different conditions, which might have been one way to account for dependence between multiple measurements on the same house. To simplify the data analysis, we chose to ignore possible dependence between measurements on the same house on different days and treat all the AER values as if they were statistically independent.

Summary Statistics. We computed summary statistics for AER and its natural logarithm LOG_AER on selected strata defined from the study, city, A/C type, and mean temperature. Cities were defined as in the original databases, except that for Los Angeles we combined all the data in the Los Angeles ozone modeling region, i.e., the counties of Los Angeles, Orange, Ventura, Riverside, and San Bernardino. A/C type was defined from the Type_of_AC2 variable, which we abbreviated as "NA" = "No A/C" and "AC" = "Central or Room A/C". The mean temperature was grouped into the following temperature bins: -15 to 0 ºC, 0 to 10 ºC, 10 to 20 ºC, 20 to 25 ºC, 25 to 30 ºC, 30 to 40 ºC (Values equal to the lower bounds are excluded from each interval). Also included were strata defined by study = "All" and/or city = "All," and/or A/C type = "All" and/or temperature bin = "All". The following summary statistics for AER and LOG_AER respectively are shown in the worksheets "SUMMARY_STATISTICS_AER" and "SUMMARY_STATISTICS_LOG_AER":

- Number of values
- Arithmetic Mean
- Arithmetic Standard Deviation
- Arithmetic Variance
- Deciles (Min, 10th, 20th … 90th percentiles, Max)

These calculations exclude all nine outliers and results are not shown for strata with 10 or fewer values, since those summary statistics are extremely unreliable.

Examination of these summary tables clearly demonstrates that the AER distributions vary greatly across cities and A/C types and temperatures, so that the selected AER distributions for the modeled cities should also depend upon the city, A/C type and temperature. For example, the mean AER for residences with A/C ranges from 0.39 for Los Angeles between 30 and 40 °C to 1.73 for New York between 20 and 25 °C. The mean AER for residences without A/C ranges from 0.46 for San Francisco between 10 and 20 °C upwards to 2.54 for Detroit between 25 and 30 °C. The need to account for the city as well as the A/C type and temperature is illustrated by the result that for residences with A/C and between 20 and 25 °C, the mean AER ranges from 0.52 for Research Triangle Park to 1.73 for New York. Statistical comparisons are described below.

Statistical Comparisons. Various statistical comparisons between the different strata are shown in the worksheets COMPARISON_STATISTICS_AER and COMPARISON_STATISTICS_LOG_AER, for the AER and its logarithm, respectively. The various strata are defined as in the Summary Statistics section, excluding the "All" cases. For each analysis, we fixed one or two of the variables Study, City, A/C type, temperature, and tested for statistically significant differences among other variables. The comparisons are listed in Table 2.

Table 2. Summary of Comparisons of Means.

Comparison Analysis Number	Comparison Variable(s) "Groups Compared"	Stratification Variable(s) (not missing in worksheet)	Total Comparisons	Cases with significantly different means (5 % level)	
				AER	Log AER
1	City	Type of A/C AND Temp. Range	12	8	9
2	Temp. Range	Study AND City	13	6	6
3	Type of A/C	Study AND City	16	6	6
4	City	Type of A/C	2	2	2
5	City	Temp. Range	6	6	6
6	Type of A/C AND Temp. Range	Study AND City	18	7	7

For example, the first set of comparisons fix the Type of A/C and the temperature range; there are twelve such combinations. For each of these twelve combinations, we compare the AER distributions across different cities. This analysis determines whether the AER distribution is appropriately defined by the A/C type and temperature range, without specifying the city. Similarly, for the sixth set of comparisons, the study and city are held fixed (18 combinations) and in each case we compare AER distributions across groups defined by the combination of the A/C type and the temperature range.

The F-Statistic comparisons compare the mean values between groups using a one way analysis of variance (ANOVA). This test assumes that the AER or log(AER) values are normally distributed with a mean that may vary with the comparison variable(s) and a constant variance. Shown in the worksheets are the F-Statistic and its p-value. P-values above 0.05 indicate cases where all the group means are not statistically significantly different at the 5 percent level. Those results are summarized in the last two columns of the above table "Summary of Comparisons of Means" which gives the number of cases where the means are significantly different. Comparison analyses 2, 3, and 6 show that for a given study and city, slightly less than half of the comparisons show significant differences in the means across temperature ranges, A/C types, or both. Comparison analyses 1, 4, and 5 show that for the majority of cases, means vary significantly across cities, whether you first stratify by temperature range, A/C type, or both.

The Kruskal-Wallis Statistic comparisons are non-parametric tests that are extensions of the more familiar Wilcoxon tests to two or more groups. The analysis is valid if the AER minus the group median has the same distribution for each group, and tests whether the group medians are equal. (The test is also consistent under weaker assumptions against more general alternatives). The P-values show similar patterns to the parametric F-test comparisons of the means. Since the logarithm is a strictly increasing function and the test is non-parametric, the Kruskal-Wallis tests give identical results for AER and Log (AER).

The Mood Statistic comparisons are non-parametric tests that compare the scale statistics for two or more groups. The scale statistic measures variation about the central value, which is a non-parametric generalization of the standard deviation. Specifically, suppose there is a total of N AER or log(AER) values, summing across all the groups. These N values are ranked from 1 to N, and the j'th highest value is given a score of $\{j - (N+1)/2\}^2$. The Mood statistic uses a one way ANOVA statistic to compare the total scores for each group. Generally, the Mood statistics show that in most cases the scale statistics are not statistically significantly different. Since the logarithm is a strictly increasing function and the test is non-parametric, the Mood tests give identical results for AER and Log (AER).

Fitting Distributions. Based on the summary statistics and the statistical comparisons, the need to fit different AER distributions to each combination of A/C type, city, and temperature is apparent. For each combination with a minimum of 11 AER values, we fitted and compared exponential, log-normal, normal, and Weibull distributions to the AER values.

The first analysis used the same stratifications as in the above "Summary Statistics" and "Statistical Comparisons" sections. Results are not reported for all strata because of the minimum data requirement of 11 values. Results for each combination of A/C type, city, and temperature (i.e., A, C, and T) are given in the worksheet FIT_STATISTICS_ACT. Each combination has four rows, one for each fitted distribution. For each distribution we report the fitted parameters (mean, standard deviation, scale, shape) and the p-value for three standard goodness-of-fit tests: Kolmogorov-Smirnov (K-S), Cramer-Von-Mises (C-M), Anderson-Darling (A-D). Each goodness-of-fit test compares the empirical distribution of the AER values to the fitted distribution. The K-S and C-M tests are different tests examining the overall fit, while the Anderson-Darling test gives more weight to the fit in the tails of the distribution. For each combination, the best-fitting of the four distributions has the highest p-value and is marked by an x in the final three columns. The mean and standard deviation (Std_Dev) are the values for the fitted distribution. The scale and shape parameters are defined by:

- Exponential: density = $\sigma^{-1} \exp(-x/\sigma)$, where shape = mean = σ
- Log-normal: density = $\{\sigma x\sqrt{(2\pi)}\}^{-1} \exp\{ -(\log x - \zeta)^2 / (2\sigma^2)\}$, where shape = σ and scale = ζ. Thus the geometric mean and geometric standard deviation are given by $\exp(\zeta)$ and $\exp(\sigma)$, respectively.
- Normal: density = $\{\sigma\sqrt{(2\pi)}\}^{-1} \exp\{ -(x - \mu)^2 / (2\sigma^2)\}$, where mean = μ and standard deviation = σ
- Weibull: density = $(c/\sigma)\ (x/\sigma)^{c-1} \exp\{-(x/\sigma)^c\}$, where shape = c and scale = σ

Generally, the log-normal distribution was the best-fitting of the four distributions, and so, for consistency, we recommend using the fitted log-normal distributions for all the cases. For the log-normal distributions, the worksheet includes the geometric mean and geometric standard deviation.

One limitation of the initial analysis was that distributions were available only for selected cities, and yet the summary statistics and comparisons demonstrate that the AER distributions depend upon the city as well as the temperature range and A/C type. As one option to address this issue, we considered modeling cities not listed in the FIT_STATISTICS_ACT worksheet by using the AER distributions across all cities and dates for a given temperature range and A/C type. Those fitted distributions are presented in the FIT_STATISTICS_AT worksheet.

Another important limitation of the initial analysis was that distributions were not fitted to all of the temperature ranges due to inadequate data. There are missing values between temperature ranges, and the temperature ranges are all bounded. To address this issue, the temperature ranges were regrouped to cover the entire range of temperatures from minus to plus infinity, although obviously the available data to fit these ranges have finite temperatures. Stratifying by A/C type, city, and the new temperature ranges produces the results in the worksheet FIT_STATISTICS_ACT2. Results are reported for five cities: Detroit, AC and NA; Houston, AC and NA; Los Angeles, AC and NA; New York, AC and NA; Research Triangle Park, AC. As noted above the DEARS study sample is likely to be representative of an older housing stock than for the Detroit metropolitan area as a whole. For this reason, the worksheet FIT_STATISTICS_ACT2 also reports results for the aggregation of the Detroit and New

York data ("Detroit or NYC," AC and NA); these results can be used for modeling Chicago, Cleveland and Detroit.

Corresponding to each of the fitted distributions in FIT_STATISTICS _ACT2, we created histograms to compare each of the fitted distributions with the empirical distributions. For these graphs, the A/C type, city, and temperature range combinations were assigned a letter code, as shown in the first column of the worksheet. The first three digits are a numerical code used for sorting the graphs, and the remaining characters code the A/C type, temperature range, and city in a fairly obvious manner. These graphs are in the attached file Graphs_ACT2.Dec 2012.rtf, which can be read directly by Word.

AER Distributions for Eight Cities. Based upon the FIT_STATISTICS_ACT2 results for the five cities and the corresponding graphs, we propose using those fitted distributions for the two cities Houston and New York. For Los Angeles, more data were available for the type of air conditioning system, and we decided to use distributions for the more detailed air conditioning type, Type_of_AC3, as described below.

For Atlanta, Boston, and Philadelphia, we propose using the distribution for one of the cities thought to have similar characteristics to the city to be modeled with respect to factors that might influence AERs. These factors include the age composition of housing stock, construction methods, and other meteorological variables not explicitly treated in the analysis, such as humidity and wind speed patterns.

As noted above the DEARS study sample is likely to be representative of an older housing stock than for the Detroit metropolitan area as a whole. Therefore, we combined the DEARS study data with the data from New York to create an aggregate distribution for application to Detroit and other cities with similar characteristics, i.e., Chicago and Cleveland.

The distributions proposed for the eight cities are as follows:

- Atlanta, GA: Use log-normal distributions for Research Triangle Park. Residences with A/C only.
- Boston, MA: Use log-normal distributions for New York.
- Chicago, IL: Use log-normal distributions for "Detroit or NYC."
- Cleveland, OH: Use log-normal distributions for "Detroit or NYC."
- Detroit, MI: Use log-normal distributions for "Detroit or NYC."
- Houston, TX: Use log-normal distributions for Houston.
- New York, NY: Use log-normal distributions for New York.
- Philadelphia, PA: Use log-normal distributions for New York.

All the above distributions are to be found in the FIT_STATISTICS_ACT2 worksheet. Since the AER data for Research Triangle Park was only available for residences with air conditioning, AER distributions for Atlanta residences without air conditioning are discussed below.

To avoid unusually extreme simulated AER values, we propose to set a minimum AER value of 0.01 and a maximum AER value of 10.

Obviously, we would be prefer to model each city using data from the same city, but this approach was chosen as a reasonable alternative, given the available AER data.

AER Distributions for Sacramento and St. Louis. For these two cities, a direct mapping to one of the five cities Detroit, Houston, Los Angeles, New York, and Research Triangle Park is not recommended because the cities are likely to be too dissimilar. Instead, we decided to use the distribution for the inland parts of Los Angeles to represent Sacramento and to use the aggregate distributions for all cities outside of California to represent St. Louis. The results are presented in the worksheet FIT_STATISTICS_ACT3 and corresponding histogram graphs Graphs_ACT3.Dec 2012.rtf. The results for the city denoted by "Sacramento" were obtained by combining all the available AER data for Sacramento, Riverside, and San Bernardino counties. The results for the city denoted by "St. Louis" were obtained by combining all non-California AER data. Thus our proposal is:

- Sacramento, CA: Use log-normal distributions for "Sacramento" in FIT_STATISTICS_ACT3
- St. Louis, MO-IL: Use log-normal distributions for "St. Louis" in FIT_STATISTICS_ACT3

To avoid unusually extreme simulated AER values, we propose to set a minimum AER value of 0.01 and a maximum AER value of 10.

AER Distributions for Baltimore and Washington DC. Baltimore and Washington DC were judged likely to have similar characteristics to Detroit, Research Triangle Park and New York City. To choose between these three cities, we compared the Murray and Burmaster AER data for Maryland with AER data from each of those cities. The Murray and Burmaster study included AER data for Baltimore and for Gaithersburg and Rockville, primarily collected in March. April, and May 1987, although there is no information on daily mean temperatures or A/C type. We collected all the March, April, and May AER data for Detroit, Research Triangle Park and New York City, and compared those three distributions with the Murray and Burmaster Maryland data for the same three months. The summary statistics for AER and Log (AER) are given in the worksheets SUM_STAT_MURRAY_AER and SUM_STAT_MURRAY_LOG_AER, using the same formats as the other summary statistics worksheets. The corresponding statistical comparisons between Detroit and Maryland, Research Triangle Park and Maryland, and between New York and Maryland, are shown in the worksheets COMP_STAT_MURRAY_AER and COMP_STAT_MURRAY_LOG_AER.

The results for the means and central values show significant differences at the 5 percent level between the New York and Maryland distributions and between the Detroit and Maryland distributions. Between Research Triangle Park and Maryland, the central values and the mean AER values are not statistically significantly different, and the differences in the mean log (AER) values are much less statistically significant than between New York and Maryland. The scale statistic comparisons are not statistically significantly different between New York and Maryland or between Detroit and

Maryland, but were statistically significantly different between Research Triangle Park and Maryland. Since matching central and mean values is generally more important than matching the scales, we propose to model both Baltimore and Washington DC residences with air conditioning using the Research Triangle Park distributions, stratified by temperature:

- Washington DC: Use log-normal distributions for Research Triangle Park in FIT_STATISTICS_ACT2. Residences with A/C only.
- Baltimore, MD: Use log-normal distributions for Research Triangle Park in FIT_STATISTICS_ACT2. Residences with A/C only.

Since the AER data for Research Triangle Park was only available for residences with air conditioning, the estimated AER distributions for Baltimore and Washington DC residences without air conditioning are discussed below.

To avoid unusually extreme simulated AER values, we propose to set a minimum AER value of 0.01 and a maximum AER value of 10.

AER Distributions for Washington DC, Baltimore MD, and Atlanta GA Residences With No A/C. For Atlanta, Baltimore, and Washington DC we have proposed to use the AER distributions for Research Triangle Park. However, all the Research Triangle Park data (from the RTP Panel study) were from houses with air conditioning, so there are no available distributions for the "No A/C" cases. For these three cities, one option is to use AER distributions fitted to all the study data for residences without A/C, stratified by temperature. These fitted distributions are given in the worksheet FIT_STATISTICS_AT2. The distributions for "No A/C" and "Central or Room A/C" are both presented for completeness, although we only propose applying the "No A/C" distributions for modeling these two cities for residences without A/C. However, since Atlanta, Baltimore, and Washington DC residences are expected to be better represented by residences outside of California, we instead propose to use the "No A/C" AER distributions aggregated across cities outside of California, which is the same as the recommended choice for the St. Louis "No A/C" AER distributions.

- Washington DC, No A/C: Use log-normal distributions for "St. Louis" in FIT_STATISTICS_ACT3. Residences without A/C only.
- Atlanta, GA, No A/C: Use log-normal distributions for "St. Louis" in FIT_STATISTICS_ACT3. Residences without A/C only.
- Baltimore, MD. No A/C: Use log-normal distributions for "St. Louis" in FIT_STATISTICS_ACT3. Residences without A/C only.

To avoid unusually extreme simulated AER values, we propose to set a minimum AER value of 0.01 and a maximum AER value of 10.

AER Distributions for Los Angeles. Los Angeles data were collected in the Avol, RIOPA, TEACH, Wilson 1984, and Wilson 1991 studies, and in most cases the data included whether or not the residence had central A/C, and whether or not the residence had room A/C. We decided to evaluate whether to stratify the Los Angeles air exchange rate distribution by this more detailed A/C type, defined by the variable Type_of_AC3, as well as by the temperature range. The Avol study included information on whether or not the residence had an A/C but not on whether or not the residence has Central A/C, Room A/C, or both. To avoid potential bias due to location we decided to exclude all the Avol data from these analyses, although we could have decided to include the 62 Avol study residences without A/C.

Summary statistics of the AER and its logarithm for Los Angeles, stratified by Type_of_AC3 and temperature range are shown in the worksheets SUM_STAT_LOS_ANGELES_AER and SUM_STAT_LOS_ANGELES_LOG_AER, respectively. The corresponding statistical comparisons between residences with Central A/C only and Room A/C only are shown in the worksheets COMPARE_LA_AER and COMPARE_LA_LOG_AER. These comparisons did not include the residences with no A/C, since the goal was to evaluate the need to further stratify the AER distribution by the specific A/C types, rather than only by the presence or absence of any A/C. For each of the temperature ranges 10-20 and 20-25, the statistical comparisons show statistically significant differences between Central A/C and Room A/C residences in the means and central values, except for the mean comparison of AER in the range 10-20 (p-value 0.06), and no statistically significant differences in the scale statistics. For the temperature range 25-30, the statistical comparisons show no statistically significant differences between Central A/C and Room A/C residences in the means and central values, and statistically significant differences in the scale statistics. For the temperature range 0-10, the statistical comparisons show no statistically significant differences between Central A/C and Room A/C residences. On this basis, we decided to stratify the Los Angeles AER distributions by the Type_of_AC3 air conditioner type variable as well as by the temperature range.

The worksheet FIT_STATISTICS_ACT4 compares exponential, log-normal, normal, and Weibull distributions fitted to each stratum of the Los Angeles data, defined by Type_of_AC3 and temperature range, using the temperature ranges "<= 20" "20-25" and ">25". Histograms of the fitted distributions are in the attached Graphs_ACT4.Dec 2012.rtf file, using the letter codes given in the first column of FIT_STATISTICS_ACT4. Since the tabulated goodness-of-fit statistics and histograms generally support the lognormal distribution, we propose to model Los Angeles using the Los Angeles data stratified by the detailed A/C type, Type_of_AC3, and by the temperature range:

- Los Angeles, CA: Use log-normal distributions for Los Angeles in FIT_STATISTICS_ACT4

A/C Type and Temperature Distributions. Since the proposed AER distribution is conditional on the A/C type and temperature range, these values also need to be simulated using APEX in order to select the appropriate AER distribution. Mean daily temperatures are one of the available APEX inputs for each modeled city, so that the temperature range can be determined for each modeled day according to the mean daily temperature. To simulate the A/C type, we obtained estimates of A/C prevalence from the American Housing Survey. Thus for each city/metropolitan area other than Los Angeles, we obtained the estimated fraction of residences with Central or Room A/C (see Table 3), which gives the probability p for selecting the A/C type "Central or Room A/C". Obviously, 1-p is the probability for "No A/C".

For comparison with Atlanta, Baltimore, and Washington DC, we have included the A/C type percentage for Charlotte, NC (representing Research Triangle Park, NC). As discussed above, we propose modeling the 96-98% of Atlanta, Baltimore and Washington DC residences with A/C using the Research Triangle Park AER distributions, and modeling the 2-4 % of Atlanta, Baltimore, and Washington DC residences without A/C using the combined study No A/C AER distributions (denoted as "St Louis" in FIT_STATISTICS_ACT3).

Table 3. Fraction of occupied residences with central or room A/C (from American Housing Survey).

CITY	SURVEY AREA & YEAR	PERCENTAGE
Atlanta	Atlanta, 2004	98.03, replace by 100*
Baltimore	Baltimore, 2007	96.70, replace by 100*
Boston	Boston, 2007	85.67
Chicago	Chicago, 2009	94.74
Cleveland	Cleveland, 2004	79.80
Denver	Denver, 2004	84.56
Detroit	Detroit, 2009	89.61
Houston	Houston, 2007	98.77
New York	New York, 2009	86.93
Philadelphia	Philadelphia, 2009	95.00
Sacramento	Sacramento, 2004	92.81
St. Louis	St. Louis, 2004	98.73
Washington DC	Washington DC, 2007	98.15, replace by 100*
Research Triangle Park	Charlotte NC, 2002	97.89

* See text

For Los Angeles, we have proposed to use different AER distributions depending upon whether or not the residence had central A/C, and whether or not the residence had room A/C. To simulate the A/C type, we obtained estimates of A/C prevalence from the American Housing Survey as shown in Table 4:

Table 4. Fraction of occupied Los Angeles CMSA residences with central or room A/C[1]

TYPE OF A/C	PERCENTAGE
No A/C	33.89
Room A/C	16.44
Central A/C	49.68

[1] American Housing Survey, Los Angeles-Long Beach, 2003, Anaheim-Santa Ana 2002, Riverside-San Bernardino-Ontario 2002

Other AER Studies

We have not used information from some additional residential and non-residential AER studies that might provide additional information or data. Indoor / outdoor ozone and PAN distributions were studied by Jakobi and Fabian (1997). Liu et al (1995) studied residential ozone and AER distributions in Toronto, Canada. Weschler and Shields (2000) describes a modeling study of ventilation and air exchange rates. Weschler (2000) includes a useful overview of residential and non-residential AER studies.

AER Distributions for Other Indoor Environments

To estimate AER distributions for non-residential, indoor environments (e.g., offices and schools), we obtained three AER data sets:

The early "Turk" data set (Turk et al, 1989) includes 40 AER measurements from offices (25 values), schools (7 values), libraries (3 values), and multi-purpose (5 values), each measured using an SF_6 tracer over two- or four-hours in different seasons of the year.

The more recent "Persily" data (Persily and Gorfain 2004; Persily et al. 2005) were derived from the US EPA Building Assessment Survey and Evaluation (BASE) study, which was conducted to assess indoor air quality, including ventilation, in a large number of randomly selected office buildings throughout the U.S. The data base consists of a total of 390 AER measurements in 96 large, mechanically ventilated offices; each office was measured up to four times over two days, Wednesday and Thursday AM and PM. The office spaces were relatively large, with at least 25 occupants, and preferably 50 to 60 occupants. AERs were measured both by a volumetric method and by a CO_2 ratio method, and included their uncertainty estimates. For these analyses, we used the recommended "Best Estimates" defined by the values with the lower estimated uncertainty; in the vast majority of cases the best estimate was from the volumetric method.

The most recent "Bennett" data (Bennett et al 2011) was a field study of 37 small and medium commercial buildings throughout California conducted in 2008 to 2010. The data base includes information on ventilation rate, temperature, and heating, ventilating, and air conditioning (HVAC) system characteristics. The study included: seven retail establishments; five restaurants; eight offices; two each of gas stations, hair salons, healthcare facilities, grocery stores, dental offices, and fitness gyms; and five other buildings. The selection of buildings was semi-randomized, providing a minimum coverage of the vast variety of SMCBs across space, age, size, and building use category. Buildings were almost evenly distributed across each of five regions of the state; north-coastal, north-inland, south-coastal, south-inland, and central-inland. Measurements were made in summer and winter, and three of the buildings were sampled twice, once in each season. Thus there were a total of 40 measured AER values. Whole building ventilation rates were determined with a tracer decay method using SF_6, a well established method for commercial buildings.

The office AER SAS databases used for these analyses are attached: turkdata.sas7bdat for the Turk study, basedata.sas7bdat for the Persily study, and bennettdata.sas7bdat for the Bennett study.

Due to the small sample size of the Turk data, the data were analyzed without stratification by building type and/or season. For the Persily data, the AER values for each office space were averaged, rather using the individual measurements, to account for the strong dependence of the AER measurements for the same office space over a relatively short period. For the Bennett data, analyses of variance of the AER values and of their logarithms confirmed the finding of the study authors (Bennett et al 2011) that the AER varied statistically significantly between restaurants and other buildings. Analyses of variance using a categorized outdoor temperature variable (15-25, 25-35, and 35 or more) also showed that the AER did not vary statistically significantly with the outdoor temperature. Although the study authors also reported significant differences depending upon whether or not doors were open, the stratification by whether or not doors are open would not be feasible for APEX modeling since that would require a model for whether or not doors are kept open (perhaps based on the outdoor temperature and some building characteristics). The study authors did not find a significant effect due to the presence or absence of HVAC systems. Therefore the Bennett data were stratified by whether or not the building is a restaurant.

Summary statistics of AER and log (AER) for the three studies are presented in Table 5.

The overall mean values are similar for the three studies, but the mean value for the restaurants in the Bennett study is almost four times the mean value for the non-restaurants. Compared to the Turk study, the standard deviations are about twice as high for the Persily data and for the Bennett study restaurants, but are much lower for the Bennett study non-restaurants. The proposed AER distributions were derived only from the most recent Bennett data study stratified into restaurants and non-restaurants.

Similarly to the analyses of the residential AER distributions, we fitted exponential, log-normal, normal, and Weibull distributions to the AER values. The results are shown in Tables 6 and 7.

Table 5. AER summary statistics for offices and other non-residential buildings.

Study	Variable	Subgroup	N	Mean	Std Dev	Min	25th %ile	Median	75th %ile	Max
Bennett	AER	All	40	1.622	1.649	0.300	0.705	1.035	1.890	9.070
Bennett	AER	Not restaurant	34	1.144	0.768	0.300	0.620	0.995	1.460	4.020
Bennett	AER	Restaurant	6	4.328	2.642	1.460	2.640	3.855	5.090	9.070
Persily	AER	All	96	1.962	2.325	0.071	0.501	1.080	2.756	13.824
Turk	AER	All	40	1.540	0.881	0.300	0.850	1.500	2.050	4.100
Bennett	Log AER	All	40	0.152	0.785	-1.204	-0.350	0.034	0.637	2.205
Bennett	Log AER	Not restaurant	34	-0.052	0.619	-1.204	-0.478	-0.005	0.378	1.391
Bennett	Log AER	Restaurant	6	1.312	0.618	0.378	0.971	1.344	1.627	2.205
Persily	Log AER	All	96	0.104	1.104	-2.642	-0.694	0.077	1.012	2.626
Turk	Log AER	All	40	0.254	0.639	-1.204	-0.164	0.405	0.715	1.411

Table 6. Best fitting restaurant AER distributions from the Bennett et al. (2011).

Scale	Shape	Mean	Std_Dev	Distribution	P-Value Kolmogorov-Smirnov	P-Value Cramer-von Mises	P-Value Anderson-Darling
4.328		4.328	4.328	Exponential	0.25	0.23	0.25
1.312	0.618	4.492	3.062	Lognormal	0.15	0.50	0.50
		4.328	2.642	Normal	0.15	0.25	0.25
4.907	1.922	4.353	2.358	Weibull		0.25	0.25

Table 7. Best fitting non-restaurant AER distributions from the Bennett et al. (2011).

Scale	Shape	Mean	Std_Dev	Distribution	P-Value Kolmogorov-Smirnov	P-Value Cramer-von Mises	P-Value Anderson-Darling
1.144		1.144	1.144	Exponential	0.00	0.00	0.00
-0.052	0.619	1.149	0.785	Lognormal	0.15	0.50	0.50
		1.144	0.768	Normal	0.01	0.01	0.01
1.290	1.654	1.153	0.716	Weibull		0.21	0.18

(For an explanation of the Kolmogorov-Smirnov, Cramer-von Mises, and Anderson-Darling P-values see the discussion of residential AER distributions above.) For restaurants, according to two of the three goodness-of-fit measures the best-fitting distribution is the log-normal; the best-fitting distribution is exponential for the Kolmogorov-Smirnov test. For non-restaurants, according to all three goodness-of-fit measures the best-fitting distribution is the log-normal. Reasonable choices for the lower and upper bounds are the observed minimum and maximum AER values.

We therefore propose the following indoor, non-residential AER distributions.

- AER distribution for indoor, non-residential, restaurant microenvironments: Lognormal, with scale and shape parameters 1.312 and 0.618, i.e., geometric mean = 3.712, geometric standard deviation = 1.855. Lower Bound = 1.46. Upper bound = 9.07.

- AER distribution for indoor, non-residential, non-restaurant microenvironments: Lognormal, with scale and shape parameters -0.052 and 0.618, i.e., geometric mean = 0.949, geometric standard deviation = 1.857. Lower Bound = 0.30. Upper bound = 4.02.

Application of the proposed distributions in APEX would require estimates of the proportions of restaurants or similar facilities among the non-residential buildings in each city.

Proximity and Penetration Factors For Outdoors, In-vehicle, and Mass Transit

For the APEX modeling of the outdoor, in-vehicle, and mass transit micro-environments, an approach using proximity and penetration factors is proposed, as follows.

Outdoors Near Road

Penetration factor = 1.

For the Proximity factor, we propose using ratio distributions developed from the Cincinnati Ozone Study (American Petroleum Institute, 1997, Appendix B; Johnson et al. 1995). The field study was conducted in the greater Cincinnati metropolitan area in August and September, 1994. Vehicle tests were conducted according to an experimental design specifying the vehicle type, road type, vehicle speed, and ventilation mode. Vehicle types were defined by the three study vehicles: a minivan, a full-size car, and a compact car. Road types were interstate highways (interstate), principal urban arterial roads (urban), and local roads (local). Nominal vehicle speeds (typically met over one minute intervals within 5 mph) were at 35 mph, 45 mph, or 55 mph. Ventilation modes were as follows:

- Vent Open: Air conditioner off. Ventilation fan at medium. Driver's window half open. Other windows closed.
- Normal A/C. Air conditioner at normal. All windows closed.
- Max A/C: Air conditioner at maximum. All windows closed.

Ozone concentrations were measured inside the vehicle, outside the vehicle, and at six fixed site monitors in the Cincinnati area.

The proximity factor can be estimated from the distributions of the ratios of the outside-vehicle ozone concentrations to the fixed-site ozone concentrations, reported in Table 8 of Johnson et al. (1995). Ratio distributions were computed by road type (local, urban, interstate, all) and by the fixed-site monitor (each of the six sites, as well as the nearest monitor to the test location). For this analysis we propose to use the ratios of outside-vehicle concentrations to the concentrations at the nearest fixed site monitor, as shown in Table 8.

Table 8. Ratio of outside-vehicle ozone to ozone at nearest fixed site[1]

Road Type[1]	Number of cases[1]	Mean[1]	Standard Deviation[1]	25th Percentile[1]	50th Percentile[1]	75th Percentile[1]	Estimated 5th Percentile[2]
Local	191	0.755	0.203	0.645	0.742	0.911	0.422
Urban	299	0.754	0.243	0.585	0.722	0.896	0.355
Interstate	241	0.364	0.165	0.232	0.369	0.484	0.093
All	731	0.626	0.278	0.417	0.623	0.808	0.170

[1] From Table 8 of Johnson et al. (1995). Data excluded if fixed-site concentration < 40 ppb.
[2] Estimated using a normal approximation as Mean – 1.64 × Standard Deviation.

For the outdoors-near- road microenvironment, we recommend using the distribution for local roads, since most of the outdoors-near-road ozone exposure will occur on local roads. The summary data from the Cincinnati Ozone Study are too limited to allow fitting of distributions, but the 25th and 75th percentiles appear to be approximately equidistant from the median (50th percentile). Therefore we propose using a normal distribution with the observed mean and standard deviation. A plausible upper bound for the proximity factor equals 1. Although the normal distribution allows small positive values and can even produce impossible, negative values (with a very low probability), the titration of ozone concentrations near a road is limited. Therefore, as an empirical approach, we recommend a lower bound of the estimated 5th percentile, as shown in the final column of the above table. Therefore in summary we propose:

- Penetration factor for outdoors, near road: 1.
- Proximity factor for outdoors, near road: Normal distribution. Mean = 0.755. Standard Deviation = 0.203. Lower Bound = 0.422. Upper Bound = 1.

Outdoors, Public Garage / Parking Lot

This micro-environment is similar to the outdoors-near-road microenvironment. We therefore recommend the same distributions as for outdoors-near-road:

- Penetration factor for outdoors, public garage / parking lot: 1.
- Proximity factor for outdoors, public garage / parking lot: Normal distribution. Mean = 0.755. Standard Deviation = 0.203. Lower Bound = 0.422. Upper Bound = 1.

Outdoors, Other

The outdoors, other ozone concentrations should be well represented by the ambient monitors. Therefore we propose:

- Penetration factor for outdoors, other: 1.
- Proximity factor for outdoors, other: 1.

In-Vehicle

For the proximity factor for in-vehicle, we also recommend using the results of the Cincinnati Ozone Study presented in Table 5. For this microenvironment, the ratios depend upon the road type, and the relative prevalences of the road types can be estimated by the proportions of vehicle miles traveled in each modeled city. The proximity factors are assumed, as before, to be normally distributed, the upper bound to be1, and the lower bound to be the estimated 5th percentile.

- Proximity factor for in-vehicle, local roads: Normal distribution. Mean = 0.755. Standard Deviation = 0.203. Lower Bound = 0.422. Upper Bound = 1.
- Proximity factor for in-vehicle, urban roads: Normal distribution. Mean = 0.754. Standard Deviation = 0.243. Lower Bound = 0.355. Upper Bound = 1.
- Proximity factor for in-vehicle, interstates: Normal distribution. Mean = 0.364. Standard Deviation = 0.165. Lower Bound = 0.093. Upper Bound = 1.

To complete the specification, the distribution of road type needs to be estimated for each city to be modeled. Vehicle miles traveled (VMT) by city (defined by the Federal-AID urbanized area) and road type were obtained from the Federal Highway Administration. For local and interstate road types, the VMT for the same DOT categories were used. For urban roads, the VMT for all other road types was summed

(Other freeways/expressways, Other principal arterial, Minor arterial, Collector). The computed VMT ratios for each city are shown in Table 9

Table 9. Vehicle Miles Traveled by City and Road Type (2008)[1].

	FRACTION VMT BY ROAD TYPE		
FEDERAL-AID URBANIZED AREA	INTERSTATE	URBAN	LOCAL
Atlanta	0.32	0.45	0.23
Baltimore	0.34	0.59	0.07
Boston	0.32	0.54	0.14
Chicago	0.31	0.57	0.12
Cleveland	0.38	0.45	0.17
Denver	0.25	0.65	0.10
Detroit	0.24	0.66	0.10
Houston	0.24	0.73	0.03
Los Angeles	0.29	0.66	0.05
New York	0.18	0.67	0.15
Philadelphia	0.23	0.65	0.12
Sacramento	0.24	0.69	0.07
St. Louis	0.37	0.45	0.18
Washington	0.30	0.62	0.08

[1]http://www.fhwa.dot.gov/policyinformation/statistics/2008/pdf/hm71.pdf

Thus to simulate the proximity factor in APEX, we propose to first select the road type according to the above probability table of road types, then select the AER distribution (normal) for that road type as defined in the last set of bullets.

For the penetration factor for in-vehicle, we recommend using the inside-vehicle to outside-vehicle ratios from the Cincinnati Ozone Study. The ratio distributions were summarized for all the data and for stratifications by vehicle type, vehicle speed, road type, traffic (light, moderate, or heavy), and ventilation. The overall results and results by ventilation type are shown in Table 10.

Table 10. Ratio of inside-vehicle ozone to outside-vehicle ozone[1].

Ventilation[1]	Number of cases[1]	Mean[1]	Standard Deviation[1]	25th Percentile[1]	50th Percentile[1]	75th Percentile[1]	Estimated 5th Percentile[2]
Vent Open	226	0.361	0.217	0.199	0.307	0.519	0.005
Normal A/C	332	0.417	0.211	0.236	0.408	0.585	0.071
Maximum A/C	254	0.093	0.088	0.016	0.071	0.149	0.000[3]
All	812	0.300	0.232	0.117	0.251	0.463	0.000[3]

1. From Table 7 of Johnson et al.(1995). Data excluded if outside-vehicle concentration < 20 ppb.
2. Estimated using a normal approximation as Mean – 1.64 × Standard Deviation
3. Negative estimate (impossible value) replaced by zero.

Although the data in Table 7 indicate that the inside-to-outside ozone ratios strongly depend upon the ventilation type, it would be very difficult to find suitable data to estimate the ventilation type distributions for each modeled city. Furthermore, since the Cincinnati Ozone Study was scripted, the ventilation conditions may not represent real-world vehicle ventilation scenarios. Therefore, we propose to use the overall average distributions.

- Penetration factor for in-vehicle: Normal distribution. Mean = 0.300. Standard Deviation = 0.232. Lower Bound = 0.000. Upper Bound = 1.

<u>Mass Transit</u>

The mass transit microenvironment is expected to be similar to the in-vehicle microenvironment. Therefore we recommend using the same APEX modeling approach:

- Proximity factor for mass transit, local roads: Normal distribution. Mean = 0.755. Standard Deviation = 0.203. Lower Bound = 0.422. Upper Bound = 1.
- Proximity factor for mass transit, urban roads: Normal distribution. Mean = 0.754. Standard Deviation = 0.243. Lower Bound = 0.355. Upper Bound = 1.
- Proximity factor for mass transit, interstates: Normal distribution. Mean = 0.364. Standard Deviation = 0.165. Lower Bound = 0.093. Upper Bound = 1.
- Road type distributions for mass transit: See Table 8.
- Penetration factor for mass transit: Normal distribution. Mean = 0.300. Standard Deviation = 0.232. Lower Bound = 0.000. Upper Bound = 1.

References

American Petroleum Institute (1997). *Sensitivity testing of pNEM/O3 exposure to changes in the model algorithms.* Health and Environmental Sciences Department.

Avol, E. L., W. C. Navioli, and S. D. Colome (1998) Modeling ozone levels in and around southern California homes. *Environ. Sci. Technol.* 32, 463-468.

Bennett, D. H, W. Fisk, M. G. Apte, X. Wu, A. Trout, D. Faulkner, D. Sulivan (2011). *Ventilation, Temperature, and HVAC Characteristics in Small and Medium Commercial Buildings (SMCBs) in California.* Journal submission.

Chilrud, S. N., D. Epstein, J. M. Ross, S. N. Sax, D. Pederson, J. D. Spengler, P. L. Kinney (2004). Elevated airborne exposures of teenagers to manganese, chromium, and iron from steel dust and New York City's subway system. *Environ. Sci. Technol.* 38, 732-737.

Colome, S.D., A. L. Wilson, Y. Tian (1993). *California Residential Indoor Air Quality Study, Volume 1, Methodology and Descriptive Statistics.* Report prepared for the Gas Research Institute, Pacific Gas & Electric Co., San Diego Gas & Electric Co., Southern California Gas Co.

Jakobi, G and Fabian, P. (1997). Indoor/outdoor concentrations of ozone and peroxyacetyl nitrate (PAN). *Int. J. Biometeorol.* 40: 162-165..

Johnson, T., A. Pakrasi, A. Wisbeth, G. Meiners, W. M. Ollison (1995). Ozone exposures within motor vehicles – results of a field study in Cincinnati, Ohio. *Proceedings 88th annual meeting and exposition of the Air & Waste Management Association, June 18-23, 1995.* San Antonio, TX. Preprint paper 95-WA84A.02.

Kinney, P. L., S. N. Chillrud, S. Ramstrom, J. Ross, J. D. Spengler (2002). Exposures to multiple air toxics in New York City. *Environ Health Perspect* 110, 539-546.

Liu, L.-J. S, P. Koutrakis, J. Leech, I. Broder, (1995) Assessment of ozone exposures in the greater metropolitan Toronto area. *J. Air Waste Manage. Assoc.* 45: 223-234

Meng, Q. Y., B. J. Turpin, L. Korn, C. P. Weisel, M. Morandi, S. Colome, J. J. Zhang, T. Stock, D. Spektor, A. Winer, L. Zhang, J. H. Lee, R. Giovanetti, W. Cui, J. Kwon, S. Alimokhtari, D. Shendell, J. Jones, C. Farrar, S. Maberti (2004). Influence of ambient (outdoor) sources on residential indoor and personal $PM_{2.5}$ concentrations: Analyses of RIOPA data. *Journal of Exposure Analysis and Environ Epidemiology*. Preprint.

Murray, D. M. and D. E. Burmaster (1995). Residential Air Exchange Rates in the United States: Empirical and Estimated Parametric Distributions by Season and Climatic Region. *Risk Analysis*, Vol. 15, No. 4, 459-465.

Persily, A. and J. Gorfain.(2004). *Analysis of ventilation data from the U.S. Environmental Protection Agency Building Assessment Survey and Evaluation (BASE) Study.* National Institute of Standards and Technology, NISTIR 7145, December 2004.

Persily, A., J. Gorfain, G. Brunner.(2005). Ventilation design and performance in U.S. office buildings. *ASHRAE Journal.* April 2005, 30-35.

Sax, S. N., D. H. Bennett, S. N. Chillrud, P. L. Kinney, J. D. Spengler (2004). Differences in source emission rates of volatile organic compounds in inner-city residences of New York City and Los Angeles. *Journal of Exposure Analysis and Environ Epidemiology.* Preprint.

Sheldon, L. (2007). The Detroit Exposure and Aerosol Research Study (DEARS). Presentation.

Turk, B. H., D. T. Grimsrud, J. T. Brown, K. L. Geisling-Sobotka, J. Harrison, R. J. Prill (1989). *Commercial building ventilation rates and particle concentrations.* ASHRAE, No. 3248.

Weschler, C. J. (2000) Ozone in indoor environments: concentration and chemistry. Indoor Air 10: 269-288.

Weschler, C. J. and Shields, H. C. (2000) The influence of ventilation on reactions among indoor pollutants: modeling and experimental observations. *Indoor Air.* 10: 92-100.

Weisel, C. P., J. J. Zhang, B. J. Turpin, M. T. Morandi, S. Colome, T. H. Stock, D. M. Spektor, L. Korn, A. Winer, S. Alimokhtari, J. Kwon, K. Mohan, R. Harrington, R. Giovanetti, W. Cui, M. Afshar, S. Maberti, D. Shendell (2004). Relationship of Indoor, Outdoor and Personal Air (RIOPA) study; study design, methods and quality assurance / control results. *Journal of Exposure Analysis and Environ Epidemiology.* Preprint.

Williams, R., J. Suggs, A. Rea, K. Leovic, A. Vette, C. Croghan, L. Sheldon, C. Rodes, J. Thornburg, A. Ejire, M. Herbst, W. Sanders Jr. (2003a). The Research Triangle Park particulate matter panel study: PM mass concentration relationships. *Atmos Env* 37, 5349-5363.

Williams, R., J. Suggs, A. Rea, L. Sheldon, C. Rodes, J. Thornburg (2003b). The Research Triangle Park particulate patter panel study: modeling ambient source contribution to personal and residential PM mass concentrations. *Atmos Env* 37, 5365-5378.

Wilson, A. L., S. D. Colome, P. E. Baker, E. W. Becker (1986). *Residential Indoor Air Quality Characterization Study of Nitrogen Dioxide, Phase I, Final Report.* Prepared for Southern California Gas Company, Los Angeles.

Wilson, A. L., S. D. Colome, Y. Tian (1994). *California Residential Indoor Air Quality Study, Volume 2, Carbon Monoxide and Air Exchange Rate: A Univariate & Multivariate Analysis.* Report prepared for the Gas Research Institute, Pacific Gas & Electric Co., San Diego Gas & Electric Co., Southern California Gas Co.

Wilson, A. L., S. D. Colome, Y. Tian, P. E. Baker, E. W. Becker, D. W. Behrens, I. H. Billick, C. A. Garrison (1996). California residential air exchange rates and residence volumes. *Journal of Exposure Analysis and Environ Epidemiology.* Vol. 6, No. 3.

Appendix 5-F

Detailed Exposure Results

Table of Contents

5F-1 Exposure Modeling Results for Base Air Quality 3
5F-2 Exposure Modeling Results for Adjusted Air Quality 16

List of Tables

Table 5F-1. Percent of **all school-age children** with O_3 exposures at or above 60, 70, and 80 ppb-8hr while at moderate or greater exertion, years 2006-2010, base air quality.... 12
Table 5F-2. Percent of **all school-age children** with O_3 exposures at or above 60, 70, and 80 ppb-8hr while at moderate or greater exertion, years 2006-2010, adjusted air quality. 24
Table 5F-3. Mean and maximum number of **all school-age children** (and associated days per O_3 season) with at least one O_3 exposure at or above 60 ppb-8hr while at moderate or greater exertion. 53
Table 5F-4. Total number of persons experiencing at least one or two 8-hour exposures in all study areas by year, base air quality and air quality adjusted to just meeting the existing 75 ppb standard. 57

List of Figures

Figure 5F-1. Percent of **all school-age children** with at least one O_3 exposure at or above 60, 70, and 80 ppb-8hr while at moderate or greater exertion, years 2006-2010, base air quality. 6
Figure 5F-2. Percent of **asthmatic school-age children** with at least one O_3 exposure at or above 60, 70, and 80 ppb-8hr while at moderate or greater exertion, years 2006-2010, base air quality 7
Figure 5F-3. Percent of **asthmatic adults** with at least one O_3 exposure at or above 60, 70, and s 80 ppb-8hr while at moderate or greater exertion, years 2006-2010, base air quality. 8
Figure 5F-4. Percent of **older adults** with at least one O_3 exposure at or above 60, 70, and 80 ppb-8hr while at moderate or greater exertion, years 2006-2010, base air quality...... 9
Figure 5F-5. Percent of **school-age children** with **multiple O_3 exposures** at or above 60 ppb-8hr per study area O_3 season, while at moderate or greater exertion, years 2006-2010, base air quality 10
Figure 5F-6. Percent of **school-age children** with **multiple O_3 exposures** at or above 70 ppb-8hr per study area O_3 season, while at moderate or greater exertion, years 2006-2010, base air quality 11

Figure 5F-7. Incremental increases in percent of **all school-age children** with at least one exposure at or above 60 ppb-8hr (top panel), 70 ppb-8hr (middle panel), or 80 ppb-8hr (bottom panel) using the **maximum percent** exposed for each study area, year 2006-2010 adjusted air quality. 17

Figure 5F-8. Incremental increases in percent of **asthmatic school-age children** with at least one exposure at or above 60 ppb-8hr (top panel), 70 ppb-8hr (middle panel), or 80 ppb-8hr (bottom panel) using the **maximum percent** exposed for each study area, year 2006-2010 adjusted air quality. 18

Figure 5F-9. Incremental increases in percent of **asthmatic adults** with at least one exposure at or above 60 ppb-8hr (top panel), 70 ppb-8hr (middle panel), or 80 ppb-8hr (bottom panel) using the **maximum percent** exposed for each study area, year 2006-2010 adjusted air quality. 19

Figure 5F-10. Incremental increases in percent of **all older adults** with at least one exposure at or above 60 ppb-8hr (top panel), 70 ppb-8hr (middle panel), or 80 ppb-8hr (bottom panel) using the **maximum percent** exposed for each study area, year 2006-2010 adjusted air quality. 20

Figure 5F-11. Incremental increases in percent of **all school-age children** with at least one exposure at or above 60 ppb-8hr (top panel), 70 ppb-8hr (middle panel), or 80 ppb-8hr (bottom panel) using the **mean percent** exposed for each study area, year 2006-2010 adjusted air quality. 21

Figure 5F-12. Incremental increases in percent of **all school-age children** with **at least two exposures** at or above 60 ppb-8hr (top panel), 70 ppb-8hr (middle panel), or 80 ppb-8hr (bottom panel) using the **maximum percent** exposed for each study area, year 2006-2010 adjusted air quality. 22

Figure 5F-13. Incremental increases in percent of **all school-age children** with at least two exposures at or above 60 ppb-8hr (top panel), 70 ppb-8hr (middle panel), or 80 ppb-8hr (bottom panel) using the **mean percent** exposed for each study area, year 2006-2010 adjusted air quality. 23

Figure 5F-14. Average number of persons with at least one 8-hour exposure at or above 60 ppb considering the existing and alternative standards, year 2006-2010 adjusted air quality. All school-age children (top left), asthmatic school-age children (top right), asthmatic adults (bottom left), older adults (bottom right). 55

Figure 5F-15. Average total number of days in an O_3 season where simulated persons experienced 8-hour exposures at or above 60 ppb considering the existing and alternative standards, year 2006-2010 adjusted air quality. All school-age children (top left), asthmatic school-age children (top right), asthmatic adults (bottom left), older adults (bottom right). 56

This appendix contains the detailed results for the primary APEX simulations performed to estimate exposures associated with base air quality (section 5F-1) and for air quality just meeting the existing and alternative standard levels (section 5F-2).

5F-1 EXPOSURE MODELING RESULTS FOR BASE AIR QUALITY

As described in the main body of the REA, comprehensive multi-panel displays of exposure results are presented for each of the study groups of interest, i.e., all school-age children (ages 5 to 18), asthmatic school-age children, asthmatic adults (ages 19 to 95), and older adults (ages 65 to 95) (**Figure 5F-1** to **Figure 5F-4**, respectively). Included in each display are the three benchmark levels (60, 70, and 80 ppb-8hr), the five years of air quality (2006-2010), for the 15 study areas. Modeled exposures in the 15 study areas and considering each benchmark level are presented on the same scale to allow for direct comparisons across the multi-panel display. The most notable patterns in the exposure results are described here using one study group (i.e., school-age children), as there is a general consistency in the year-to-year variability within each study area across all four study groups. Any deviation from the observed pattern will be discussed for the subsequent study group. **Table 5F-1** is also provided and contains the complete exposure output for all study areas and years for school-age children.

Figure 5F-1 presents the percent of school-age children experiencing at least one O_3 exposure at or above the selected benchmark levels while at moderate or greater exertion. Consistent with the previously discussed observations regarding year-to-year variability in ambient concentrations (Chapter 4), most study areas have the greatest percent of school-age children experiencing concentrations at or above the three benchmark levels during 2006 or 2007 along with having the lowest percent of school-age children exposed during 2009. Three Western U.S. study areas, Dallas, Los Angeles, and Sacramento, differ slightly from this pattern in that they exhibit a minimum percent of school-age children exposed during 2010, while in Houston and Chicago the minimum exposures occur during year 2008. In general, between 20 to 40% of school-age children experience at least one O_3 exposure at or above 60 ppb-8hr, 10 to 20% experience at least one O_3 exposure at or above 70 ppb-8hr, and 0 to 10% experience at least one O_3 exposure at or above 80 ppb-8hr, all while at moderate or greater exertion and considering the base air quality (2006-2010).

The percent of asthmatic school-age children experiencing at least one O_3 exposure at or above the selected benchmark levels while at moderate or greater exertion (**Figure 5F-2**) is virtually indistinguishable from that of all school-age children (**Figure 5F-1**) regarding both the year-to-year pattern and percent of persons exposed. This is the result of having both simulated study groups use an identical time-location-activity diary pool to construct each simulated individual's time series of activities performed and locations visited. Different however would be the relative number of asthmatic school-age children exposed in each study area if compared with non-asthmatic school-age children, as the asthma prevalence rates vary by U.S. location (REA, Table 5-2) though on average are about 10% of all school-age children.

As mentioned above, the overall year-to-year pattern of exposure for asthmatic adults is similar to that observed for school-age children, though the percent of asthmatic adults experiencing exposures at or above the health effect benchmark levels is lower by a factor of about three or more (**Figure 5F-3**). Having a lower percent of asthmatic adults exposed is expected given that outdoor time expenditure is an important determinant of O_3 exposure (REA section 5.3.2) and that adults spend less time outdoors than children (REA section 5.3.1). In general, between 5 to 10% of asthmatic adults experience at least one O_3 exposure at or above 60 ppb-8hr, 0 to 5% experience at least one O_3 exposure at or above 70 ppb-8hr, and 0 to 2% experience at least one O_3 exposure at or above 80 ppb-8hr, all while at moderate or greater.

While the percent of asthmatic adults exposed is much lower, the number of asthmatic adults at or above the exposure benchmarks is generally just below that estimated number of asthmatic school-age children. As an example, for year 2006 in Atlanta, approximately 44% of asthmatic school-age children (or about 37,000) were estimated to experience at least one exposure at or above 60 ppb-8hr. Though a much smaller percent of asthmatic adults were estimated to experience a similar exposure for the same year (i.e., about 16%), this is equivalent to nearly 31,000 asthmatic adults exposed, at least one time, to an 8-hr average O_3 concentration at or above 60 ppb.

The percent of older adults (ages 65 to 95) experiencing exposures at or above the selected benchmark levels (**Figure 5F-4**) is lower by a fewer percentage points when compared with the results for asthmatic adults. Again, older adults, on average, would tend to spend less time outdoors when compared with both adults and children (REA section 5.3.1), in addition to fewer older adults performing activities at moderate or greater exertion for extended periods of time, thus leading to fewer older adults exposed to concentrations of concern. In general, less than 10% of older adults experience at least one O_3 exposure at or above 60 ppb-8hr, less than 5% experience at least one O_3 exposure at or above 70 ppb-8hr, and about 2% or less experience at least one O_3 exposure at or above 80 ppb-8hr, all while at moderate or greater exertion considering base air quality.

Given the similar year-to-year patterns of the single and multiple exposure occurrences and when considering any of the four study groups, we present the graphic multi-day exposure results here considering school-age children only. All multi-day exposure results are provided in **Table 5F-1**. **Figure 5F-5** illustrates the percent of school-age children having multiple exposures at or above 60 ppb-8hr for each of the 15 study areas, considering base air quality (2006-2010). Depending on the year and study area, about 10 to 25% of school-age children could experience at least two exposures above the 60 ppb benchmark during the ozone season, while about 5 to 10% could experience at least four. Most study areas and years are estimated to have fewer than 5% of school-age children experience six or more exposures above 60 ppb-8hr considering the base air quality. When considering the multi-day exposures for school-age children at or above the 70 ppb benchmark (**Figure 5F-6**), about 2 to 10% of school-age children could experience at least two exposures during the ozone season, while four or more exposures were generally limited to fewer than 4% of school-age children.

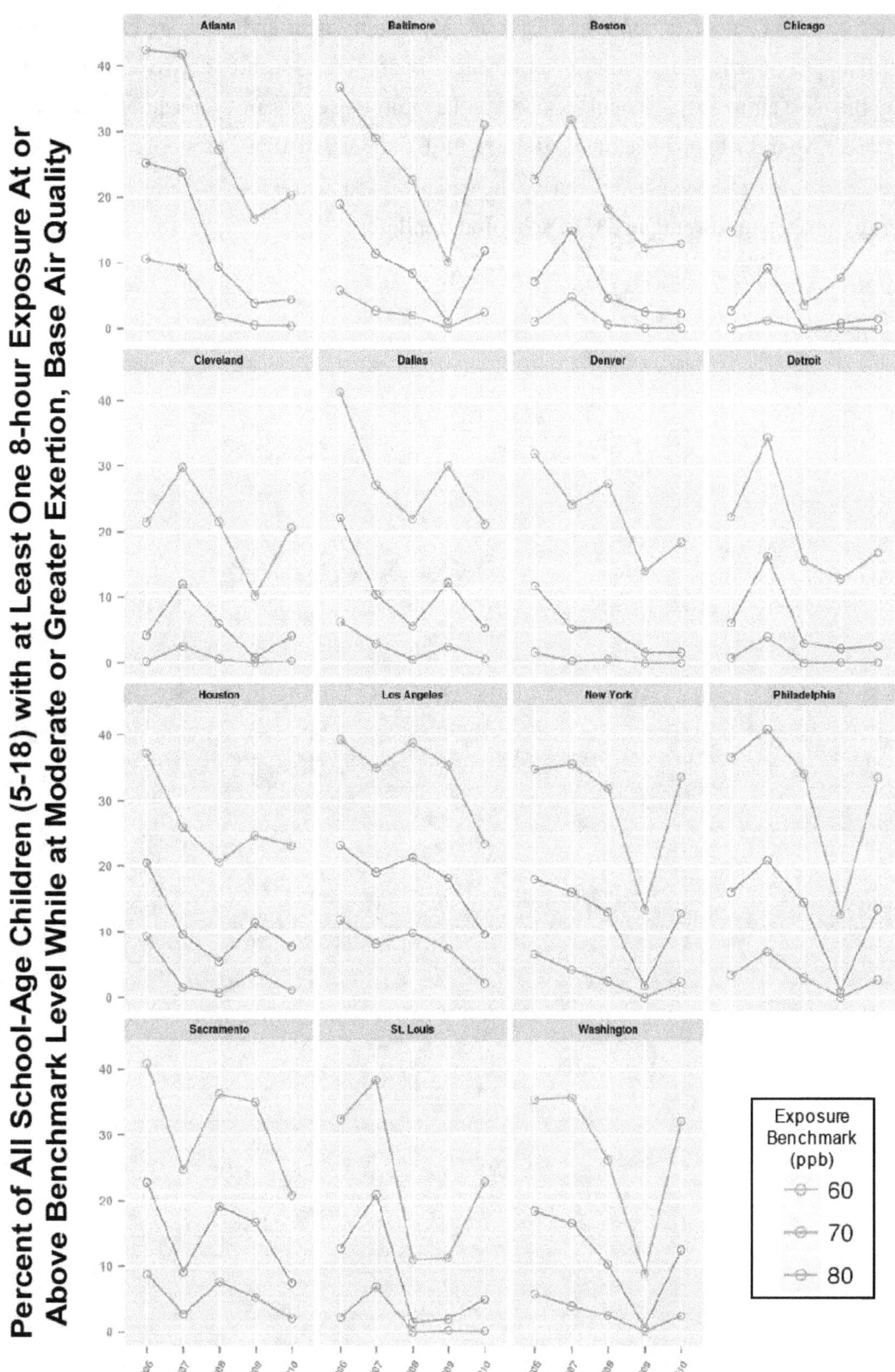

Figure 5F-1. Percent of **all school-age children** with at least one O_3 exposure at or above 60, 70, and 80 ppb-8hr while at moderate or greater exertion, years 2006-2010, base air quality.

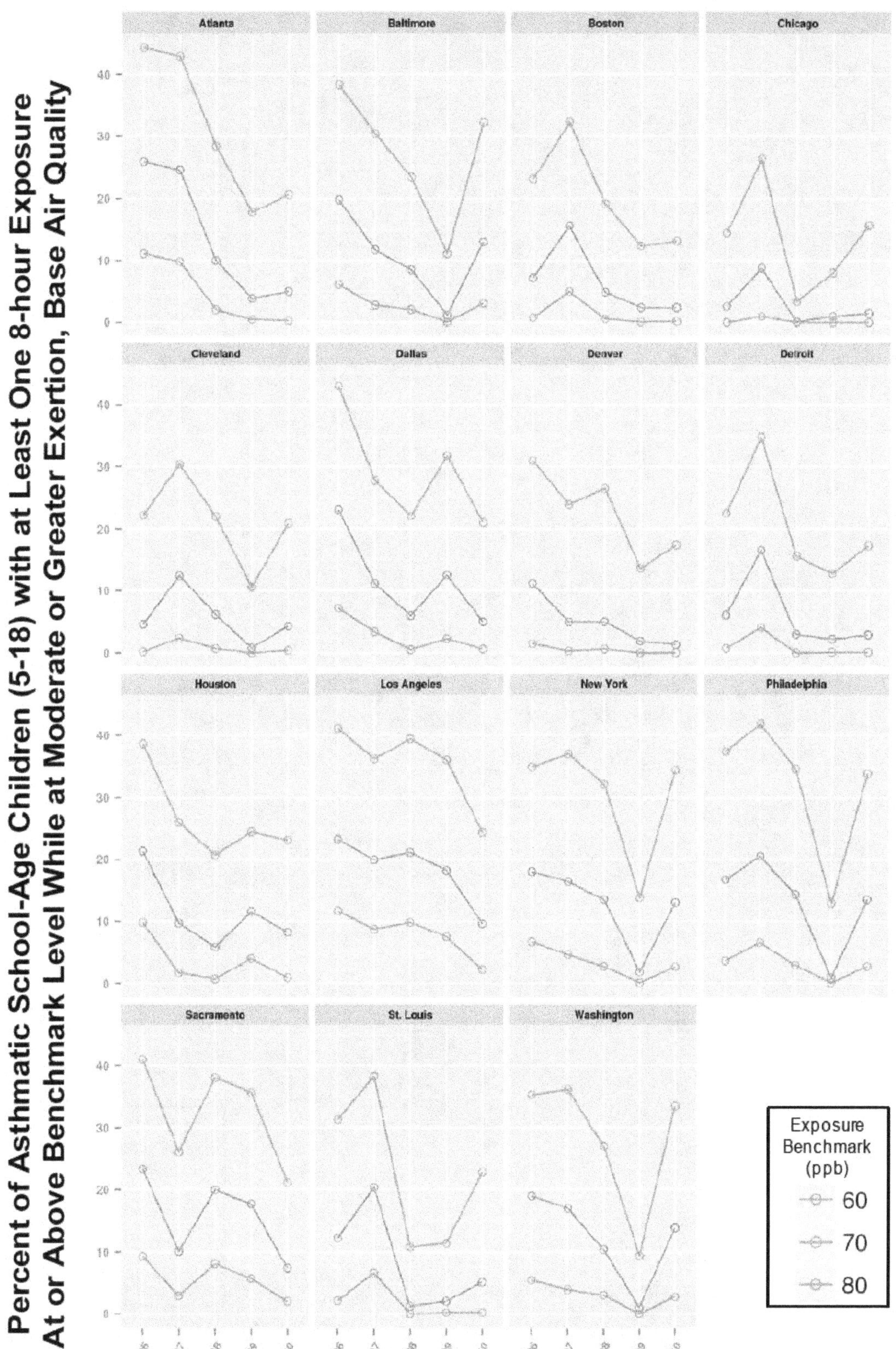

Figure 5F-2. Percent of **asthmatic school-age children** with at least one O_3 exposure at or above 60, 70, and 80 ppb-8hr while at moderate or greater exertion, years 2006-2010, base air quality.

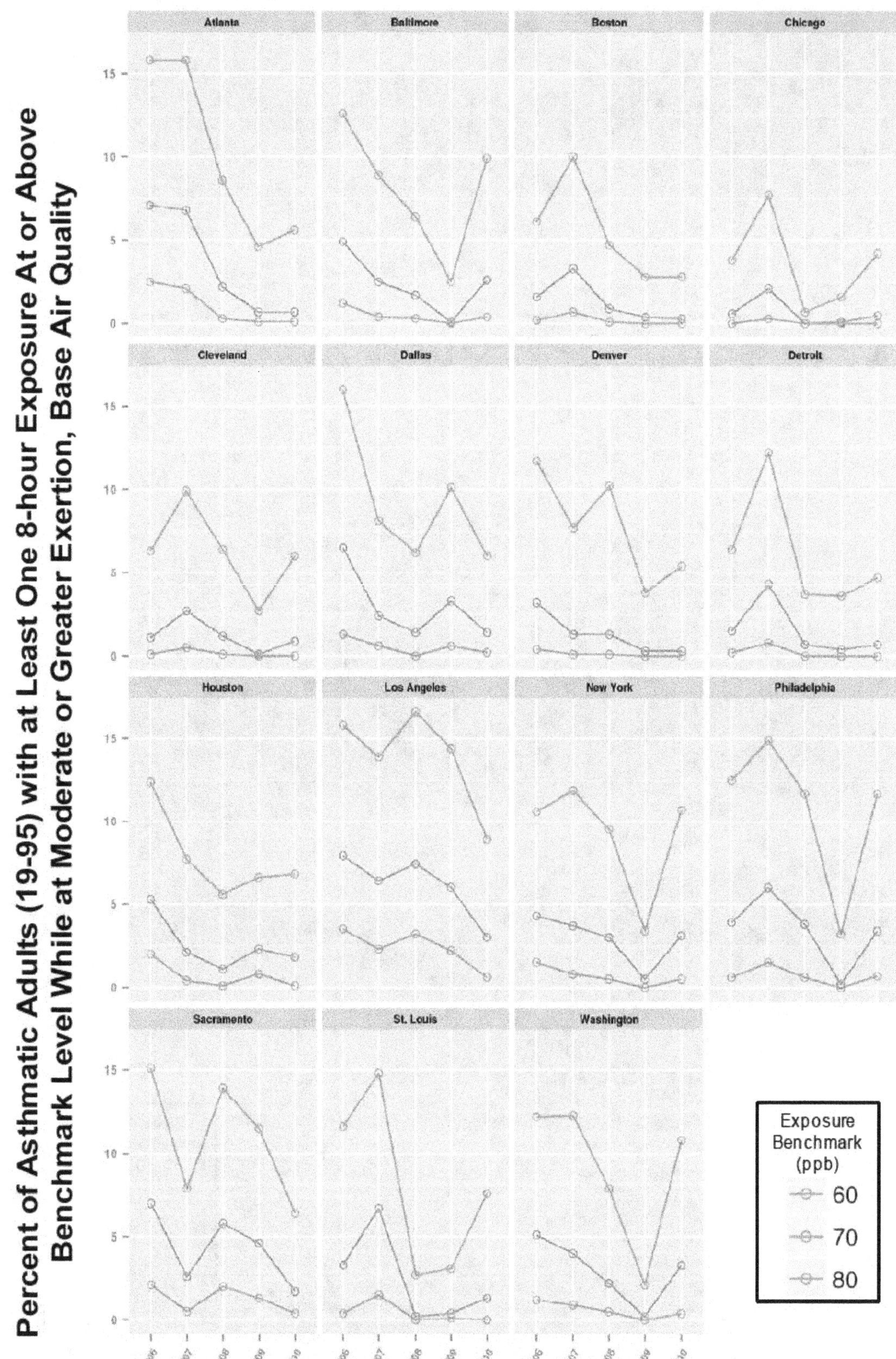

Figure 5F-3. Percent of **asthmatic adults** with at least one O_3 exposure at or above 60, 70, and s 80 ppb-8hr while at moderate or greater exertion, years 2006-2010, base air quality.

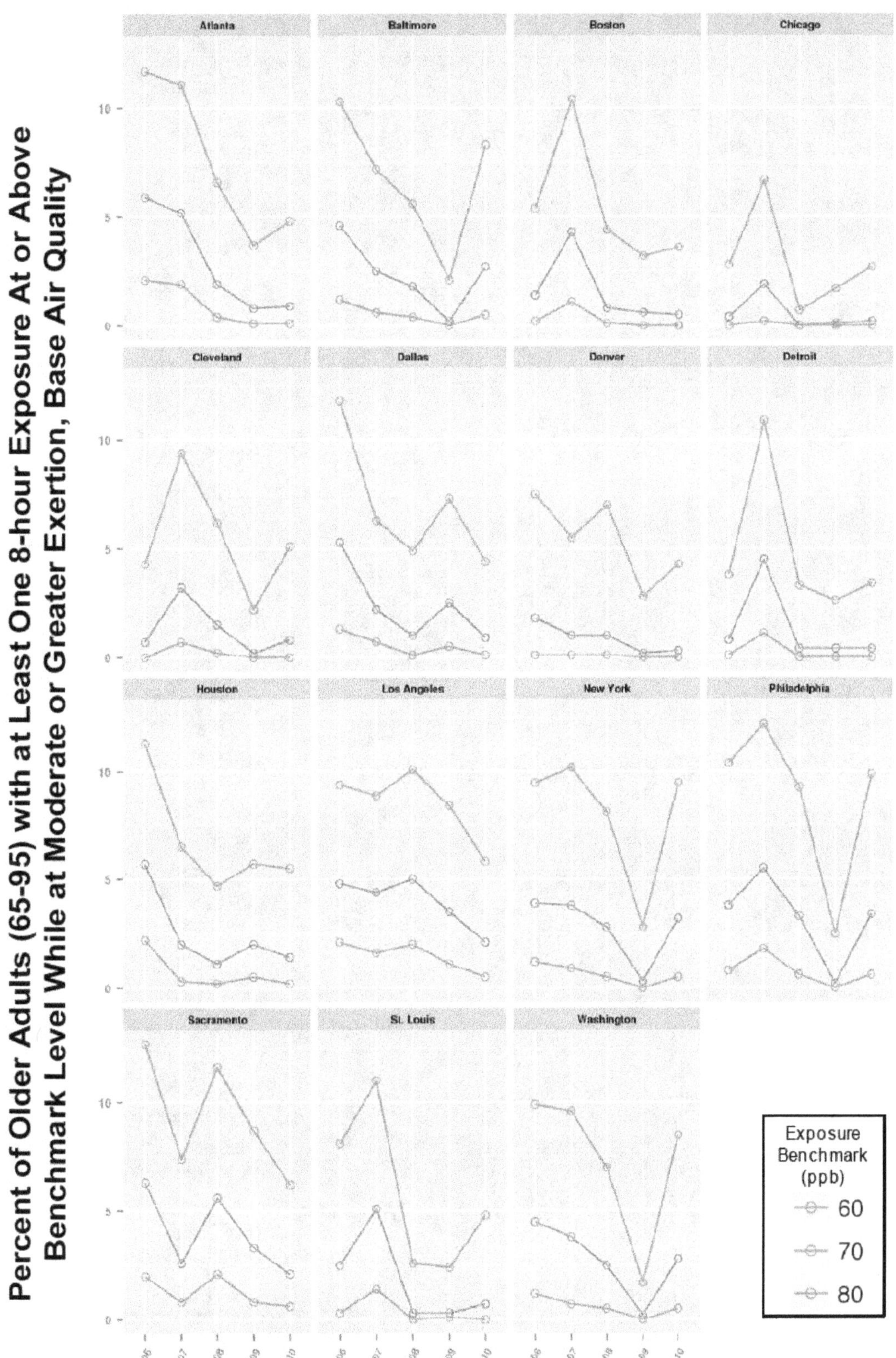

Figure 5F-4. Percent of **older adults** with at least one O_3 exposure at or above 60, 70, and 80 ppb-8hr while at moderate or greater exertion, years 2006-2010, base air quality.

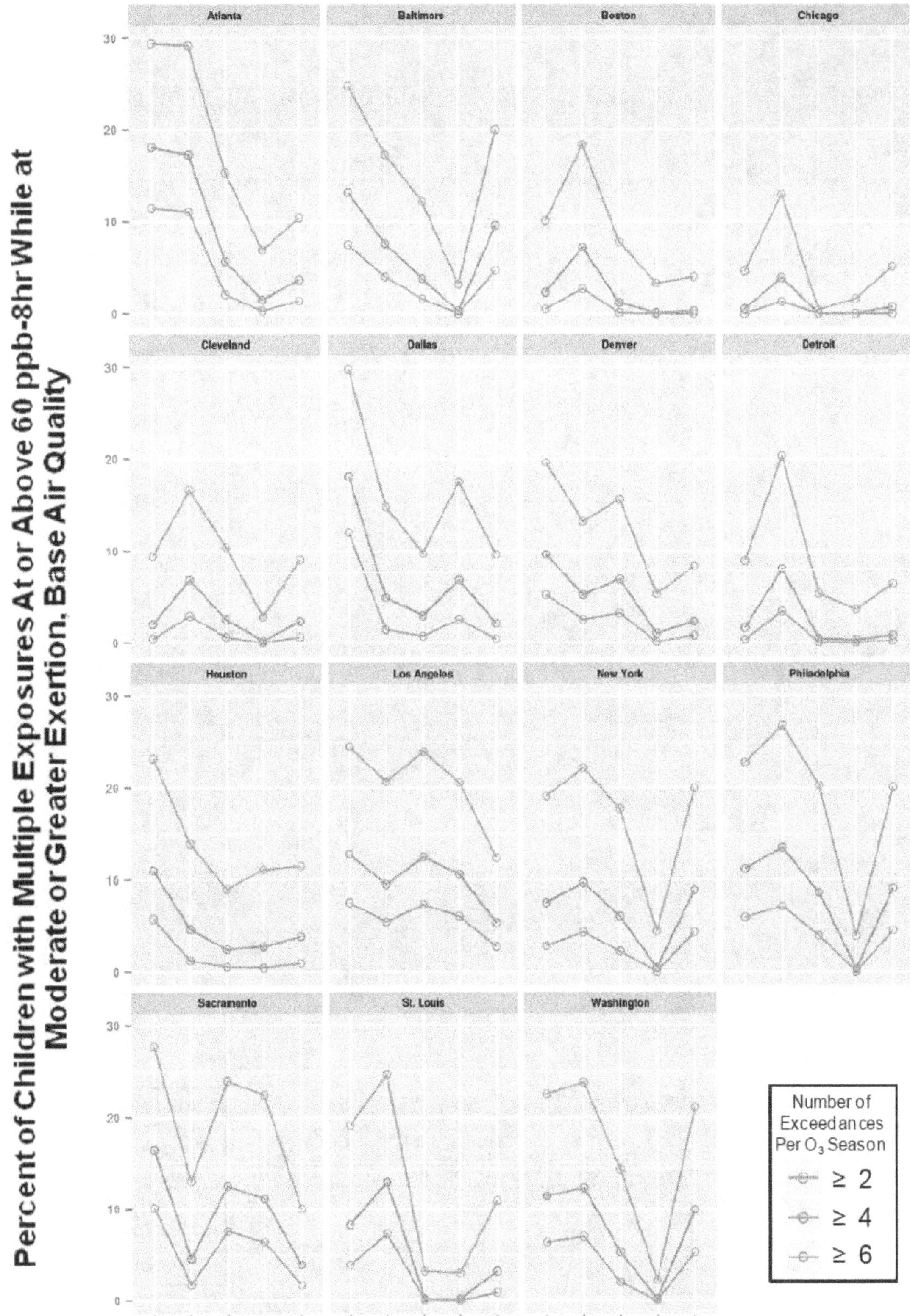

Figure 5F-5. Percent of **school-age children** with **multiple O_3 exposures** at or above 60 ppb-8hr per study area O_3 season, while at moderate or greater exertion, years 2006-2010, base air quality.

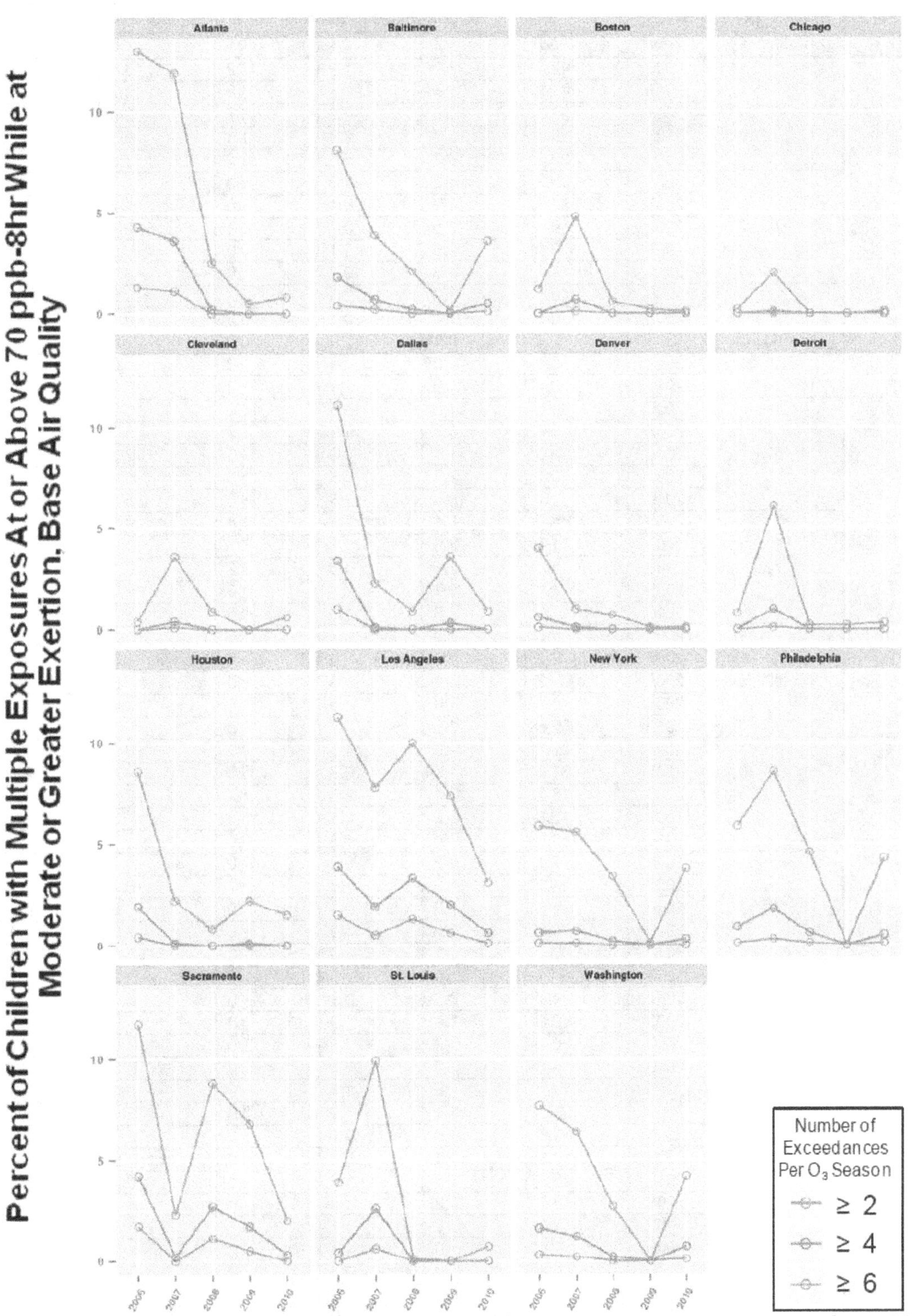

Figure 5F-6. Percent of **school-age children** with **multiple O_3 exposures** at or above 70 ppb-8hr per study area O_3 season, while at moderate or greater exertion, years 2006-2010, base air quality.

Table 5F-1. Percent of **all school-age children** with O_3 exposures at or above 60, 70, and 80 ppb-8hr while at moderate or greater exertion, years 2006-2010, base air quality.

Study Area	Exposure Benchmark (ppb-8hr)	Year	% of school-age children experiencing multiple exposures per O_3 season at or above benchmarks, base air quality					
			≥1	≥2	≥3	≥4	≥5	≥6
Atlanta	60	2006	42.4	28.9	21.9	17.2	13.8	11
		2007	41.8	28.5	21.5	16.8	13.3	10.5
		2008	27.3	15	9.2	5.7	3.6	2.3
		2009	16.7	6.7	2.9	1.3	0.5	0.2
		2010	20.3	10.1	5.6	3.2	1.8	1.1
	70	2006	25.2	13	7.3	4.3	2.4	1.3
		2007	23.8	11.9	6.4	3.6	2	1.1
		2008	9.4	2.5	0.7	0.2	0.1	0
		2009	3.8	0.5	0.1	0	0	0
		2010	4.4	0.8	0.2	0	0	0
	80	2006	10.6	2.7	0.8	0.2	0	0
		2007	9.3	2.4	0.7	0.2	0.1	0
		2008	1.8	0.1	0	0	0	0
		2009	0.5	0	0	0	0	0
		2010	0.4	0	0	0	0	0
Baltimore	60	2006	36.8	23.5	16.6	12.2	9	6.7
		2007	29	16.5	10.9	7.2	5.1	3.7
		2008	22.6	11.1	6	3.6	2.2	1.3
		2009	10.1	2.7	0.8	0.3	0.1	0
		2010	31	18.6	12.5	8.6	5.8	4
	70	2006	19	8.1	3.8	1.8	0.8	0.4
		2007	11.4	3.9	1.7	0.7	0.3	0.2
		2008	8.4	2.1	0.6	0.2	0.1	0
		2009	1.1	0.1	0	0	0	0
		2010	11.8	3.6	1.3	0.5	0.2	0.1
	80	2006	5.8	0.9	0.1	0	0	0
		2007	2.7	0.4	0	0	0	0
		2008	2	0.1	0	0	0	0
		2009	0.1	0	0	0	0	0
		2010	2.5	0.2	0	0	0	0
Boston	60	2006	22.7	9.6	4.5	2.1	0.9	0.4
		2007	31.8	17.5	10.8	6.8	4.2	2.6
		2008	18.3	7.1	2.8	1.2	0.5	0.2
		2009	12.3	3.2	0.8	0.1	0	0
		2010	12.9	4	1.3	0.4	0.1	0
	70	2006	7.1	1.2	0.2	0	0	0
		2007	14.9	4.8	1.8	0.7	0.2	0.1
		2008	4.6	0.6	0.1	0	0	0
		2009	2.6	0.2	0	0	0	0
		2010	2.3	0.1	0	0	0	0
	80	2006	1	0	0	0	0	0
		2007	4.8	0.6	0.1	0	0	0
		2008	0.6	0	0	0	0	0
		2009	0.1	0	0	0	0	0
		2010	0.1	0	0	0	0	0
Chicago	60	2006	14.7	4.5	1.4	0.5	0.1	0
		2007	26.5	13.1	7.1	3.9	2.1	1.2
		2008	3.6	0.4	0.1	0	0	0
		2009	7.8	1.4	0.3	0	0	0
		2010	15.6	5	1.7	0.5	0.1	0
	70	2006	2.7	0.2	0	0	0	0
		2007	9.2	2	0.4	0.1	0	0
		2008	0	0	0	0	0	0
		2009	0.8	0	0	0	0	0
		2010	1.5	0.1	0	0	0	0
	80	2006	0.1	0	0	0	0	0
		2007	1.2	0	0	0	0	0
		2008	0	0	0	0	0	0
		2009	0.1	0	0	0	0	0
		2010	0	0	0	0	0	0

Study Area	Exposure Benchmark (ppb-8hr)	Year	% of school-age children experiencing multiple exposures per O_3 season at or above benchmarks, base air quality					
			≥1	≥2	≥3	≥4	≥5	≥6
Cleveland	60	2006	21.4	9.1	4.2	1.8	0.8	0.3
		2007	29.7	16.4	10.2	6.4	3.9	2.5
		2008	21.5	9.7	4.7	2.2	1.1	0.5
		2009	10.2	2.6	0.8	0.2	0.1	0
		2010	20.6	8.9	4.3	2.1	1	0.4
	70	2006	4.2	0.4	0.1	0	0	0
		2007	12	3.6	1.2	0.4	0.2	0.1
		2008	6	0.9	0.1	0	0	0
		2009	0.8	0	0	0	0	0
		2010	4.1	0.6	0.1	0	0	0
	80	2006	0.2	0	0	0	0	0
		2007	2.5	0.2	0	0	0	0
		2008	0.6	0	0	0	0	0
		2009	0	0	0	0	0	0
		2010	0.3	0	0	0	0	0
Dallas	60	2006	41.3	28.4	21.7	17.2	13.8	11
		2007	27.1	14.1	8	4.5	2.5	1.3
		2008	21.9	10.3	5.4	2.8	1.5	0.7
		2009	30	17	10.6	6.7	4.2	2.6
		2010	21.1	9.6	4.6	2.2	1	0.4
	70	2006	22.1	11.1	6.1	3.4	1.9	1
		2007	10.4	2.3	0.5	0.1	0	0
		2008	5.6	0.9	0.2	0	0	0
		2009	12.2	3.6	1.1	0.3	0.1	0
		2010	5.3	0.9	0.2	0	0	0
	80	2006	6.3	1.3	0.4	0.1	0	0
		2007	2.9	0.1	0	0	0	0
		2008	0.5	0	0	0	0	0
		2009	2.5	0.2	0	0	0	0
		2010	0.6	0	0	0	0	0
Denver	60	2006	31.9	19.6	13.3	9.5	7	5.2
		2007	24.1	13	8.1	5.3	3.4	2.2
		2008	27.3	15.7	10.2	6.8	4.6	3.1
		2009	14	5.1	2	0.8	0.3	0.1
		2010	18.4	8.6	4.4	2.4	1.2	0.6
	70	2006	11.7	4	1.5	0.6	0.2	0.1
		2007	5.2	1	0.3	0.1	0	0
		2008	5.3	0.7	0.1	0	0	0
		2009	1.6	0.1	0	0	0	0
		2010	1.6	0.1	0	0	0	0
	80	2006	1.6	0.1	0	0	0	0
		2007	0.3	0	0	0	0	0
		2008	0.6	0	0	0	0	0
		2009	0	0	0	0	0	0
		2010	0	0	0	0	0	0
Detroit	60	2006	22.3	9.1	4.1	1.7	0.7	0.3
		2007	34.4	20.1	13.1	8.5	5.6	3.6
		2008	15.6	5.3	1.8	0.6	0.2	0
		2009	12.8	3.6	1	0.3	0.1	0
		2010	16.8	6.3	2.6	1	0.4	0.2
	70	2006	6.1	0.8	0.1	0	0	0
		2007	16.3	6.1	2.5	1	0.4	0.1
		2008	3	0.2	0	0	0	0
		2009	2.2	0.2	0	0	0	0
		2010	2.6	0.3	0	0	0	0
	80	2006	0.8	0	0	0	0	0
		2007	4	0.6	0.1	0	0	0
		2008	0	0	0	0	0	0
		2009	0.1	0	0	0	0	0
		2010	0.1	0	0	0	0	0

Study Area	Exposure Benchmark (ppb-8hr)	Year	% of school-age children experiencing multiple exposures per O_3 season at or above benchmarks, base air quality					
			≥1	≥2	≥3	≥4	≥5	≥6
		2006	37.2	23	15.7	11	7.8	5.5
		2007	25.9	13.4	7.5	4.2	2.4	1.2
	60	2008	20.6	8.9	4.3	2	0.9	0.4
		2009	24.7	11	5.1	2.3	1	0.4
		2010	23.1	11.1	6	3.2	1.6	0.8
		2006	20.5	8.6	4	1.9	0.9	0.4
		2007	9.6	2.2	0.6	0.1	0	0
Houston	70	2008	5.5	0.8	0.1	0	0	0
		2009	11.4	2.2	0.4	0.1	0	0
		2010	7.8	1.5	0.3	0	0	0
		2006	9.3	2.3	0.6	0.1	0	0
		2007	1.5	0.1	0	0	0	0
	80	2008	0.7	0	0	0	0	0
		2009	3.8	0.2	0	0	0	0
		2010	1.1	0	0	0	0	0
		2006	39.3	24.1	16.9	12.6	9.6	7.4
		2007	35	20.1	13.1	9.4	6.9	5.2
	60	2008	38.8	23.6	16.5	12.2	9.2	7.2
		2009	35.4	20.4	13.8	10	7.5	5.7
		2010	23.4	12.2	7.6	5.1	3.7	2.7
		2006	23.2	11.3	6.4	3.9	2.4	1.5
		2007	19.1	7.8	3.8	1.9	1	0.5
Los Angeles	70	2008	21.3	10	5.6	3.3	2.1	1.3
		2009	18.2	7.4	3.8	2	1.1	0.6
		2010	9.6	3.1	1.3	0.6	0.3	0.1
		2006	11.8	3.9	1.5	0.5	0.2	0.1
		2007	8.2	1.8	0.4	0.1	0	0
	80	2008	9.8	3	1.1	0.5	0.2	0.1
		2009	7.4	1.5	0.4	0.1	0	0
		2010	2.2	0.3	0.1	0	0	0
		2006	34.8	19.1	11.4	7	4.1	2.5
		2007	35.6	21.1	13.6	9	6.1	4
	60	2008	31.9	17.2	9.8	5.7	3.3	1.8
		2009	13.5	4	1.2	0.3	0.1	0
		2010	33.6	19.4	12.5	8.3	5.5	3.6
		2006	18	5.9	2	0.6	0.2	0.1
		2007	16.1	5.6	2	0.7	0.2	0.1
New York	70	2008	13	3.4	0.9	0.2	0	0
		2009	1.8	0.1	0	0	0	0
		2010	12.8	3.8	1.2	0.3	0.1	0
		2006	6.6	0.8	0.1	0	0	0
		2007	4.2	0.5	0.1	0	0	0
	80	2008	2.5	0.2	0	0	0	0
		2009	0	0	0	0	0	0
		2010	2.4	0.2	0	0	0	0
		2006	36.6	22.4	15.4	10.8	7.6	5.4
		2007	40.9	26	18.1	13.2	9.8	7.1
	60	2008	34.1	19.8	12.7	8.4	5.5	3.7
		2009	12.7	3.7	1.1	0.3	0.1	0
		2010	33.5	19.8	13.1	8.9	6.2	4.3
		2006	16.1	5.9	2.3	0.9	0.4	0.1
		2007	20.9	8.6	3.9	1.8	0.7	0.3
Philadelphia	70	2008	14.5	4.6	1.6	0.6	0.2	0.1
		2009	1	0	0	0	0	0
		2010	13.5	4.3	1.5	0.5	0.2	0.1
		2006	3.4	0.4	0	0	0	0
		2007	7	1.1	0.2	0	0	0
	80	2008	3	0.3	0	0	0	0
		2009	0	0	0	0	0	0
		2010	2.7	0.3	0	0	0	0

Study Area	Exposure Benchmark (ppb-8hr)	Year	% of school-age children experiencing multiple exposures per O_3 season at or above benchmarks, base air quality					
			≥1	≥2	≥3	≥4	≥5	≥6
		2006	40.9	26.9	20	15.3	12	9.5
		2007	24.7	12.4	7.2	4.3	2.7	1.7
	60	2008	36.3	22.8	16.1	11.7	8.9	6.8
		2009	35	21.4	14.5	10.4	7.6	5.7
		2010	20.8	9.9	5.6	3.5	2.3	1.4
		2006	22.8	11.7	6.8	4.2	2.7	1.7
		2007	9.2	2.3	0.7	0.2	0.1	0
Sacramento	70	2008	19.1	8.8	4.7	2.7	1.7	1.1
		2009	16.7	6.8	3.2	1.7	0.8	0.5
		2010	7.4	2	0.7	0.3	0.1	0
		2006	8.8	2.7	0.9	0.3	0.2	0
		2007	2.7	0.2	0	0	0	0
	80	2008	7.6	2.3	0.9	0.4	0.1	0.1
		2009	5.2	1	0.3	0.1	0	0
		2010	2	0.3	0	0	0	0
		2006	32.4	19.7	13.2	9	6	4
		2007	38.4	24.8	17.8	13	9.9	7.4
	60	2008	11	3	0.9	0.2	0.1	0
		2009	11.3	2.8	0.8	0.2	0	0
		2010	22.9	11	5.8	3.1	1.6	0.9
		2006	12.7	3.9	1.3	0.4	0.1	0
		2007	20.9	9.9	5	2.6	1.3	0.6
St. Louis	70	2008	1.4	0.1	0	0	0	0
		2009	1.9	0	0	0	0	0
		2010	4.9	0.7	0.1	0	0	0
		2006	2.2	0.1	0	0	0	0
		2007	6.8	1.4	0.3	0.1	0	0
	80	2008	0	0	0	0	0	0
		2009	0.2	0	0	0	0	0
		2010	0.1	0	0	0	0	0
		2006	35.4	22.1	15.3	11	8	5.8
		2007	35.7	22.8	16.1	11.7	8.8	6.5
	60	2008	26.2	14.1	8.4	5	2.9	1.7
		2009	8.9	2	0.5	0.1	0	0
		2010	32.1	19.8	13.5	9.6	6.9	4.9
		2006	18.4	7.7	3.5	1.6	0.7	0.3
		2007	16.6	6.4	2.7	1.2	0.6	0.2
Washington DC	70	2008	10.3	2.7	0.7	0.2	0	0
		2009	0.7	0	0	0	0	0
		2010	12.5	4.2	1.6	0.7	0.3	0.1
		2006	5.7	1	0.1	0	0	0
		2007	3.9	0.5	0.1	0	0	0
	80	2008	2.5	0.2	0	0	0	0
		2009	0.1	0	0	0	0	0
		2010	2.4	0.3	0	0	0	0

5F-2 EXPOSURE MODELING RESULTS FOR ADJUSTED AIR QUALITY

In this section, we present the exposures estimated when considering the air quality adjusted to just meeting the existing O_3 NAAQS standard, as well as when considering potential alternative standard levels (55, 60, 65, 70 ppb 8-hr) of the existing standard. We note that one study area (Chicago) O_3 ambient monitor design values were below that of the existing standard during the 2008-2010, therefore APEX simulations could not be performed for that 3-year period. We could not simulate just meeting a standard level of 60 ppb-8hr or below in the New York study area, thus APEX simulations for these air quality scenarios could not be performed for the New York study area.

First are presented three-paneled figures for each of the four exposure study groups of interest (i.e. school-age children, asthmatic school-age children, asthmatic adults, older adults), one panel of which was briefly summarized at the end of Chapter 5 in the key observation section (all school-age children, 60 ppb-8hr benchmark). Presented for each of the three exposure benchmarks (60 ppb-8hr, 70 ppb-8hr, 80 ppb-8hr) are the highest estimated percent exposed while at moderate or greater exertion in each study area, considering just meeting the existing and alternative standards (**Figure 5F-7** to **Figure 5F-10**).

Exposures for the all school-age children study group were additionally characterized by calculating the mean percent (averaged over the study years) experiencing at least one exposure at or above each of the three benchmarks (60 ppb-8hr, 70 ppb-8hr, 80 ppb-8hr) while at moderate or greater exertion (**Figure 5F-11**). Further, the maximum (**Figure 5F-12**) and mean (**Figure 5F-13**) percent of all school-age children experiencing at least two exposures at or above the three health effect benchmark levels are presented. Following these figures, **Table 5F-2** provides the complete exposure output for all study areas, years, benchmark levels, and adjusted air quality scenarios for all school-age children, the study group containing the greatest percent and number of persons exposed in the REA.

And finally, the mean and maximum number of all school-age children and associated person days with at least one exposure at or above each of the benchmark levels is provided in **Table 5F-3**, by study area and air quality scenario. A similar but more visually pleasing presentation is given in **Figure 5F-14** and **Figure 5F-15**, providing the average number (and person-days, respectively) of all four exposure study study groups experiencing at least one 8-hr average exposure at or above 60 ppb across the 15 study areas considering each of the adjusted air quality scenarios. **Table 5F-4** contains the total number of persons experiencing at least one or two 8-hour exposures in all study areas by year, base air quality and air quality adjusted to just meeting the existing 75 ppb standard.

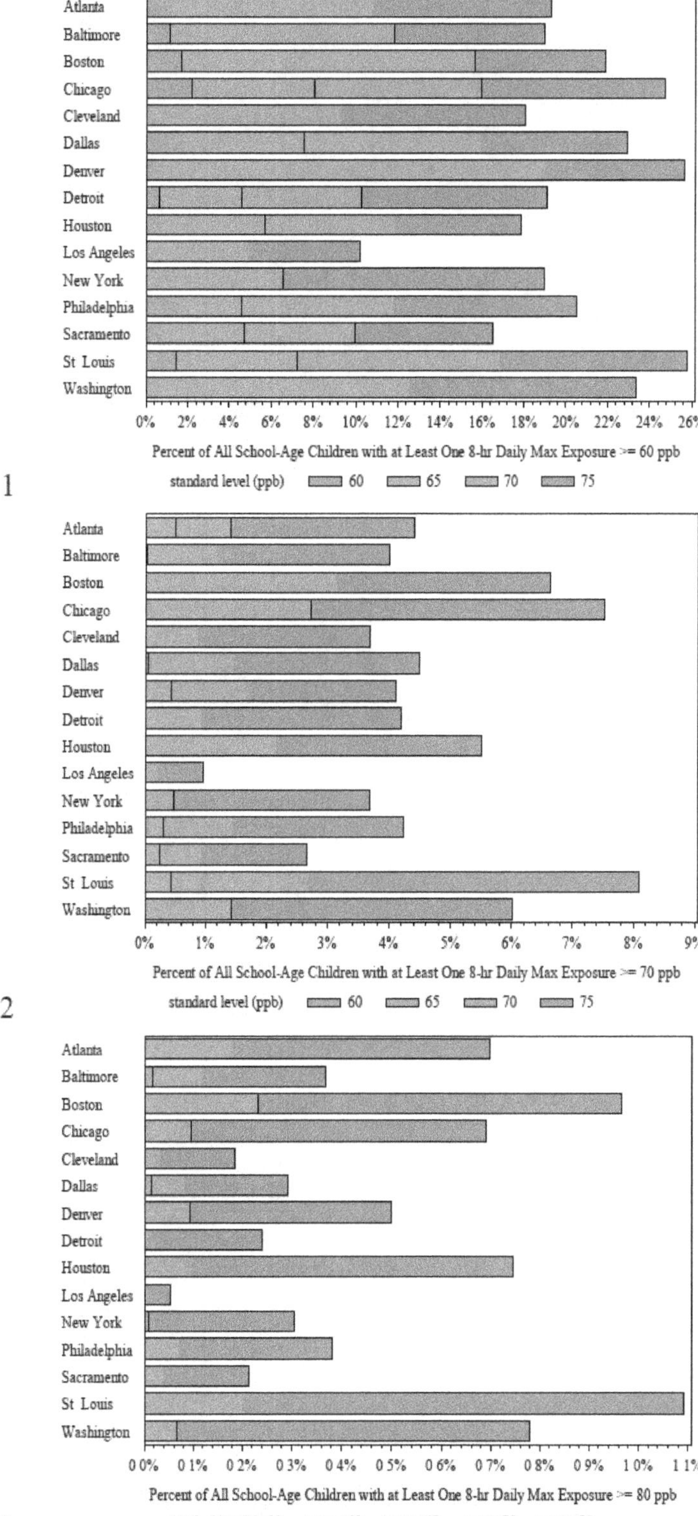

Figure 5F-7. Incremental increases in percent of **all school-age children** with at least one exposure at or above 60 ppb-8hr (top panel), 70 ppb-8hr (middle panel), or 80 ppb-8hr (bottom panel) using the **maximum percent** exposed for each study area, year 2006-2010 adjusted air quality.

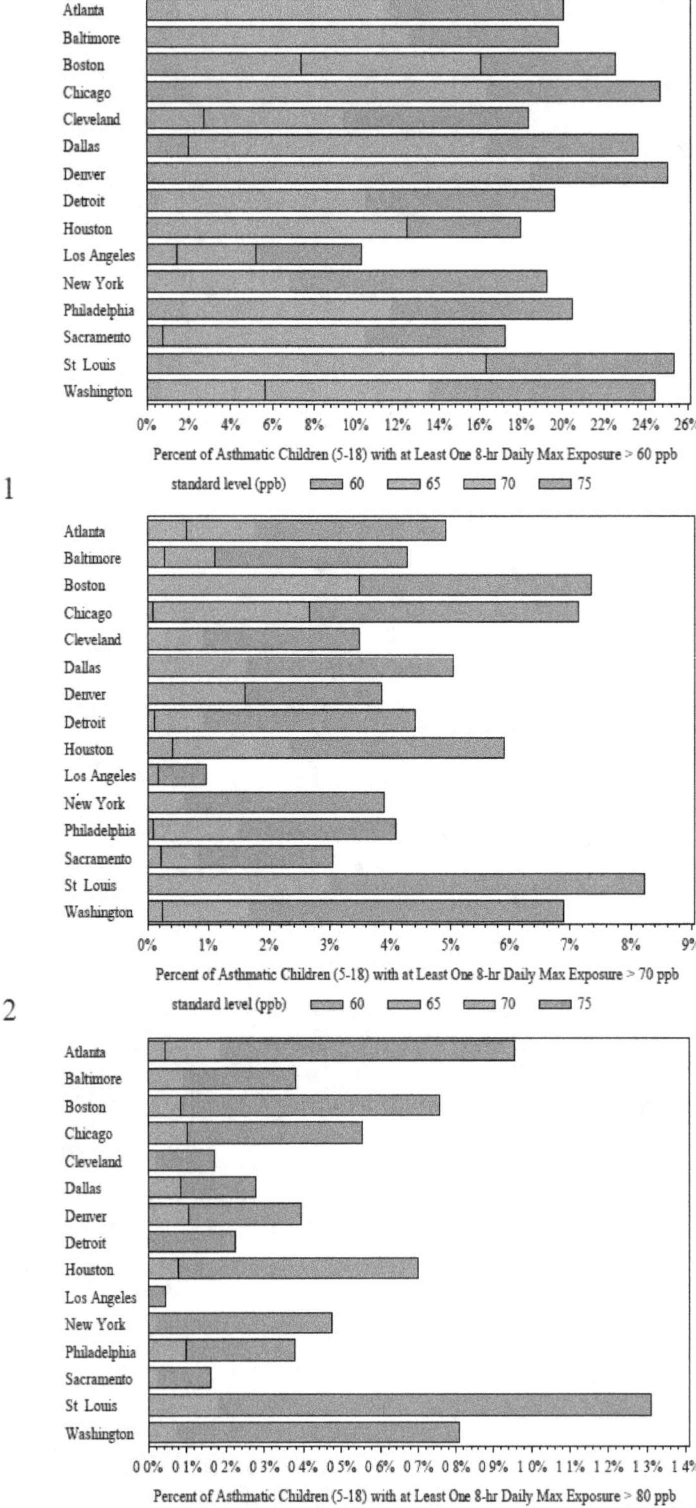

Figure 5F-8. Incremental increases in percent of **asthmatic school-age children** with at least one exposure at or above 60 ppb-8hr (top panel), 70 ppb-8hr (middle panel), or 80 ppb-8hr (bottom panel) using the **maximum percent** exposed for each study area, year 2006-2010 adjusted air quality.

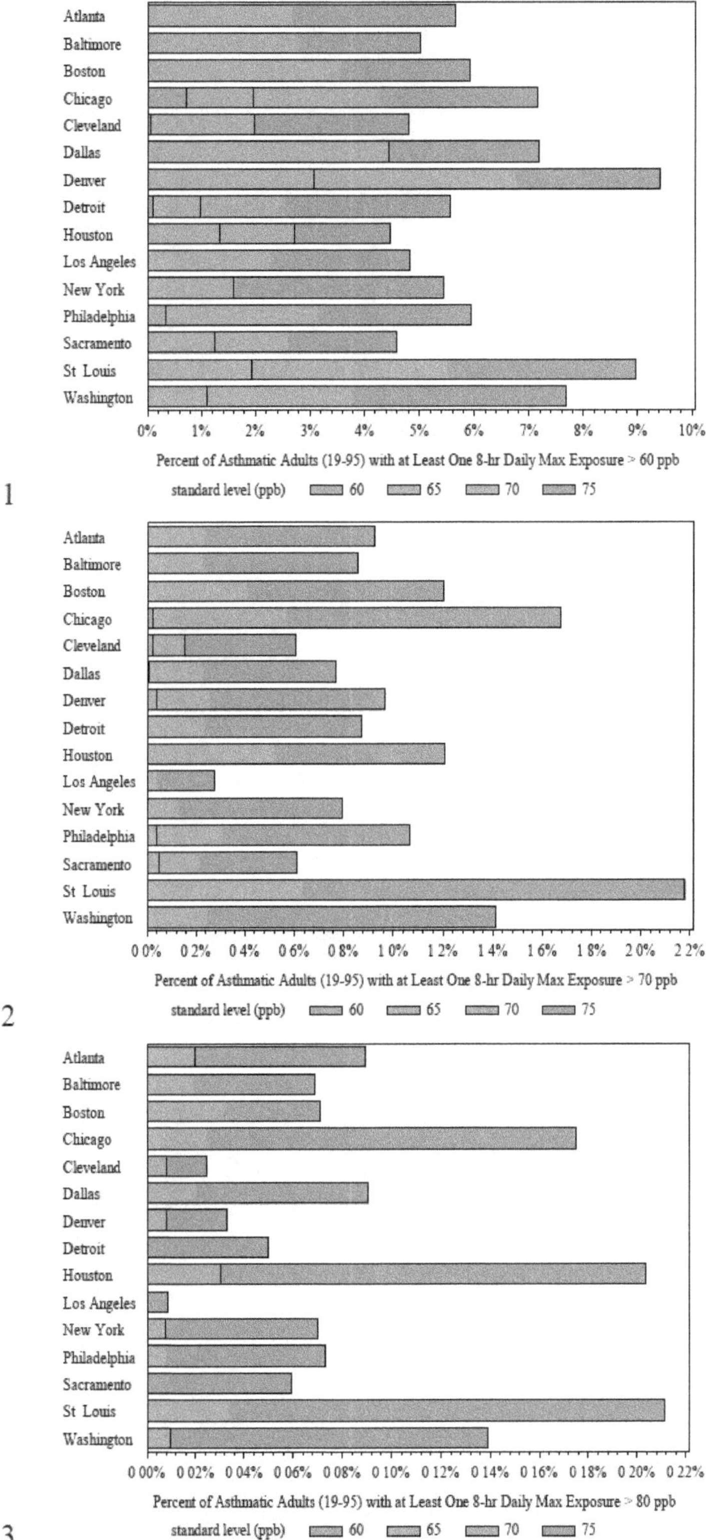

Figure 5F-9. Incremental increases in percent of **asthmatic adults** with at least one exposure at or above 60 ppb-8hr (top panel), 70 ppb-8hr (middle panel), or 80 ppb-8hr (bottom panel) using the **maximum percent** exposed for each study area, year 2006-2010 adjusted air quality.

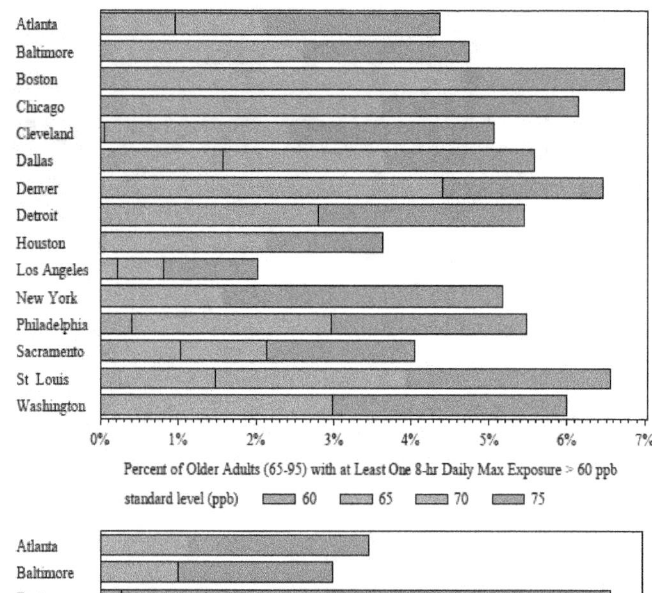

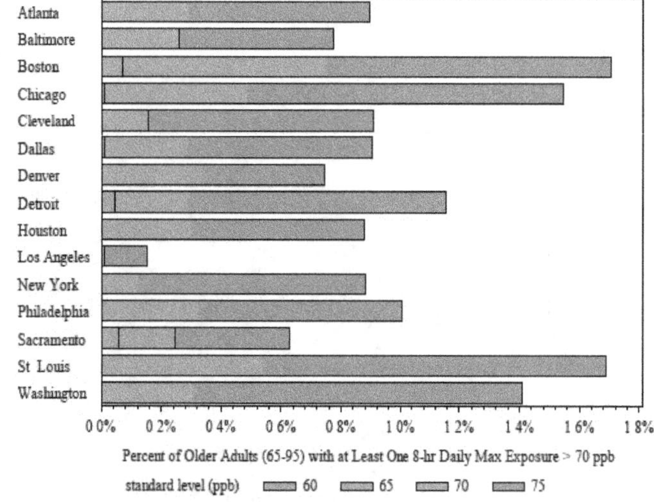

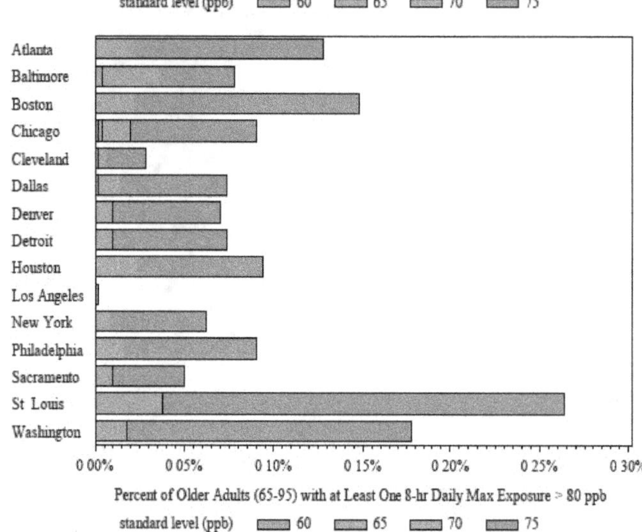

Figure 5F-10. Incremental increases in percent of **all older adults** with at least one exposure at or above 60 ppb-8hr (top panel), 70 ppb-8hr (middle panel), or 80 ppb-8hr (bottom panel) using the **maximum percent** exposed for each study area, year 2006-2010 adjusted air quality.

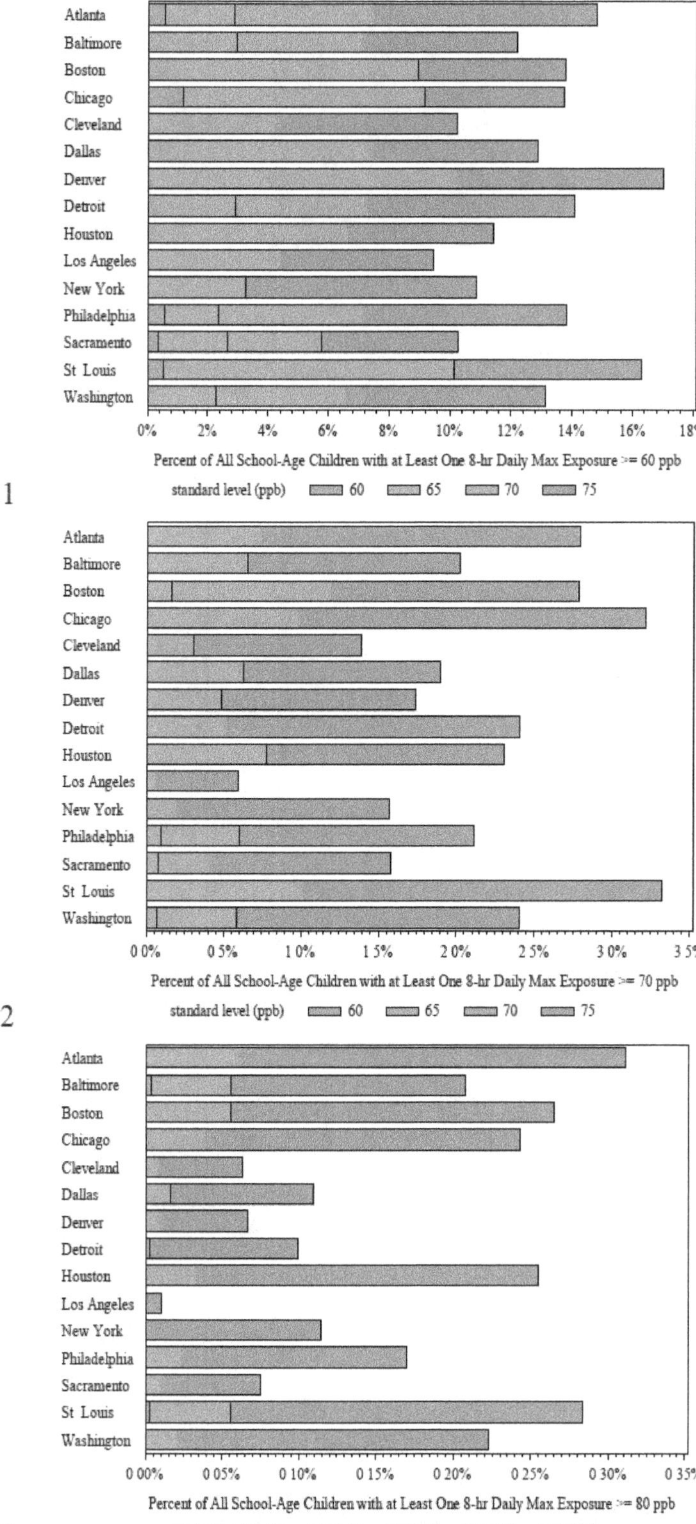

Figure 5F-11. Incremental increases in percent of **all school-age children** with at least one exposure at or above 60 ppb-8hr (top panel), 70 ppb-8hr (middle panel), or 80 ppb-8hr (bottom panel) using the **mean percent** exposed for each study area, year 2006-2010 adjusted air quality.

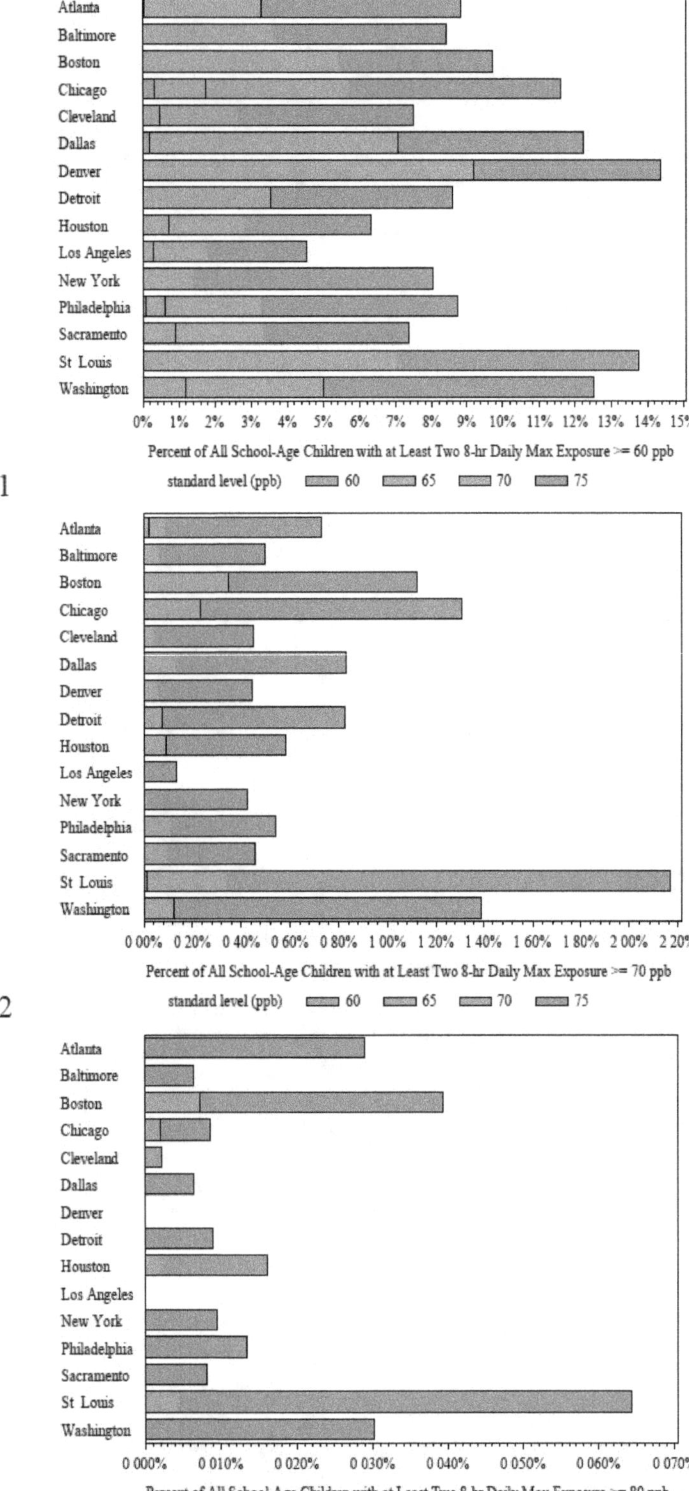

Figure 5F-12. Incremental increases in percent of **all school-age children** with **at least two exposures** at or above 60 ppb-8hr (top panel), 70 ppb-8hr (middle panel), or 80 ppb-8hr (bottom panel) using the **maximum percent** exposed for each study area, year 2006-2010 adjusted air quality.

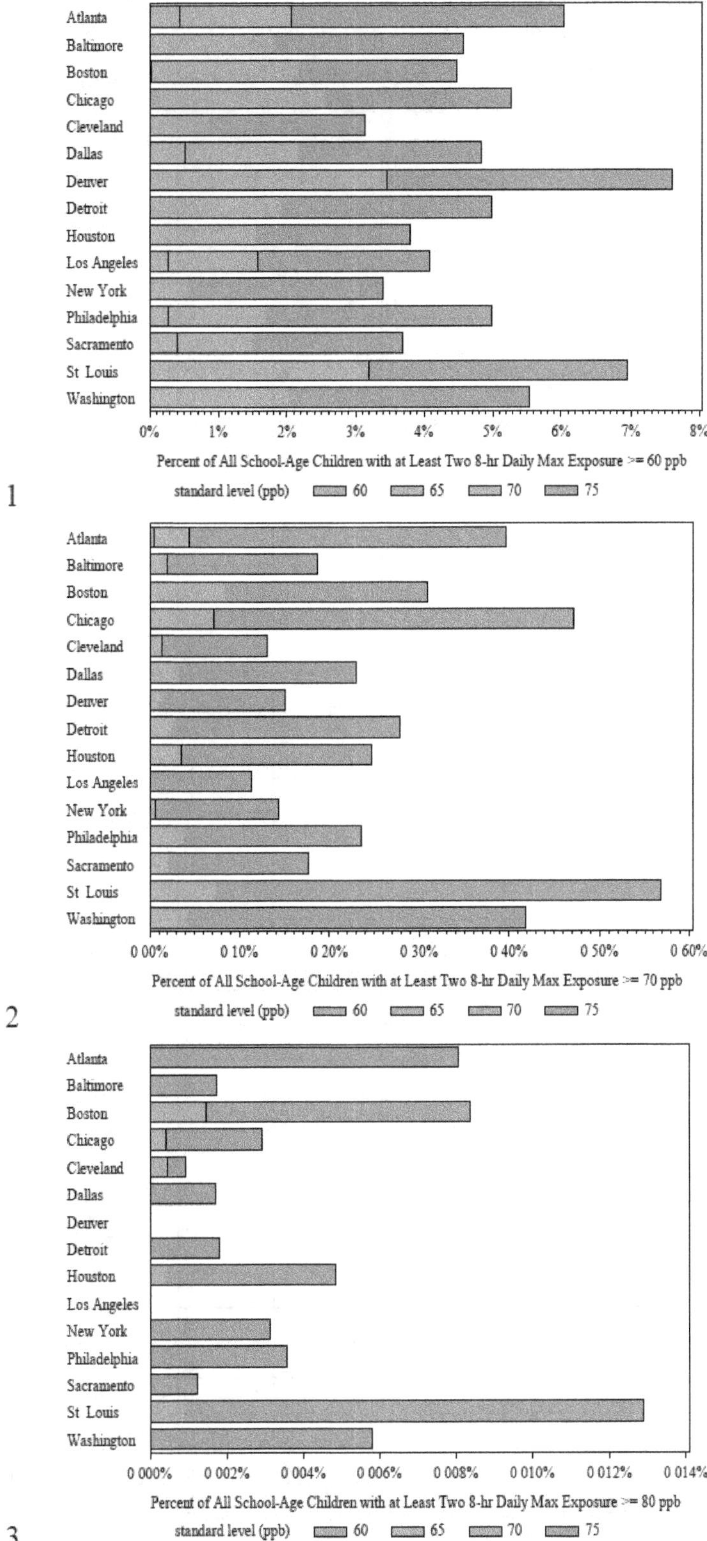

Figure 5F-13. Incremental increases in percent of **all school-age children** with at least two exposures at or above 60 ppb-8hr (top panel), 70 ppb-8hr (middle panel), or 80 ppb-8hr (bottom panel) using the **mean percent** exposed for each study area, year 2006-2010 adjusted air quality.

Table 5F-2. Percent of **all school-age children** with O_3 exposures at or above 60, 70, and 80 ppb-8hr while at moderate or greater exertion, years 2006-2010, adjusted air quality.

Study Area	Exposure Benchmark (ppb-8hr)	Air Quality Scenario[1]	Year	% of all school-age children experiencing multiple exposures per O_3 season at or above benchmarks, adjusted air quality					
				≥1	≥2	≥3	≥4	≥5	≥6
Atlanta	60	55(06-08)	2006	0.3	0	0	0	0	0
Atlanta	60	55(06-08)	2007	0.2	0	0	0	0	0
Atlanta	60	55(06-08)	2008	0	0	0	0	0	0
Atlanta	60	55(08-10)	2008	0	0	0	0	0	0
Atlanta	60	55(08-10)	2009	0	0	0	0	0	0
Atlanta	60	55(08-10)	2010	0	0	0	0	0	0
Atlanta	70	55(06-08)	2006	0	0	0	0	0	0
Atlanta	70	55(06-08)	2007	0	0	0	0	0	0
Atlanta	70	55(06-08)	2008	0	0	0	0	0	0
Atlanta	70	55(08-10)	2008	0	0	0	0	0	0
Atlanta	70	55(08-10)	2009	0	0	0	0	0	0
Atlanta	70	55(08-10)	2010	0	0	0	0	0	0
Atlanta	80	55(06-08)	2006	0	0	0	0	0	0
Atlanta	80	55(06-08)	2007	0	0	0	0	0	0
Atlanta	80	55(06-08)	2008	0	0	0	0	0	0
Atlanta	80	55(08-10)	2008	0	0	0	0	0	0
Atlanta	80	55(08-10)	2009	0	0	0	0	0	0
Atlanta	80	55(08-10)	2010	0	0	0	0	0	0
Atlanta	60	60(06-08)	2006	1.2	0.1	0	0	0	0
Atlanta	60	60(06-08)	2007	0.8	0	0	0	0	0
Atlanta	60	60(06-08)	2008	0	0	0	0	0	0
Atlanta	60	60(08-10)	2008	0.8	0	0	0	0	0
Atlanta	60	60(08-10)	2009	0.2	0	0	0	0	0
Atlanta	60	60(08-10)	2010	0.3	0	0	0	0	0
Atlanta	70	60(06-08)	2006	0.1	0	0	0	0	0
Atlanta	70	60(06-08)	2007	0	0	0	0	0	0
Atlanta	70	60(06-08)	2008	0	0	0	0	0	0
Atlanta	70	60(08-10)	2008	0	0	0	0	0	0
Atlanta	70	60(08-10)	2009	0	0	0	0	0	0
Atlanta	70	60(08-10)	2010	0	0	0	0	0	0
Atlanta	80	60(06-08)	2006	0	0	0	0	0	0
Atlanta	80	60(06-08)	2007	0	0	0	0	0	0
Atlanta	80	60(06-08)	2008	0	0	0	0	0	0
Atlanta	80	60(08-10)	2008	0	0	0	0	0	0
Atlanta	80	60(08-10)	2009	0	0	0	0	0	0
Atlanta	80	60(08-10)	2010	0	0	0	0	0	0
Atlanta	60	65(06-08)	2006	4.8	0.8	0.2	0	0	0
Atlanta	60	65(06-08)	2007	4.2	0.7	0.2	0	0	0
Atlanta	60	65(06-08)	2008	0.5	0	0	0	0	0
Atlanta	60	65(08-10)	2008	3.3	0.6	0.1	0	0	0
Atlanta	60	65(08-10)	2009	1.4	0.1	0	0	0	0
Atlanta	60	65(08-10)	2010	2.2	0.3	0	0	0	0
Atlanta	70	65(06-08)	2006	0.5	0	0	0	0	0
Atlanta	70	65(06-08)	2007	0.2	0	0	0	0	0

Study Area	Exposure Benchmark (ppb-8hr)	Air Quality Scenario[1]	Year	% of all school-age children experiencing multiple exposures per O_3 season at or above benchmarks, adjusted air quality					
				≥1	≥2	≥3	≥4	≥5	≥6
Atlanta	70	65(06-08)	2008	0	0	0	0	0	0
Atlanta	70	65(08-10)	2008	0.2	0	0	0	0	0
Atlanta	70	65(08-10)	2009	0	0	0	0	0	0
Atlanta	70	65(08-10)	2010	0	0	0	0	0	0
Atlanta	80	65(06-08)	2006	0	0	0	0	0	0
Atlanta	80	65(06-08)	2007	0	0	0	0	0	0
Atlanta	80	65(06-08)	2008	0	0	0	0	0	0
Atlanta	80	65(08-10)	2008	0	0	0	0	0	0
Atlanta	80	65(08-10)	2009	0	0	0	0	0	0
Atlanta	80	65(08-10)	2010	0	0	0	0	0	0
Atlanta	60	70(06-08)	2006	10.8	3.3	1.2	0.4	0.1	0
Atlanta	60	70(06-08)	2007	9.4	2.8	0.9	0.3	0.1	0
Atlanta	60	70(06-08)	2008	1.8	0.2	0	0	0	0
Atlanta	60	70(08-10)	2008	10	3.1	1.1	0.4	0.1	0.1
Atlanta	60	70(08-10)	2009	4.4	0.8	0.2	0	0	0
Atlanta	60	70(08-10)	2010	6.9	1.8	0.5	0.2	0.1	0
Atlanta	70	70(06-08)	2006	1.4	0.1	0	0	0	0
Atlanta	70	70(06-08)	2007	1	0.1	0	0	0	0
Atlanta	70	70(06-08)	2008	0.1	0	0	0	0	0
Atlanta	70	70(08-10)	2008	1.2	0.1	0	0	0	0
Atlanta	70	70(08-10)	2009	0.4	0	0	0	0	0
Atlanta	70	70(08-10)	2010	0.3	0	0	0	0	0
Atlanta	80	70(06-08)	2006	0.2	0	0	0	0	0
Atlanta	80	70(06-08)	2007	0.1	0	0	0	0	0
Atlanta	80	70(06-08)	2008	0	0	0	0	0	0
Atlanta	80	70(08-10)	2008	0.1	0	0	0	0	0
Atlanta	80	70(08-10)	2009	0	0	0	0	0	0
Atlanta	80	70(08-10)	2010	0	0	0	0	0	0
Atlanta	60	75(06-08)	2006	19.3	8.9	4.5	2.4	1.4	0.7
Atlanta	60	75(06-08)	2007	17.7	7.9	3.9	2.1	1.1	0.5
Atlanta	60	75(06-08)	2008	5.4	1.2	0.3	0.1	0	0
Atlanta	60	75(08-10)	2008	19.3	8.7	4.5	2.4	1.3	0.7
Atlanta	60	75(08-10)	2009	10.6	3	1	0.4	0.1	0
Atlanta	60	75(08-10)	2010	14	5.5	2.5	1.2	0.6	0.3
Atlanta	70	75(06-08)	2006	4.4	0.7	0.1	0	0	0
Atlanta	70	75(06-08)	2007	3.7	0.6	0.1	0	0	0
Atlanta	70	75(06-08)	2008	0.4	0	0	0	0	0
Atlanta	70	75(08-10)	2008	4.2	0.7	0.1	0	0	0
Atlanta	70	75(08-10)	2009	1.6	0.1	0	0	0	0
Atlanta	70	75(08-10)	2010	1.9	0.2	0	0	0	0
Atlanta	80	75(06-08)	2006	0.7	0	0	0	0	0
Atlanta	80	75(06-08)	2007	0.4	0	0	0	0	0
Atlanta	80	75(06-08)	2008	0	0	0	0	0	0
Atlanta	80	75(08-10)	2008	0.5	0	0	0	0	0
Atlanta	80	75(08-10)	2009	0.1	0	0	0	0	0
Atlanta	80	75(08-10)	2010	0.1	0	0	0	0	0

Study Area	Exposure Benchmark (ppb-8hr)	Air Quality Scenario[1]	Year	% of all school-age children experiencing multiple exposures per O_3 season at or above benchmarks, adjusted air quality					
				≥1	≥2	≥3	≥4	≥5	≥6
Baltimore	60	55(06-08)	2006	0.1	0	0	0	0	0
Baltimore	60	55(06-08)	2007	0	0	0	0	0	0
Baltimore	60	55(06-08)	2008	0	0	0	0	0	0
Baltimore	60	55(08-10)	2008	0	0	0	0	0	0
Baltimore	60	55(08-10)	2009	0	0	0	0	0	0
Baltimore	60	55(08-10)	2010	0.2	0	0	0	0	0
Baltimore	70	55(06-08)	2006	0	0	0	0	0	0
Baltimore	70	55(06-08)	2007	0	0	0	0	0	0
Baltimore	70	55(06-08)	2008	0	0	0	0	0	0
Baltimore	70	55(08-10)	2008	0	0	0	0	0	0
Baltimore	70	55(08-10)	2009	0	0	0	0	0	0
Baltimore	70	55(08-10)	2010	0	0	0	0	0	0
Baltimore	80	55(06-08)	2006	0	0	0	0	0	0
Baltimore	80	55(06-08)	2007	0	0	0	0	0	0
Baltimore	80	55(06-08)	2008	0	0	0	0	0	0
Baltimore	80	55(08-10)	2008	0	0	0	0	0	0
Baltimore	80	55(08-10)	2009	0	0	0	0	0	0
Baltimore	80	55(08-10)	2010	0	0	0	0	0	0
Baltimore	60	60(06-08)	2006	1	0.1	0	0	0	0
Baltimore	60	60(06-08)	2007	0.4	0	0	0	0	0
Baltimore	60	60(06-08)	2008	0.3	0	0	0	0	0
Baltimore	60	60(08-10)	2008	0.6	0	0	0	0	0
Baltimore	60	60(08-10)	2009	0	0	0	0	0	0
Baltimore	60	60(08-10)	2010	1.2	0	0	0	0	0
Baltimore	70	60(06-08)	2006	0	0	0	0	0	0
Baltimore	70	60(06-08)	2007	0	0	0	0	0	0
Baltimore	70	60(06-08)	2008	0	0	0	0	0	0
Baltimore	70	60(08-10)	2008	0	0	0	0	0	0
Baltimore	70	60(08-10)	2009	0	0	0	0	0	0
Baltimore	70	60(08-10)	2010	0	0	0	0	0	0
Baltimore	80	60(06-08)	2006	0	0	0	0	0	0
Baltimore	80	60(06-08)	2007	0	0	0	0	0	0
Baltimore	80	60(06-08)	2008	0	0	0	0	0	0
Baltimore	80	60(08-10)	2008	0	0	0	0	0	0
Baltimore	80	60(08-10)	2009	0	0	0	0	0	0
Baltimore	80	60(08-10)	2010	0	0	0	0	0	0
Baltimore	60	65(06-08)	2006	5.4	0.9	0.2	0	0	0
Baltimore	60	65(06-08)	2007	3.3	0.4	0	0	0	0
Baltimore	60	65(06-08)	2008	2.1	0.1	0	0	0	0
Baltimore	60	65(08-10)	2008	2.5	0.2	0	0	0	0
Baltimore	60	65(08-10)	2009	0.1	0	0	0	0	0
Baltimore	60	65(08-10)	2010	3.8	0.5	0.1	0	0	0
Baltimore	70	65(06-08)	2006	0.3	0	0	0	0	0
Baltimore	70	65(06-08)	2007	0.2	0	0	0	0	0
Baltimore	70	65(06-08)	2008	0.1	0	0	0	0	0
Baltimore	70	65(08-10)	2008	0.1	0	0	0	0	0

Study Area	Exposure Benchmark (ppb-8hr)	Air Quality Scenario[1]	Year	% of all school-age children experiencing multiple exposures per O_3 season at or above benchmarks, adjusted air quality					
				≥1	≥2	≥3	≥4	≥5	≥6
Baltimore	70	65(08-10)	2009	0	0	0	0	0	0
Baltimore	70	65(08-10)	2010	0.3	0	0	0	0	0
Baltimore	80	65(06-08)	2006	0	0	0	0	0	0
Baltimore	80	65(06-08)	2007	0	0	0	0	0	0
Baltimore	80	65(06-08)	2008	0	0	0	0	0	0
Baltimore	80	65(08-10)	2008	0	0	0	0	0	0
Baltimore	80	65(08-10)	2009	0	0	0	0	0	0
Baltimore	80	65(08-10)	2010	0	0	0	0	0	0
Baltimore	60	70(06-08)	2006	11.8	3.7	1.2	0.5	0.2	0.1
Baltimore	60	70(06-08)	2007	7.7	1.8	0.5	0.2	0.1	0
Baltimore	60	70(06-08)	2008	5.3	0.8	0.2	0	0	0
Baltimore	60	70(08-10)	2008	6.3	1.2	0.3	0.1	0	0
Baltimore	60	70(08-10)	2009	0.6	0	0	0	0	0
Baltimore	60	70(08-10)	2010	9.5	2.5	0.7	0.2	0.1	0
Baltimore	70	70(06-08)	2006	1.2	0.1	0	0	0	0
Baltimore	70	70(06-08)	2007	0.6	0	0	0	0	0
Baltimore	70	70(06-08)	2008	0.4	0	0	0	0	0
Baltimore	70	70(08-10)	2008	0.6	0	0	0	0	0
Baltimore	70	70(08-10)	2009	0	0	0	0	0	0
Baltimore	70	70(08-10)	2010	1	0	0	0	0	0
Baltimore	80	70(06-08)	2006	0.1	0	0	0	0	0
Baltimore	80	70(06-08)	2007	0.1	0	0	0	0	0
Baltimore	80	70(06-08)	2008	0	0	0	0	0	0
Baltimore	80	70(08-10)	2008	0	0	0	0	0	0
Baltimore	80	70(08-10)	2009	0	0	0	0	0	0
Baltimore	80	70(08-10)	2010	0.1	0	0	0	0	0
Baltimore	60	75(06-08)	2006	19	8.4	4.1	2.1	1.1	0.5
Baltimore	60	75(06-08)	2007	13.6	5.1	2.1	0.9	0.4	0.2
Baltimore	60	75(06-08)	2008	9.7	2.5	0.8	0.3	0.1	0
Baltimore	60	75(08-10)	2008	10.7	3	1	0.3	0.1	0
Baltimore	60	75(08-10)	2009	1.9	0.2	0	0	0	0
Baltimore	60	75(08-10)	2010	16.2	6.4	2.8	1.3	0.6	0.2
Baltimore	70	75(06-08)	2006	4	0.5	0.1	0	0	0
Baltimore	70	75(06-08)	2007	2	0.2	0	0	0	0
Baltimore	70	75(06-08)	2008	1.4	0.1	0	0	0	0
Baltimore	70	75(08-10)	2008	1.7	0.1	0	0	0	0
Baltimore	70	75(08-10)	2009	0.1	0	0	0	0	0
Baltimore	70	75(08-10)	2010	2.6	0.2	0	0	0	0
Baltimore	80	75(06-08)	2006	0.3	0	0	0	0	0
Baltimore	80	75(06-08)	2007	0.2	0	0	0	0	0
Baltimore	80	75(06-08)	2008	0.1	0	0	0	0	0
Baltimore	80	75(08-10)	2008	0.1	0	0	0	0	0
Baltimore	80	75(08-10)	2009	0	0	0	0	0	0
Baltimore	80	75(08-10)	2010	0.4	0	0	0	0	0
Boston	60	55(06-08)	2006	0	0	0	0	0	0
Boston	60	55(06-08)	2007	0.2	0	0	0	0	0

Study Area	Exposure Benchmark (ppb-8hr)	Air Quality Scenario[1]	Year	% of all school-age children experiencing multiple exposures per O_3 season at or above benchmarks, adjusted air quality					
				≥1	≥2	≥3	≥4	≥5	≥6
Boston	60	55(06-08)	2008	0	0	0	0	0	0
Boston	60	55(08-10)	2008	0	0	0	0	0	0
Boston	60	55(08-10)	2009	0.1	0	0	0	0	0
Boston	60	55(08-10)	2010	0	0	0	0	0	0
Boston	70	55(06-08)	2006	0	0	0	0	0	0
Boston	70	55(06-08)	2007	0	0	0	0	0	0
Boston	70	55(06-08)	2008	0	0	0	0	0	0
Boston	70	55(08-10)	2008	0	0	0	0	0	0
Boston	70	55(08-10)	2009	0	0	0	0	0	0
Boston	70	55(08-10)	2010	0	0	0	0	0	0
Boston	80	55(06-08)	2006	0	0	0	0	0	0
Boston	80	55(06-08)	2007	0	0	0	0	0	0
Boston	80	55(06-08)	2008	0	0	0	0	0	0
Boston	80	55(08-10)	2008	0	0	0	0	0	0
Boston	80	55(08-10)	2009	0	0	0	0	0	0
Boston	80	55(08-10)	2010	0	0	0	0	0	0
Boston	60	60(06-08)	2006	0.1	0	0	0	0	0
Boston	60	60(06-08)	2007	1.7	0.1	0	0	0	0
Boston	60	60(06-08)	2008	0	0	0	0	0	0
Boston	60	60(08-10)	2008	0.3	0	0	0	0	0
Boston	60	60(08-10)	2009	1.3	0.1	0	0	0	0
Boston	60	60(08-10)	2010	0.4	0	0	0	0	0
Boston	70	60(06-08)	2006	0	0	0	0	0	0
Boston	70	60(06-08)	2007	0	0	0	0	0	0
Boston	70	60(06-08)	2008	0	0	0	0	0	0
Boston	70	60(08-10)	2008	0	0	0	0	0	0
Boston	70	60(08-10)	2009	0	0	0	0	0	0
Boston	70	60(08-10)	2010	0	0	0	0	0	0
Boston	80	60(06-08)	2006	0	0	0	0	0	0
Boston	80	60(06-08)	2007	0	0	0	0	0	0
Boston	80	60(06-08)	2008	0	0	0	0	0	0
Boston	80	60(08-10)	2008	0	0	0	0	0	0
Boston	80	60(08-10)	2009	0	0	0	0	0	0
Boston	80	60(08-10)	2010	0	0	0	0	0	0
Boston	60	65(06-08)	2006	1.2	0.1	0	0	0	0
Boston	60	65(06-08)	2007	6.7	1.1	0.2	0	0	0
Boston	60	65(06-08)	2008	0.8	0	0	0	0	0
Boston	60	65(08-10)	2008	2.9	0.3	0	0	0	0
Boston	60	65(08-10)	2009	4.6	0.5	0	0	0	0
Boston	60	65(08-10)	2010	2.5	0.2	0	0	0	0
Boston	70	65(06-08)	2006	0	0	0	0	0	0
Boston	70	65(06-08)	2007	0.5	0	0	0	0	0
Boston	70	65(06-08)	2008	0	0	0	0	0	0
Boston	70	65(08-10)	2008	0.1	0	0	0	0	0
Boston	70	65(08-10)	2009	0.2	0	0	0	0	0
Boston	70	65(08-10)	2010	0	0	0	0	0	0

Study Area	Exposure Benchmark (ppb-8hr)	Air Quality Scenario[1]	Year	% of all school-age children experiencing multiple exposures per O_3 season at or above benchmarks, adjusted air quality					
				≥1	≥2	≥3	≥4	≥5	≥6
Boston	80	65(06-08)	2006	0	0	0	0	0	0
Boston	80	65(06-08)	2007	0	0	0	0	0	0
Boston	80	65(06-08)	2008	0	0	0	0	0	0
Boston	80	65(08-10)	2008	0	0	0	0	0	0
Boston	80	65(08-10)	2009	0	0	0	0	0	0
Boston	80	65(08-10)	2010	0	0	0	0	0	0
Boston	60	70(06-08)	2006	6.7	1.2	0.2	0	0	0
Boston	60	70(06-08)	2007	15.7	5.5	2	0.8	0.3	0.1
Boston	60	70(06-08)	2008	4.7	0.7	0.1	0	0	0
Boston	60	70(08-10)	2008	9.1	2.1	0.5	0.1	0	0
Boston	60	70(08-10)	2009	9	1.8	0.2	0	0	0
Boston	60	70(08-10)	2010	6.7	1.1	0.2	0	0	0
Boston	70	70(06-08)	2006	0.3	0	0	0	0	0
Boston	70	70(06-08)	2007	3.2	0.4	0	0	0	0
Boston	70	70(06-08)	2008	0.2	0	0	0	0	0
Boston	70	70(08-10)	2008	0.8	0	0	0	0	0
Boston	70	70(08-10)	2009	1.3	0	0	0	0	0
Boston	70	70(08-10)	2010	0.7	0	0	0	0	0
Boston	80	70(06-08)	2006	0	0	0	0	0	0
Boston	80	70(06-08)	2007	0.2	0	0	0	0	0
Boston	80	70(06-08)	2008	0	0	0	0	0	0
Boston	80	70(08-10)	2008	0.1	0	0	0	0	0
Boston	80	70(08-10)	2009	0	0	0	0	0	0
Boston	80	70(08-10)	2010	0	0	0	0	0	0
Boston	60	75(06-08)	2006	11.9	3.1	0.9	0.2	0	0
Boston	60	75(06-08)	2007	21.9	9.7	4.6	2.3	1.1	0.5
Boston	60	75(06-08)	2008	9.1	2.1	0.5	0.1	0	0
Boston	60	75(08-10)	2008	15.9	5.5	2.1	0.8	0.3	0.1
Boston	60	75(08-10)	2009	11.4	2.8	0.6	0.1	0	0
Boston	60	75(08-10)	2010	11.2	3	0.9	0.2	0.1	0
Boston	70	75(06-08)	2006	1.4	0.1	0	0	0	0
Boston	70	75(06-08)	2007	6.6	1.1	0.2	0	0	0
Boston	70	75(06-08)	2008	0.8	0	0	0	0	0
Boston	70	75(08-10)	2008	3.2	0.3	0	0	0	0
Boston	70	75(08-10)	2009	2.1	0.1	0	0	0	0
Boston	70	75(08-10)	2010	1.7	0.1	0	0	0	0
Boston	80	75(06-08)	2006	0	0	0	0	0	0
Boston	80	75(06-08)	2007	1	0	0	0	0	0
Boston	80	75(06-08)	2008	0.1	0	0	0	0	0
Boston	80	75(08-10)	2008	0.4	0	0	0	0	0
Boston	80	75(08-10)	2009	0.1	0	0	0	0	0
Boston	80	75(08-10)	2010	0.1	0	0	0	0	0
Chicago	60	55(06-08)	2006	0.1	0	0	0	0	0
Chicago	60	55(06-08)	2007	0.3	0	0	0	0	0
Chicago	60	55(06-08)	2008	0	0	0	0	0	0
Chicago	60	55(08-10)	2008	0	0	0	0	0	0

Study Area	Exposure Benchmark (ppb-8hr)	Air Quality Scenario[1]	Year	% of all school-age children experiencing multiple exposures per O_3 season at or above benchmarks, adjusted air quality					
				≥1	≥2	≥3	≥4	≥5	≥6
Chicago	60	55(08-10)	2009	0.3	0	0	0	0	0
Chicago	60	55(08-10)	2010	0.3	0	0	0	0	0
Chicago	70	55(06-08)	2006	0	0	0	0	0	0
Chicago	70	55(06-08)	2007	0	0	0	0	0	0
Chicago	70	55(06-08)	2008	0	0	0	0	0	0
Chicago	70	55(08-10)	2008	0	0	0	0	0	0
Chicago	70	55(08-10)	2009	0	0	0	0	0	0
Chicago	70	55(08-10)	2010	0	0	0	0	0	0
Chicago	80	55(06-08)	2006	0	0	0	0	0	0
Chicago	80	55(06-08)	2007	0	0	0	0	0	0
Chicago	80	55(06-08)	2008	0	0	0	0	0	0
Chicago	80	55(08-10)	2008	0	0	0	0	0	0
Chicago	80	55(08-10)	2009	0	0	0	0	0	0
Chicago	80	55(08-10)	2010	0	0	0	0	0	0
Chicago	60	60(06-08)	2006	0.7	0	0	0	0	0
Chicago	60	60(06-08)	2007	2.2	0.3	0	0	0	0
Chicago	60	60(06-08)	2008	0	0	0	0	0	0
Chicago	60	60(08-10)	2008	0.1	0	0	0	0	0
Chicago	60	60(08-10)	2009	1.1	0.1	0	0	0	0
Chicago	60	60(08-10)	2010	2	0.3	0.1	0	0	0
Chicago	70	60(06-08)	2006	0	0	0	0	0	0
Chicago	70	60(06-08)	2007	0.1	0	0	0	0	0
Chicago	70	60(06-08)	2008	0	0	0	0	0	0
Chicago	70	60(08-10)	2008	0	0	0	0	0	0
Chicago	70	60(08-10)	2009	0.1	0	0	0	0	0
Chicago	70	60(08-10)	2010	0	0	0	0	0	0
Chicago	80	60(06-08)	2006	0	0	0	0	0	0
Chicago	80	60(06-08)	2007	0	0	0	0	0	0
Chicago	80	60(06-08)	2008	0	0	0	0	0	0
Chicago	80	60(08-10)	2008	0	0	0	0	0	0
Chicago	80	60(08-10)	2009	0	0	0	0	0	0
Chicago	80	60(08-10)	2010	0	0	0	0	0	0
Chicago	60	65(06-08)	2006	2.9	0.3	0	0	0	0
Chicago	60	65(06-08)	2007	8.1	1.8	0.4	0.1	0	0
Chicago	60	65(06-08)	2008	0.2	0	0	0	0	0
Chicago	60	65(08-10)	2008	0.8	0	0	0	0	0
Chicago	60	65(08-10)	2009	2.9	0.4	0.1	0	0	0
Chicago	60	65(08-10)	2010	6.5	1.5	0.4	0.1	0	0
Chicago	70	65(06-08)	2006	0.1	0	0	0	0	0
Chicago	70	65(06-08)	2007	0.4	0	0	0	0	0
Chicago	70	65(06-08)	2008	0	0	0	0	0	0
Chicago	70	65(08-10)	2008	0	0	0	0	0	0
Chicago	70	65(08-10)	2009	0.3	0	0	0	0	0
Chicago	70	65(08-10)	2010	0.2	0	0	0	0	0
Chicago	80	65(06-08)	2006	0	0	0	0	0	0
Chicago	80	65(06-08)	2007	0	0	0	0	0	0

Study Area	Exposure Benchmark (ppb-8hr)	Air Quality Scenario[1]	Year	% of all school-age children experiencing multiple exposures per O_3 season at or above benchmarks, adjusted air quality					
				≥1	≥2	≥3	≥4	≥5	≥6
Chicago	80	65(06-08)	2008	0	0	0	0	0	0
Chicago	80	65(08-10)	2008	0	0	0	0	0	0
Chicago	80	65(08-10)	2009	0	0	0	0	0	0
Chicago	80	65(08-10)	2010	0	0	0	0	0	0
Chicago	60	70(06-08)	2006	7.7	1.6	0.4	0.1	0	0
Chicago	60	70(06-08)	2007	16	5.7	2.1	0.8	0.3	0.1
Chicago	60	70(06-08)	2008	1.1	0.1	0	0	0	0
Chicago	60	70(08-10)	2008	2.9	0.3	0	0	0	0
Chicago	60	70(08-10)	2009	6.6	1.2	0.2	0	0	0
Chicago	60	70(08-10)	2010	13.6	4.1	1.3	0.4	0.1	0
Chicago	70	70(06-08)	2006	0.5	0	0	0	0	0
Chicago	70	70(06-08)	2007	2.7	0.2	0	0	0	0
Chicago	70	70(06-08)	2008	0	0	0	0	0	0
Chicago	70	70(08-10)	2008	0	0	0	0	0	0
Chicago	70	70(08-10)	2009	0.6	0	0	0	0	0
Chicago	70	70(08-10)	2010	1	0.1	0	0	0	0
Chicago	80	70(06-08)	2006	0	0	0	0	0	0
Chicago	80	70(06-08)	2007	0.1	0	0	0	0	0
Chicago	80	70(06-08)	2008	0	0	0	0	0	0
Chicago	80	70(08-10)	2008	0	0	0	0	0	0
Chicago	80	70(08-10)	2009	0.1	0	0	0	0	0
Chicago	80	70(08-10)	2010	0	0	0	0	0	0
Chicago	60	75(06-08)	2006	13.5	3.9	1.2	0.4	0.1	0
Chicago	60	75(06-08)	2007	24.7	11.6	6	3.1	1.6	0.8
Chicago	60	75(06-08)	2008	3	0.3	0	0	0	0
Chicago	70	75(06-08)	2006	2.1	0.1	0	0	0	0
Chicago	70	75(06-08)	2007	7.5	1.3	0.2	0	0	0
Chicago	70	75(06-08)	2008	0	0	0	0	0	0
Chicago	80	75(06-08)	2006	0	0	0	0	0	0
Chicago	80	75(06-08)	2007	0.7	0	0	0	0	0
Chicago	80	75(06-08)	2008	0	0	0	0	0	0
Cleveland	60	55(06-08)	2006	0	0	0	0	0	0
Cleveland	60	55(06-08)	2007	0	0	0	0	0	0
Cleveland	60	55(06-08)	2008	0	0	0	0	0	0
Cleveland	60	55(08-10)	2008	0	0	0	0	0	0
Cleveland	60	55(08-10)	2009	0	0	0	0	0	0
Cleveland	60	55(08-10)	2010	0	0	0	0	0	0
Cleveland	70	55(06-08)	2006	0	0	0	0	0	0
Cleveland	70	55(06-08)	2007	0	0	0	0	0	0
Cleveland	70	55(06-08)	2008	0	0	0	0	0	0
Cleveland	70	55(08-10)	2008	0	0	0	0	0	0
Cleveland	70	55(08-10)	2009	0	0	0	0	0	0
Cleveland	70	55(08-10)	2010	0	0	0	0	0	0
Cleveland	80	55(06-08)	2006	0	0	0	0	0	0
Cleveland	80	55(06-08)	2007	0	0	0	0	0	0
Cleveland	80	55(06-08)	2008	0	0	0	0	0	0

Study Area	Exposure Benchmark (ppb-8hr)	Air Quality Scenario[1]	Year	% of all school-age children experiencing multiple exposures per O_3 season at or above benchmarks, adjusted air quality					
				≥1	≥2	≥3	≥4	≥5	≥6
Cleveland	80	55(08-10)	2008	0	0	0	0	0	0
Cleveland	80	55(08-10)	2009	0	0	0	0	0	0
Cleveland	80	55(08-10)	2010	0	0	0	0	0	0
Cleveland	60	60(06-08)	2006	0.1	0	0	0	0	0
Cleveland	60	60(06-08)	2007	0.4	0	0	0	0	0
Cleveland	60	60(06-08)	2008	0	0	0	0	0	0
Cleveland	60	60(08-10)	2008	0	0	0	0	0	0
Cleveland	60	60(08-10)	2009	0.2	0	0	0	0	0
Cleveland	60	60(08-10)	2010	0.1	0	0	0	0	0
Cleveland	70	60(06-08)	2006	0	0	0	0	0	0
Cleveland	70	60(06-08)	2007	0	0	0	0	0	0
Cleveland	70	60(06-08)	2008	0	0	0	0	0	0
Cleveland	70	60(08-10)	2008	0	0	0	0	0	0
Cleveland	70	60(08-10)	2009	0	0	0	0	0	0
Cleveland	70	60(08-10)	2010	0	0	0	0	0	0
Cleveland	80	60(06-08)	2006	0	0	0	0	0	0
Cleveland	80	60(06-08)	2007	0	0	0	0	0	0
Cleveland	80	60(06-08)	2008	0	0	0	0	0	0
Cleveland	80	60(08-10)	2008	0	0	0	0	0	0
Cleveland	80	60(08-10)	2009	0	0	0	0	0	0
Cleveland	80	60(08-10)	2010	0	0	0	0	0	0
Cleveland	60	65(06-08)	2006	0.5	0	0	0	0	0
Cleveland	60	65(06-08)	2007	3	0.5	0.1	0	0	0
Cleveland	60	65(06-08)	2008	1.2	0.1	0	0	0	0
Cleveland	60	65(08-10)	2008	0.6	0	0	0	0	0
Cleveland	60	65(08-10)	2009	0.7	0	0	0	0	0
Cleveland	60	65(08-10)	2010	0.5	0	0	0	0	0
Cleveland	70	65(06-08)	2006	0	0	0	0	0	0
Cleveland	70	65(06-08)	2007	0.2	0	0	0	0	0
Cleveland	70	65(06-08)	2008	0	0	0	0	0	0
Cleveland	70	65(08-10)	2008	0	0	0	0	0	0
Cleveland	70	65(08-10)	2009	0	0	0	0	0	0
Cleveland	70	65(08-10)	2010	0	0	0	0	0	0
Cleveland	80	65(06-08)	2006	0	0	0	0	0	0
Cleveland	80	65(06-08)	2007	0	0	0	0	0	0
Cleveland	80	65(06-08)	2008	0	0	0	0	0	0
Cleveland	80	65(08-10)	2008	0	0	0	0	0	0
Cleveland	80	65(08-10)	2009	0	0	0	0	0	0
Cleveland	80	65(08-10)	2010	0	0	0	0	0	0
Cleveland	60	70(06-08)	2006	3.1	0.4	0.1	0	0	0
Cleveland	60	70(06-08)	2007	9.3	2.6	0.7	0.2	0.1	0
Cleveland	60	70(06-08)	2008	5.3	0.9	0.2	0	0	0
Cleveland	60	70(08-10)	2008	4.1	0.6	0.1	0	0	0
Cleveland	60	70(08-10)	2009	1.9	0.2	0	0	0	0
Cleveland	60	70(08-10)	2010	2.2	0.3	0.1	0	0	0
Cleveland	70	70(06-08)	2006	0.1	0	0	0	0	0

Study Area	Exposure Benchmark (ppb-8hr)	Air Quality Scenario[1]	Year	% of all school-age children experiencing multiple exposures per O_3 season at or above benchmarks, adjusted air quality					
				≥1	≥2	≥3	≥4	≥5	≥6
Cleveland	70	70(06-08)	2007	0.9	0	0	0	0	0
Cleveland	70	70(06-08)	2008	0.4	0	0	0	0	0
Cleveland	70	70(08-10)	2008	0.2	0	0	0	0	0
Cleveland	70	70(08-10)	2009	0.2	0	0	0	0	0
Cleveland	70	70(08-10)	2010	0.1	0	0	0	0	0
Cleveland	80	70(06-08)	2006	0	0	0	0	0	0
Cleveland	80	70(06-08)	2007	0	0	0	0	0	0
Cleveland	80	70(06-08)	2008	0	0	0	0	0	0
Cleveland	80	70(08-10)	2008	0	0	0	0	0	0
Cleveland	80	70(08-10)	2009	0	0	0	0	0	0
Cleveland	80	70(08-10)	2010	0	0	0	0	0	0
Cleveland	60	75(06-08)	2006	9.3	2.2	0.6	0.2	0	0
Cleveland	60	75(06-08)	2007	18	7.5	3.4	1.6	0.7	0.3
Cleveland	60	75(06-08)	2008	11.7	3.4	1.2	0.4	0.1	0
Cleveland	60	75(08-10)	2008	10.6	2.9	1	0.4	0.1	0
Cleveland	60	75(08-10)	2009	4.5	0.9	0.2	0.1	0	0
Cleveland	60	75(08-10)	2010	8.2	2	0.6	0.2	0.1	0
Cleveland	70	75(06-08)	2006	0.6	0	0	0	0	0
Cleveland	70	75(06-08)	2007	3.7	0.5	0.1	0	0	0
Cleveland	70	75(06-08)	2008	1.8	0.2	0	0	0	0
Cleveland	70	75(08-10)	2008	1.5	0.1	0	0	0	0
Cleveland	70	75(08-10)	2009	0.5	0	0	0	0	0
Cleveland	70	75(08-10)	2010	0.5	0	0	0	0	0
Cleveland	80	75(06-08)	2006	0	0	0	0	0	0
Cleveland	80	75(06-08)	2007	0.2	0	0	0	0	0
Cleveland	80	75(06-08)	2008	0.1	0	0	0	0	0
Cleveland	80	75(08-10)	2008	0	0	0	0	0	0
Cleveland	80	75(08-10)	2009	0	0	0	0	0	0
Cleveland	80	75(08-10)	2010	0	0	0	0	0	0
Dallas	60	55(06-08)	2006	0.1	0	0	0	0	0
Dallas	60	55(06-08)	2007	0.3	0	0	0	0	0
Dallas	60	55(06-08)	2008	0	0	0	0	0	0
Dallas	60	55(08-10)	2008	0	0	0	0	0	0
Dallas	60	55(08-10)	2009	0	0	0	0	0	0
Dallas	60	55(08-10)	2010	0	0	0	0	0	0
Dallas	70	55(06-08)	2006	0	0	0	0	0	0
Dallas	70	55(06-08)	2007	0	0	0	0	0	0
Dallas	70	55(06-08)	2008	0	0	0	0	0	0
Dallas	70	55(08-10)	2008	0	0	0	0	0	0
Dallas	70	55(08-10)	2009	0	0	0	0	0	0
Dallas	70	55(08-10)	2010	0	0	0	0	0	0
Dallas	80	55(06-08)	2006	0	0	0	0	0	0
Dallas	80	55(06-08)	2007	0	0	0	0	0	0
Dallas	80	55(06-08)	2008	0	0	0	0	0	0
Dallas	80	55(08-10)	2008	0	0	0	0	0	0
Dallas	80	55(08-10)	2009	0	0	0	0	0	0

Study Area	Exposure Benchmark (ppb-8hr)	Air Quality Scenario[1]	Year	% of all school-age children experiencing multiple exposures per O_3 season at or above benchmarks, adjusted air quality					
				≥1	≥2	≥3	≥4	≥5	≥6
Dallas	80	55(08-10)	2010	0	0	0	0	0	0
Dallas	60	60(06-08)	2006	1.9	0.2	0	0	0	0
Dallas	60	60(06-08)	2007	1.1	0	0	0	0	0
Dallas	60	60(06-08)	2008	0.1	0	0	0	0	0
Dallas	60	60(08-10)	2008	0.1	0	0	0	0	0
Dallas	60	60(08-10)	2009	0.7	0	0	0	0	0
Dallas	60	60(08-10)	2010	0.1	0	0	0	0	0
Dallas	70	60(06-08)	2006	0	0	0	0	0	0
Dallas	70	60(06-08)	2007	0.1	0	0	0	0	0
Dallas	70	60(06-08)	2008	0	0	0	0	0	0
Dallas	70	60(08-10)	2008	0	0	0	0	0	0
Dallas	70	60(08-10)	2009	0	0	0	0	0	0
Dallas	70	60(08-10)	2010	0	0	0	0	0	0
Dallas	80	60(06-08)	2006	0	0	0	0	0	0
Dallas	80	60(06-08)	2007	0	0	0	0	0	0
Dallas	80	60(06-08)	2008	0	0	0	0	0	0
Dallas	80	60(08-10)	2008	0	0	0	0	0	0
Dallas	80	60(08-10)	2009	0	0	0	0	0	0
Dallas	80	60(08-10)	2010	0	0	0	0	0	0
Dallas	60	65(06-08)	2006	7.6	2	0.7	0.2	0.1	0
Dallas	60	65(06-08)	2007	2.6	0.1	0	0	0	0
Dallas	60	65(06-08)	2008	0.7	0	0	0	0	0
Dallas	60	65(08-10)	2008	0.7	0	0	0	0	0
Dallas	60	65(08-10)	2009	3.1	0.4	0	0	0	0
Dallas	60	65(08-10)	2010	1	0.1	0	0	0	0
Dallas	70	65(06-08)	2006	0.2	0	0	0	0	0
Dallas	70	65(06-08)	2007	0.3	0	0	0	0	0
Dallas	70	65(06-08)	2008	0	0	0	0	0	0
Dallas	70	65(08-10)	2008	0	0	0	0	0	0
Dallas	70	65(08-10)	2009	0.1	0	0	0	0	0
Dallas	70	65(08-10)	2010	0	0	0	0	0	0
Dallas	80	65(06-08)	2006	0	0	0	0	0	0
Dallas	80	65(06-08)	2007	0	0	0	0	0	0
Dallas	80	65(06-08)	2008	0	0	0	0	0	0
Dallas	80	65(08-10)	2008	0	0	0	0	0	0
Dallas	80	65(08-10)	2009	0	0	0	0	0	0
Dallas	80	65(08-10)	2010	0	0	0	0	0	0
Dallas	60	70(06-08)	2006	16	7.1	3.4	1.7	0.8	0.4
Dallas	60	70(06-08)	2007	6.4	1	0.1	0	0	0
Dallas	60	70(06-08)	2008	3	0.4	0.1	0	0	0
Dallas	60	70(08-10)	2008	3	0.4	0.1	0	0	0
Dallas	60	70(08-10)	2009	8.4	1.9	0.5	0.1	0	0
Dallas	60	70(08-10)	2010	3.7	0.5	0.1	0	0	0
Dallas	70	70(06-08)	2006	1.5	0.1	0	0	0	0
Dallas	70	70(06-08)	2007	1	0	0	0	0	0
Dallas	70	70(06-08)	2008	0.1	0	0	0	0	0

Study Area	Exposure Benchmark (ppb-8hr)	Air Quality Scenario[1]	Year	% of all school-age children experiencing multiple exposures per O_3 season at or above benchmarks, adjusted air quality					
				≥1	≥2	≥3	≥4	≥5	≥6
Dallas	70	70(08-10)	2008	0.1	0	0	0	0	0
Dallas	70	70(08-10)	2009	0.6	0	0	0	0	0
Dallas	70	70(08-10)	2010	0.1	0	0	0	0	0
Dallas	80	70(06-08)	2006	0	0	0	0	0	0
Dallas	80	70(06-08)	2007	0.1	0	0	0	0	0
Dallas	80	70(06-08)	2008	0	0	0	0	0	0
Dallas	80	70(08-10)	2008	0	0	0	0	0	0
Dallas	80	70(08-10)	2009	0	0	0	0	0	0
Dallas	80	70(08-10)	2010	0	0	0	0	0	0
Dallas	60	75(06-08)	2006	22.9	12.2	7.3	4.4	2.7	1.6
Dallas	60	75(06-08)	2007	10.9	2.8	0.8	0.2	0	0
Dallas	60	75(06-08)	2008	6.7	1.4	0.4	0.1	0	0
Dallas	60	75(08-10)	2008	7.9	1.9	0.6	0.2	0	0
Dallas	60	75(08-10)	2009	14.9	5.5	2.2	0.8	0.3	0.1
Dallas	60	75(08-10)	2010	8.3	2	0.5	0.2	0	0
Dallas	70	75(06-08)	2006	4.5	0.8	0.2	0.1	0	0
Dallas	70	75(06-08)	2007	1.9	0.1	0	0	0	0
Dallas	70	75(06-08)	2008	0.3	0	0	0	0	0
Dallas	70	75(08-10)	2008	0.5	0	0	0	0	0
Dallas	70	75(08-10)	2009	2.2	0.2	0	0	0	0
Dallas	70	75(08-10)	2010	0.6	0	0	0	0	0
Dallas	80	75(06-08)	2006	0.2	0	0	0	0	0
Dallas	80	75(06-08)	2007	0.3	0	0	0	0	0
Dallas	80	75(06-08)	2008	0	0	0	0	0	0
Dallas	80	75(08-10)	2008	0	0	0	0	0	0
Dallas	80	75(08-10)	2009	0.1	0	0	0	0	0
Dallas	80	75(08-10)	2010	0	0	0	0	0	0
Denver	60	55(06-08)	2006	0	0	0	0	0	0
Denver	60	55(06-08)	2007	0	0	0	0	0	0
Denver	60	55(06-08)	2008	0	0	0	0	0	0
Denver	60	55(08-10)	2008	0	0	0	0	0	0
Denver	60	55(08-10)	2009	0	0	0	0	0	0
Denver	60	55(08-10)	2010	0	0	0	0	0	0
Denver	70	55(06-08)	2006	0	0	0	0	0	0
Denver	70	55(06-08)	2007	0	0	0	0	0	0
Denver	70	55(06-08)	2008	0	0	0	0	0	0
Denver	70	55(08-10)	2008	0	0	0	0	0	0
Denver	70	55(08-10)	2009	0	0	0	0	0	0
Denver	70	55(08-10)	2010	0	0	0	0	0	0
Denver	80	55(06-08)	2006	0	0	0	0	0	0
Denver	80	55(06-08)	2007	0	0	0	0	0	0
Denver	80	55(06-08)	2008	0	0	0	0	0	0
Denver	80	55(08-10)	2008	0	0	0	0	0	0
Denver	80	55(08-10)	2009	0	0	0	0	0	0
Denver	80	55(08-10)	2010	0	0	0	0	0	0
Denver	60	60(06-08)	2006	0.5	0	0	0	0	0

Study Area	Exposure Benchmark (ppb-8hr)	Air Quality Scenario[1]	Year	% of all school-age children experiencing multiple exposures per O_3 season at or above benchmarks, adjusted air quality					
				≥1	≥2	≥3	≥4	≥5	≥6
Denver	60	60(06-08)	2007	0.1	0	0	0	0	0
Denver	60	60(06-08)	2008	0.4	0	0	0	0	0
Denver	60	60(08-10)	2008	0.4	0	0	0	0	0
Denver	60	60(08-10)	2009	0	0	0	0	0	0
Denver	60	60(08-10)	2010	0	0	0	0	0	0
Denver	70	60(06-08)	2006	0	0	0	0	0	0
Denver	70	60(06-08)	2007	0	0	0	0	0	0
Denver	70	60(06-08)	2008	0	0	0	0	0	0
Denver	70	60(08-10)	2008	0	0	0	0	0	0
Denver	70	60(08-10)	2009	0	0	0	0	0	0
Denver	70	60(08-10)	2010	0	0	0	0	0	0
Denver	80	60(06-08)	2006	0	0	0	0	0	0
Denver	80	60(06-08)	2007	0	0	0	0	0	0
Denver	80	60(06-08)	2008	0	0	0	0	0	0
Denver	80	60(08-10)	2008	0	0	0	0	0	0
Denver	80	60(08-10)	2009	0	0	0	0	0	0
Denver	80	60(08-10)	2010	0	0	0	0	0	0
Denver	60	65(06-08)	2006	4.4	0.8	0.2	0	0	0
Denver	60	65(06-08)	2007	2	0.3	0.1	0	0	0
Denver	60	65(06-08)	2008	2.8	0.3	0.1	0	0	0
Denver	60	65(08-10)	2008	9.5	2.8	1	0.4	0.2	0.1
Denver	60	65(08-10)	2009	3.1	0.4	0.1	0	0	0
Denver	60	65(08-10)	2010	3.3	0.7	0.2	0.1	0	0
Denver	70	65(06-08)	2006	0	0	0	0	0	0
Denver	70	65(06-08)	2007	0	0	0	0	0	0
Denver	70	65(06-08)	2008	0	0	0	0	0	0
Denver	70	65(08-10)	2008	0.4	0	0	0	0	0
Denver	70	65(08-10)	2009	0	0	0	0	0	0
Denver	70	65(08-10)	2010	0	0	0	0	0	0
Denver	80	65(06-08)	2006	0	0	0	0	0	0
Denver	80	65(06-08)	2007	0	0	0	0	0	0
Denver	80	65(06-08)	2008	0	0	0	0	0	0
Denver	80	65(08-10)	2008	0	0	0	0	0	0
Denver	80	65(08-10)	2009	0	0	0	0	0	0
Denver	80	65(08-10)	2010	0	0	0	0	0	0
Denver	60	70(06-08)	2006	12.9	4.4	1.8	0.8	0.4	0.2
Denver	60	70(06-08)	2007	7	2	0.7	0.3	0.1	0
Denver	60	70(06-08)	2008	8.9	2.5	0.9	0.3	0.1	0.1
Denver	60	70(08-10)	2008	18.9	9.2	5	2.8	1.5	0.9
Denver	60	70(08-10)	2009	7.8	2	0.6	0.2	0.1	0
Denver	60	70(08-10)	2010	9.5	3.1	1.2	0.4	0.2	0.1
Denver	70	70(06-08)	2006	0.7	0	0	0	0	0
Denver	70	70(06-08)	2007	0.2	0	0	0	0	0
Denver	70	70(06-08)	2008	0.4	0	0	0	0	0
Denver	70	70(08-10)	2008	1.7	0.1	0	0	0	0
Denver	70	70(08-10)	2009	0.4	0	0	0	0	0

Study Area	Exposure Benchmark (ppb-8hr)	Air Quality Scenario[1]	Year	% of all school-age children experiencing multiple exposures per O_3 season at or above benchmarks, adjusted air quality					
				≥1	≥2	≥3	≥4	≥5	≥6
Denver	70	70(08-10)	2010	0.2	0	0	0	0	0
Denver	80	70(06-08)	2006	0	0	0	0	0	0
Denver	80	70(06-08)	2007	0	0	0	0	0	0
Denver	80	70(06-08)	2008	0	0	0	0	0	0
Denver	80	70(08-10)	2008	0.1	0	0	0	0	0
Denver	80	70(08-10)	2009	0	0	0	0	0	0
Denver	80	70(08-10)	2010	0	0	0	0	0	0
Denver	60	75(06-08)	2006	21.3	10.4	5.7	3.1	1.8	1
Denver	60	75(06-08)	2007	13.8	5.4	2.5	1.3	0.7	0.4
Denver	60	75(06-08)	2008	16.7	7.6	3.8	1.9	1	0.6
Denver	60	75(08-10)	2008	25.6	14.4	9.1	5.9	3.8	2.5
Denver	60	75(08-10)	2009	12.5	4.2	1.5	0.6	0.2	0.1
Denver	60	75(08-10)	2010	16.3	7.1	3.4	1.7	0.8	0.4
Denver	70	75(06-08)	2006	2.9	0.3	0	0	0	0
Denver	70	75(06-08)	2007	0.9	0.1	0	0	0	0
Denver	70	75(06-08)	2008	1.2	0	0	0	0	0
Denver	70	75(08-10)	2008	4.1	0.4	0	0	0	0
Denver	70	75(08-10)	2009	1.2	0.1	0	0	0	0
Denver	70	75(08-10)	2010	1	0.1	0	0	0	0
Denver	80	75(06-08)	2006	0.1	0	0	0	0	0
Denver	80	75(06-08)	2007	0	0	0	0	0	0
Denver	80	75(06-08)	2008	0	0	0	0	0	0
Denver	80	75(08-10)	2008	0.5	0	0	0	0	0
Denver	80	75(08-10)	2009	0	0	0	0	0	0
Denver	80	75(08-10)	2010	0	0	0	0	0	0
Detroit	60	55(06-08)	2006	0	0	0	0	0	0
Detroit	60	55(06-08)	2007	0	0	0	0	0	0
Detroit	60	55(06-08)	2008	0	0	0	0	0	0
Detroit	60	55(08-10)	2008	0	0	0	0	0	0
Detroit	60	55(08-10)	2009	0	0	0	0	0	0
Detroit	60	55(08-10)	2010	0	0	0	0	0	0
Detroit	70	55(06-08)	2006	0	0	0	0	0	0
Detroit	70	55(06-08)	2007	0	0	0	0	0	0
Detroit	70	55(06-08)	2008	0	0	0	0	0	0
Detroit	70	55(08-10)	2008	0	0	0	0	0	0
Detroit	70	55(08-10)	2009	0	0	0	0	0	0
Detroit	70	55(08-10)	2010	0	0	0	0	0	0
Detroit	80	55(06-08)	2006	0	0	0	0	0	0
Detroit	80	55(06-08)	2007	0	0	0	0	0	0
Detroit	80	55(06-08)	2008	0	0	0	0	0	0
Detroit	80	55(08-10)	2008	0	0	0	0	0	0
Detroit	80	55(08-10)	2009	0	0	0	0	0	0
Detroit	80	55(08-10)	2010	0	0	0	0	0	0
Detroit	60	60(06-08)	2006	0.2	0	0	0	0	0
Detroit	60	60(06-08)	2007	0.7	0	0	0	0	0
Detroit	60	60(06-08)	2008	0	0	0	0	0	0

Study Area	Exposure Benchmark (ppb-8hr)	Air Quality Scenario[1]	Year	% of all school-age children experiencing multiple exposures per O_3 season at or above benchmarks, adjusted air quality					
				≥1	≥2	≥3	≥4	≥5	≥6
Detroit	60	60(08-10)	2008	0.4	0	0	0	0	0
Detroit	60	60(08-10)	2009	0.3	0	0	0	0	0
Detroit	60	60(08-10)	2010	0.2	0	0	0	0	0
Detroit	70	60(06-08)	2006	0	0	0	0	0	0
Detroit	70	60(06-08)	2007	0	0	0	0	0	0
Detroit	70	60(06-08)	2008	0	0	0	0	0	0
Detroit	70	60(08-10)	2008	0	0	0	0	0	0
Detroit	70	60(08-10)	2009	0	0	0	0	0	0
Detroit	70	60(08-10)	2010	0	0	0	0	0	0
Detroit	80	60(06-08)	2006	0	0	0	0	0	0
Detroit	80	60(06-08)	2007	0	0	0	0	0	0
Detroit	80	60(06-08)	2008	0	0	0	0	0	0
Detroit	80	60(08-10)	2008	0	0	0	0	0	0
Detroit	80	60(08-10)	2009	0	0	0	0	0	0
Detroit	80	60(08-10)	2010	0	0	0	0	0	0
Detroit	60	65(06-08)	2006	2	0.1	0	0	0	0
Detroit	60	65(06-08)	2007	4.6	1.1	0.3	0.1	0	0
Detroit	60	65(06-08)	2008	0.7	0	0	0	0	0
Detroit	60	65(08-10)	2008	3.6	0.5	0.1	0	0	0
Detroit	60	65(08-10)	2009	2.8	0.3	0	0	0	0
Detroit	60	65(08-10)	2010	3.1	0.4	0.1	0	0	0
Detroit	70	65(06-08)	2006	0	0	0	0	0	0
Detroit	70	65(06-08)	2007	0.2	0	0	0	0	0
Detroit	70	65(06-08)	2008	0	0	0	0	0	0
Detroit	70	65(08-10)	2008	0	0	0	0	0	0
Detroit	70	65(08-10)	2009	0.1	0	0	0	0	0
Detroit	70	65(08-10)	2010	0	0	0	0	0	0
Detroit	80	65(06-08)	2006	0	0	0	0	0	0
Detroit	80	65(06-08)	2007	0	0	0	0	0	0
Detroit	80	65(06-08)	2008	0	0	0	0	0	0
Detroit	80	65(08-10)	2008	0	0	0	0	0	0
Detroit	80	65(08-10)	2009	0	0	0	0	0	0
Detroit	80	65(08-10)	2010	0	0	0	0	0	0
Detroit	60	70(06-08)	2006	4.9	0.8	0.1	0	0	0
Detroit	60	70(06-08)	2007	10.3	3.6	1.4	0.5	0.2	0.1
Detroit	60	70(06-08)	2008	2.5	0.2	0	0	0	0
Detroit	60	70(08-10)	2008	9	2.3	0.6	0.1	0	0
Detroit	60	70(08-10)	2009	6.9	1.4	0.3	0.1	0	0
Detroit	60	70(08-10)	2010	8.7	2.5	0.8	0.2	0.1	0
Detroit	70	70(06-08)	2006	0.2	0	0	0	0	0
Detroit	70	70(06-08)	2007	0.9	0.1	0	0	0	0
Detroit	70	70(06-08)	2008	0	0	0	0	0	0
Detroit	70	70(08-10)	2008	0.9	0	0	0	0	0
Detroit	70	70(08-10)	2009	0.5	0	0	0	0	0
Detroit	70	70(08-10)	2010	0.5	0	0	0	0	0
Detroit	80	70(06-08)	2006	0	0	0	0	0	0

Study Area	Exposure Benchmark (ppb-8hr)	Air Quality Scenario[1]	Year	% of all school-age children experiencing multiple exposures per O_3 season at or above benchmarks, adjusted air quality					
				≥1	≥2	≥3	≥4	≥5	≥6
Detroit	80	70(06-08)	2007	0	0	0	0	0	0
Detroit	80	70(06-08)	2008	0	0	0	0	0	0
Detroit	80	70(08-10)	2008	0	0	0	0	0	0
Detroit	80	70(08-10)	2009	0	0	0	0	0	0
Detroit	80	70(08-10)	2010	0	0	0	0	0	0
Detroit	60	75(06-08)	2006	10.6	3	0.9	0.3	0.1	0
Detroit	60	75(06-08)	2007	19.1	8.6	4.3	2.3	1.3	0.7
Detroit	60	75(06-08)	2008	6.7	1.4	0.3	0.1	0	0
Detroit	60	75(08-10)	2008	15.6	5.3	1.8	0.6	0.2	0
Detroit	60	75(08-10)	2009	12.8	3.6	1	0.3	0.1	0
Detroit	60	75(08-10)	2010	16.8	6.3	2.6	1	0.4	0.2
Detroit	70	75(06-08)	2006	1.4	0	0	0	0	0
Detroit	70	75(06-08)	2007	4.2	0.8	0.2	0	0	0
Detroit	70	75(06-08)	2008	0.3	0	0	0	0	0
Detroit	70	75(08-10)	2008	3	0.2	0	0	0	0
Detroit	70	75(08-10)	2009	2.2	0.2	0	0	0	0
Detroit	70	75(08-10)	2010	2.6	0.3	0	0	0	0
Detroit	80	75(06-08)	2006	0	0	0	0	0	0
Detroit	80	75(06-08)	2007	0.2	0	0	0	0	0
Detroit	80	75(06-08)	2008	0	0	0	0	0	0
Detroit	80	75(08-10)	2008	0	0	0	0	0	0
Detroit	80	75(08-10)	2009	0.1	0	0	0	0	0
Detroit	80	75(08-10)	2010	0.1	0	0	0	0	0
Houston	60	55(06-08)	2006	0	0	0	0	0	0
Houston	60	55(06-08)	2007	0	0	0	0	0	0
Houston	60	55(06-08)	2008	0	0	0	0	0	0
Houston	60	55(08-10)	2008	0	0	0	0	0	0
Houston	60	55(08-10)	2009	0	0	0	0	0	0
Houston	60	55(08-10)	2010	0	0	0	0	0	0
Houston	70	55(06-08)	2006	0	0	0	0	0	0
Houston	70	55(06-08)	2007	0	0	0	0	0	0
Houston	70	55(06-08)	2008	0	0	0	0	0	0
Houston	70	55(08-10)	2008	0	0	0	0	0	0
Houston	70	55(08-10)	2009	0	0	0	0	0	0
Houston	70	55(08-10)	2010	0	0	0	0	0	0
Houston	80	55(06-08)	2006	0	0	0	0	0	0
Houston	80	55(06-08)	2007	0	0	0	0	0	0
Houston	80	55(06-08)	2008	0	0	0	0	0	0
Houston	80	55(08-10)	2008	0	0	0	0	0	0
Houston	80	55(08-10)	2009	0	0	0	0	0	0
Houston	80	55(08-10)	2010	0	0	0	0	0	0
Houston	60	60(06-08)	2006	0.3	0	0	0	0	0
Houston	60	60(06-08)	2007	0	0	0	0	0	0
Houston	60	60(06-08)	2008	0.1	0	0	0	0	0
Houston	60	60(08-10)	2008	0.2	0	0	0	0	0
Houston	60	60(08-10)	2009	0.7	0	0	0	0	0

Study Area	Exposure Benchmark (ppb-8hr)	Air Quality Scenario[1]	Year	% of all school-age children experiencing multiple exposures per O_3 season at or above benchmarks, adjusted air quality					
				≥1	≥2	≥3	≥4	≥5	≥6
Houston	60	60(08-10)	2010	0.2	0	0	0	0	0
Houston	70	60(06-08)	2006	0	0	0	0	0	0
Houston	70	60(06-08)	2007	0	0	0	0	0	0
Houston	70	60(06-08)	2008	0	0	0	0	0	0
Houston	70	60(08-10)	2008	0	0	0	0	0	0
Houston	70	60(08-10)	2009	0	0	0	0	0	0
Houston	70	60(08-10)	2010	0	0	0	0	0	0
Houston	80	60(06-08)	2006	0	0	0	0	0	0
Houston	80	60(06-08)	2007	0	0	0	0	0	0
Houston	80	60(06-08)	2008	0	0	0	0	0	0
Houston	80	60(08-10)	2008	0	0	0	0	0	0
Houston	80	60(08-10)	2009	0	0	0	0	0	0
Houston	80	60(08-10)	2010	0	0	0	0	0	0
Houston	60	65(06-08)	2006	3	0.6	0.1	0	0	0
Houston	60	65(06-08)	2007	0.2	0	0	0	0	0
Houston	60	65(06-08)	2008	0.3	0	0	0	0	0
Houston	60	65(08-10)	2008	2.1	0.2	0	0	0	0
Houston	60	65(08-10)	2009	5.7	0.7	0.1	0	0	0
Houston	60	65(08-10)	2010	3.3	0.3	0	0	0	0
Houston	70	65(06-08)	2006	0.2	0	0	0	0	0
Houston	70	65(06-08)	2007	0	0	0	0	0	0
Houston	70	65(06-08)	2008	0	0	0	0	0	0
Houston	70	65(08-10)	2008	0	0	0	0	0	0
Houston	70	65(08-10)	2009	0.4	0	0	0	0	0
Houston	70	65(08-10)	2010	0	0	0	0	0	0
Houston	80	65(06-08)	2006	0	0	0	0	0	0
Houston	80	65(06-08)	2007	0	0	0	0	0	0
Houston	80	65(06-08)	2008	0	0	0	0	0	0
Houston	80	65(08-10)	2008	0	0	0	0	0	0
Houston	80	65(08-10)	2009	0	0	0	0	0	0
Houston	80	65(08-10)	2010	0	0	0	0	0	0
Houston	60	70(06-08)	2006	7	1.9	0.6	0.2	0.1	0
Houston	60	70(06-08)	2007	1.4	0.1	0	0	0	0
Houston	60	70(06-08)	2008	0.8	0	0	0	0	0
Houston	60	70(08-10)	2008	6.9	1.4	0.3	0	0	0
Houston	60	70(08-10)	2009	11.9	2.9	0.8	0.2	0	0
Houston	60	70(08-10)	2010	9	2.2	0.6	0.2	0	0
Houston	70	70(06-08)	2006	1	0.1	0	0	0	0
Houston	70	70(06-08)	2007	0	0	0	0	0	0
Houston	70	70(06-08)	2008	0	0	0	0	0	0
Houston	70	70(08-10)	2008	0.5	0	0	0	0	0
Houston	70	70(08-10)	2009	2.1	0.1	0	0	0	0
Houston	70	70(08-10)	2010	0.6	0	0	0	0	0
Houston	80	70(06-08)	2006	0.1	0	0	0	0	0
Houston	80	70(06-08)	2007	0	0	0	0	0	0
Houston	80	70(06-08)	2008	0	0	0	0	0	0

Study Area	Exposure Benchmark (ppb-8hr)	Air Quality Scenario[1]	Year	% of all school-age children experiencing multiple exposures per O_3 season at or above benchmarks, adjusted air quality					
				≥1	≥2	≥3	≥4	≥5	≥6
Houston	80	70(08-10)	2008	0	0	0	0	0	0
Houston	80	70(08-10)	2009	0.1	0	0	0	0	0
Houston	80	70(08-10)	2010	0	0	0	0	0	0
Houston	60	75(06-08)	2006	12	4.1	1.7	0.7	0.3	0.1
Houston	60	75(06-08)	2007	4.1	0.7	0.2	0	0	0
Houston	60	75(06-08)	2008	2.5	0.2	0	0	0	0
Houston	60	75(08-10)	2008	13.2	4.4	1.5	0.5	0.2	0.1
Houston	60	75(08-10)	2009	17.8	6.3	2.4	0.9	0.3	0.1
Houston	60	75(08-10)	2010	15.3	5.6	2.2	0.9	0.3	0.1
Houston	70	75(06-08)	2006	2.5	0.4	0.1	0	0	0
Houston	70	75(06-08)	2007	0.1	0	0	0	0	0
Houston	70	75(06-08)	2008	0.1	0	0	0	0	0
Houston	70	75(08-10)	2008	1.9	0.1	0	0	0	0
Houston	70	75(08-10)	2009	5.5	0.6	0.1	0	0	0
Houston	70	75(08-10)	2010	2.4	0.2	0	0	0	0
Houston	80	75(06-08)	2006	0.3	0	0	0	0	0
Houston	80	75(06-08)	2007	0	0	0	0	0	0
Houston	80	75(06-08)	2008	0	0	0	0	0	0
Houston	80	75(08-10)	2008	0.2	0	0	0	0	0
Houston	80	75(08-10)	2009	0.7	0	0	0	0	0
Houston	80	75(08-10)	2010	0.1	0	0	0	0	0
Los Angeles	60	55(06-08)	2006	0	0	0	0	0	0
Los Angeles	60	55(06-08)	2007	0	0	0	0	0	0
Los Angeles	60	55(06-08)	2008	0	0	0	0	0	0
Los Angeles	60	55(08-10)	2008	0	0	0	0	0	0
Los Angeles	60	55(08-10)	2009	0	0	0	0	0	0
Los Angeles	60	55(08-10)	2010	0	0	0	0	0	0
Los Angeles	70	55(06-08)	2006	0	0	0	0	0	0
Los Angeles	70	55(06-08)	2007	0	0	0	0	0	0
Los Angeles	70	55(06-08)	2008	0	0	0	0	0	0
Los Angeles	70	55(08-10)	2008	0	0	0	0	0	0
Los Angeles	70	55(08-10)	2009	0	0	0	0	0	0
Los Angeles	70	55(08-10)	2010	0	0	0	0	0	0
Los Angeles	80	55(06-08)	2006	0	0	0	0	0	0
Los Angeles	80	55(06-08)	2007	0	0	0	0	0	0
Los Angeles	80	55(06-08)	2008	0	0	0	0	0	0
Los Angeles	80	55(08-10)	2008	0	0	0	0	0	0
Los Angeles	80	55(08-10)	2009	0	0	0	0	0	0
Los Angeles	80	55(08-10)	2010	0	0	0	0	0	0
Los Angeles	60	60(06-08)	2006	0	0	0	0	0	0
Los Angeles	60	60(06-08)	2007	0.2	0	0	0	0	0
Los Angeles	60	60(06-08)	2008	0	0	0	0	0	0
Los Angeles	60	60(08-10)	2008	0	0	0	0	0	0
Los Angeles	60	60(08-10)	2009	0	0	0	0	0	0
Los Angeles	60	60(08-10)	2010	0	0	0	0	0	0
Los Angeles	70	60(06-08)	2006	0	0	0	0	0	0

Study Area	Exposure Benchmark (ppb-8hr)	Air Quality Scenario[1]	Year	% of all school-age children experiencing multiple exposures per O_3 season at or above benchmarks, adjusted air quality					
				≥1	≥2	≥3	≥4	≥5	≥6
Los Angeles	70	60(06-08)	2007	0	0	0	0	0	0
Los Angeles	70	60(06-08)	2008	0	0	0	0	0	0
Los Angeles	70	60(08-10)	2008	0	0	0	0	0	0
Los Angeles	70	60(08-10)	2009	0	0	0	0	0	0
Los Angeles	70	60(08-10)	2010	0	0	0	0	0	0
Los Angeles	80	60(06-08)	2006	0	0	0	0	0	0
Los Angeles	80	60(06-08)	2007	0	0	0	0	0	0
Los Angeles	80	60(06-08)	2008	0	0	0	0	0	0
Los Angeles	80	60(08-10)	2008	0	0	0	0	0	0
Los Angeles	80	60(08-10)	2009	0	0	0	0	0	0
Los Angeles	80	60(08-10)	2010	0	0	0	0	0	0
Los Angeles	60	65(06-08)	2006	0.9	0.2	0.1	0	0	0
Los Angeles	60	65(06-08)	2007	1.5	0.3	0.1	0	0	0
Los Angeles	60	65(06-08)	2008	0.9	0.3	0.1	0	0	0
Los Angeles	60	65(08-10)	2008	1.2	0.3	0.1	0.1	0	0
Los Angeles	60	65(08-10)	2009	1.2	0.3	0.1	0.1	0	0
Los Angeles	60	65(08-10)	2010	0.8	0.2	0.1	0	0	0
Los Angeles	70	65(06-08)	2006	0	0	0	0	0	0
Los Angeles	70	65(06-08)	2007	0	0	0	0	0	0
Los Angeles	70	65(06-08)	2008	0	0	0	0	0	0
Los Angeles	70	65(08-10)	2008	0	0	0	0	0	0
Los Angeles	70	65(08-10)	2009	0	0	0	0	0	0
Los Angeles	70	65(08-10)	2010	0	0	0	0	0	0
Los Angeles	80	65(06-08)	2006	0	0	0	0	0	0
Los Angeles	80	65(06-08)	2007	0	0	0	0	0	0
Los Angeles	80	65(06-08)	2008	0	0	0	0	0	0
Los Angeles	80	65(08-10)	2008	0	0	0	0	0	0
Los Angeles	80	65(08-10)	2009	0	0	0	0	0	0
Los Angeles	80	65(08-10)	2010	0	0	0	0	0	0
Los Angeles	60	70(06-08)	2006	4.4	1.6	0.8	0.5	0.3	0.2
Los Angeles	60	70(06-08)	2007	4.9	1.6	0.8	0.5	0.3	0.2
Los Angeles	60	70(06-08)	2008	4.2	1.5	0.8	0.4	0.3	0.2
Los Angeles	60	70(08-10)	2008	4.9	1.8	0.9	0.6	0.3	0.2
Los Angeles	60	70(08-10)	2009	5	1.8	0.9	0.5	0.3	0.2
Los Angeles	60	70(08-10)	2010	3.2	1.3	0.7	0.4	0.2	0.1
Los Angeles	70	70(06-08)	2006	0	0	0	0	0	0
Los Angeles	70	70(06-08)	2007	0.2	0	0	0	0	0
Los Angeles	70	70(06-08)	2008	0	0	0	0	0	0
Los Angeles	70	70(08-10)	2008	0	0	0	0	0	0
Los Angeles	70	70(08-10)	2009	0	0	0	0	0	0
Los Angeles	70	70(08-10)	2010	0	0	0	0	0	0
Los Angeles	80	70(06-08)	2006	0	0	0	0	0	0
Los Angeles	80	70(06-08)	2007	0	0	0	0	0	0
Los Angeles	80	70(06-08)	2008	0	0	0	0	0	0
Los Angeles	80	70(08-10)	2008	0	0	0	0	0	0
Los Angeles	80	70(08-10)	2009	0	0	0	0	0	0

Study Area	Exposure Benchmark (ppb-8hr)	Air Quality Scenario[1]	Year	% of all school-age children experiencing multiple exposures per O_3 season at or above benchmarks, adjusted air quality					
				≥1	≥2	≥3	≥4	≥5	≥6
Los Angeles	80	70(08-10)	2010	0	0	0	0	0	0
Los Angeles	60	75(06-08)	2006	10.2	4.5	2.5	1.6	1.1	0.7
Los Angeles	60	75(06-08)	2007	10.2	4.3	2.4	1.4	0.9	0.7
Los Angeles	60	75(06-08)	2008	9.9	4.3	2.4	1.5	1	0.7
Los Angeles	60	75(08-10)	2008	10.2	4.5	2.5	1.6	1.1	0.7
Los Angeles	60	75(08-10)	2009	10	4.4	2.5	1.5	1	0.7
Los Angeles	60	75(08-10)	2010	6.9	2.9	1.7	1	0.7	0.4
Los Angeles	70	75(06-08)	2006	0.5	0.1	0	0	0	0
Los Angeles	70	75(06-08)	2007	1	0.1	0	0	0	0
Los Angeles	70	75(06-08)	2008	0.5	0.1	0	0	0	0
Los Angeles	70	75(08-10)	2008	0.6	0.1	0.1	0	0	0
Los Angeles	70	75(08-10)	2009	0.5	0.1	0	0	0	0
Los Angeles	70	75(08-10)	2010	0.4	0.1	0	0	0	0
Los Angeles	80	75(06-08)	2006	0	0	0	0	0	0
Los Angeles	80	75(06-08)	2007	0.1	0	0	0	0	0
Los Angeles	80	75(06-08)	2008	0	0	0	0	0	0
Los Angeles	80	75(08-10)	2008	0	0	0	0	0	0
Los Angeles	80	75(08-10)	2009	0	0	0	0	0	0
Los Angeles	80	75(08-10)	2010	0	0	0	0	0	0
New York	60	65(06-08)	2006	0	0	0	0	0	0
New York	60	65(06-08)	2007	0	0	0	0	0	0
New York	60	65(06-08)	2008	0	0	0	0	0	0
New York	60	65(08-10)	2008	0.1	0	0	0	0	0
New York	60	65(08-10)	2009	0	0	0	0	0	0
New York	60	65(08-10)	2010	0.1	0	0	0	0	0
New York	70	65(06-08)	2006	0	0	0	0	0	0
New York	70	65(06-08)	2007	0	0	0	0	0	0
New York	70	65(06-08)	2008	0	0	0	0	0	0
New York	70	65(08-10)	2008	0	0	0	0	0	0
New York	70	65(08-10)	2009	0	0	0	0	0	0
New York	70	65(08-10)	2010	0	0	0	0	0	0
New York	80	65(06-08)	2006	0	0	0	0	0	0
New York	80	65(06-08)	2007	0	0	0	0	0	0
New York	80	65(06-08)	2008	0	0	0	0	0	0
New York	80	65(08-10)	2008	0	0	0	0	0	0
New York	80	65(08-10)	2009	0	0	0	0	0	0
New York	80	65(08-10)	2010	0	0	0	0	0	0
New York	60	70(06-08)	2006	2.6	0.2	0	0	0	0
New York	60	70(06-08)	2007	2	0.3	0.1	0	0	0
New York	60	70(06-08)	2008	2.3	0.3	0.1	0	0	0
New York	60	70(08-10)	2008	5.7	1.1	0.3	0.1	0	0
New York	60	70(08-10)	2009	1.2	0.1	0	0	0	0
New York	60	70(08-10)	2010	6.6	1.4	0.3	0.1	0	0
New York	70	70(06-08)	2006	0.2	0	0	0	0	0
New York	70	70(06-08)	2007	0	0	0	0	0	0
New York	70	70(06-08)	2008	0.1	0	0	0	0	0

Study Area	Exposure Benchmark (ppb-8hr)	Air Quality Scenario[1]	Year	% of all school-age children experiencing multiple exposures per O_3 season at or above benchmarks, adjusted air quality					
				≥1	≥2	≥3	≥4	≥5	≥6
New York	70	70(08-10)	2008	0.4	0	0	0	0	0
New York	70	70(08-10)	2009	0	0	0	0	0	0
New York	70	70(08-10)	2010	0.5	0	0	0	0	0
New York	80	70(06-08)	2006	0	0	0	0	0	0
New York	80	70(06-08)	2007	0	0	0	0	0	0
New York	80	70(06-08)	2008	0	0	0	0	0	0
New York	80	70(08-10)	2008	0	0	0	0	0	0
New York	80	70(08-10)	2009	0	0	0	0	0	0
New York	80	70(08-10)	2010	0	0	0	0	0	0
New York	60	75(06-08)	2006	9.1	1.9	0.4	0.1	0	0
New York	60	75(06-08)	2007	8.5	2.2	0.7	0.2	0.1	0
New York	60	75(06-08)	2008	7.8	1.7	0.5	0.2	0.1	0
New York	60	75(08-10)	2008	17.2	6.3	2.4	1	0.4	0.2
New York	60	75(08-10)	2009	5.4	0.9	0.2	0	0	0
New York	60	75(08-10)	2010	19	8	3.6	1.8	0.9	0.4
New York	70	75(06-08)	2006	1.5	0.1	0	0	0	0
New York	70	75(06-08)	2007	0.4	0	0	0	0	0
New York	70	75(06-08)	2008	0.7	0	0	0	0	0
New York	70	75(08-10)	2008	3.4	0.4	0.1	0	0	0
New York	70	75(08-10)	2009	0.3	0	0	0	0	0
New York	70	75(08-10)	2010	3.7	0.4	0	0	0	0
New York	80	75(06-08)	2006	0.1	0	0	0	0	0
New York	80	75(06-08)	2007	0	0	0	0	0	0
New York	80	75(06-08)	2008	0	0	0	0	0	0
New York	80	75(08-10)	2008	0.3	0	0	0	0	0
New York	80	75(08-10)	2009	0	0	0	0	0	0
New York	80	75(08-10)	2010	0.3	0	0	0	0	0
Philadelphia	60	55(06-08)	2006	0	0	0	0	0	0
Philadelphia	60	55(06-08)	2007	0	0	0	0	0	0
Philadelphia	60	55(06-08)	2008	0	0	0	0	0	0
Philadelphia	60	55(08-10)	2008	0	0	0	0	0	0
Philadelphia	60	55(08-10)	2009	0	0	0	0	0	0
Philadelphia	60	55(08-10)	2010	0.2	0	0	0	0	0
Philadelphia	70	55(06-08)	2006	0	0	0	0	0	0
Philadelphia	70	55(06-08)	2007	0	0	0	0	0	0
Philadelphia	70	55(06-08)	2008	0	0	0	0	0	0
Philadelphia	70	55(08-10)	2008	0	0	0	0	0	0
Philadelphia	70	55(08-10)	2009	0	0	0	0	0	0
Philadelphia	70	55(08-10)	2010	0	0	0	0	0	0
Philadelphia	80	55(06-08)	2006	0	0	0	0	0	0
Philadelphia	80	55(06-08)	2007	0	0	0	0	0	0
Philadelphia	80	55(06-08)	2008	0	0	0	0	0	0
Philadelphia	80	55(08-10)	2008	0	0	0	0	0	0
Philadelphia	80	55(08-10)	2009	0	0	0	0	0	0
Philadelphia	80	55(08-10)	2010	0	0	0	0	0	0
Philadelphia	60	60(06-08)	2006	0.1	0	0	0	0	0

Study Area	Exposure Benchmark (ppb-8hr)	Air Quality Scenario[1]	Year	% of all school-age children experiencing multiple exposures per O_3 season at or above benchmarks, adjusted air quality					
				≥1	≥2	≥3	≥4	≥5	≥6
Philadelphia	60	60(06-08)	2007	0.7	0	0	0	0	0
Philadelphia	60	60(06-08)	2008	0.1	0	0	0	0	0
Philadelphia	60	60(08-10)	2008	1	0	0	0	0	0
Philadelphia	60	60(08-10)	2009	0	0	0	0	0	0
Philadelphia	60	60(08-10)	2010	1.7	0.1	0	0	0	0
Philadelphia	70	60(06-08)	2006	0	0	0	0	0	0
Philadelphia	70	60(06-08)	2007	0	0	0	0	0	0
Philadelphia	70	60(06-08)	2008	0	0	0	0	0	0
Philadelphia	70	60(08-10)	2008	0	0	0	0	0	0
Philadelphia	70	60(08-10)	2009	0	0	0	0	0	0
Philadelphia	70	60(08-10)	2010	0.1	0	0	0	0	0
Philadelphia	80	60(06-08)	2006	0	0	0	0	0	0
Philadelphia	80	60(06-08)	2007	0	0	0	0	0	0
Philadelphia	80	60(06-08)	2008	0	0	0	0	0	0
Philadelphia	80	60(08-10)	2008	0	0	0	0	0	0
Philadelphia	80	60(08-10)	2009	0	0	0	0	0	0
Philadelphia	80	60(08-10)	2010	0	0	0	0	0	0
Philadelphia	60	65(06-08)	2006	1.3	0.1	0	0	0	0
Philadelphia	60	65(06-08)	2007	3.3	0.4	0.1	0	0	0
Philadelphia	60	65(06-08)	2008	1	0	0	0	0	0
Philadelphia	60	65(08-10)	2008	4.3	0.6	0.1	0	0	0
Philadelphia	60	65(08-10)	2009	0	0	0	0	0	0
Philadelphia	60	65(08-10)	2010	4.6	0.6	0.1	0	0	0
Philadelphia	70	65(06-08)	2006	0	0	0	0	0	0
Philadelphia	70	65(06-08)	2007	0.1	0	0	0	0	0
Philadelphia	70	65(06-08)	2008	0	0	0	0	0	0
Philadelphia	70	65(08-10)	2008	0.1	0	0	0	0	0
Philadelphia	70	65(08-10)	2009	0	0	0	0	0	0
Philadelphia	70	65(08-10)	2010	0.3	0	0	0	0	0
Philadelphia	80	65(06-08)	2006	0	0	0	0	0	0
Philadelphia	80	65(06-08)	2007	0	0	0	0	0	0
Philadelphia	80	65(06-08)	2008	0	0	0	0	0	0
Philadelphia	80	65(08-10)	2008	0	0	0	0	0	0
Philadelphia	80	65(08-10)	2009	0	0	0	0	0	0
Philadelphia	80	65(08-10)	2010	0	0	0	0	0	0
Philadelphia	60	70(06-08)	2006	5.3	0.9	0.2	0	0	0
Philadelphia	60	70(06-08)	2007	9.9	2.4	0.7	0.2	0.1	0
Philadelphia	60	70(06-08)	2008	4.3	0.6	0.1	0	0	0
Philadelphia	60	70(08-10)	2008	11.8	3.2	1.1	0.3	0.1	0
Philadelphia	60	70(08-10)	2009	0.8	0	0	0	0	0
Philadelphia	60	70(08-10)	2010	11.6	3.3	1.1	0.4	0.1	0.1
Philadelphia	70	70(06-08)	2006	0.1	0	0	0	0	0
Philadelphia	70	70(06-08)	2007	0.9	0.1	0	0	0	0
Philadelphia	70	70(06-08)	2008	0.1	0	0	0	0	0
Philadelphia	70	70(08-10)	2008	1.1	0.1	0	0	0	0
Philadelphia	70	70(08-10)	2009	0	0	0	0	0	0

Study Area	Exposure Benchmark (ppb-8hr)	Air Quality Scenario[1]	Year	% of all school-age children experiencing multiple exposures per O_3 season at or above benchmarks, adjusted air quality					
				≥1	≥2	≥3	≥4	≥5	≥6
Philadelphia	70	70(08-10)	2010	1.5	0.1	0	0	0	0
Philadelphia	80	70(06-08)	2006	0	0	0	0	0	0
Philadelphia	80	70(06-08)	2007	0	0	0	0	0	0
Philadelphia	80	70(06-08)	2008	0	0	0	0	0	0
Philadelphia	80	70(08-10)	2008	0	0	0	0	0	0
Philadelphia	80	70(08-10)	2009	0	0	0	0	0	0
Philadelphia	80	70(08-10)	2010	0.1	0	0	0	0	0
Philadelphia	60	75(06-08)	2006	12.1	3.7	1.3	0.4	0.2	0.1
Philadelphia	60	75(06-08)	2007	17.5	6.6	2.7	1.2	0.6	0.2
Philadelphia	60	75(06-08)	2008	10	2.4	0.7	0.2	0.1	0
Philadelphia	60	75(08-10)	2008	20.5	8.6	3.9	1.9	0.8	0.4
Philadelphia	60	75(08-10)	2009	4	0.4	0	0	0	0
Philadelphia	60	75(08-10)	2010	20.2	8.7	4.3	2	1	0.5
Philadelphia	70	75(06-08)	2006	0.8	0	0	0	0	0
Philadelphia	70	75(06-08)	2007	3	0.3	0.1	0	0	0
Philadelphia	70	75(06-08)	2008	0.8	0	0	0	0	0
Philadelphia	70	75(08-10)	2008	4.1	0.5	0.1	0	0	0
Philadelphia	70	75(08-10)	2009	0	0	0	0	0	0
Philadelphia	70	75(08-10)	2010	4.2	0.5	0.1	0	0	0
Philadelphia	80	75(06-08)	2006	0	0	0	0	0	0
Philadelphia	80	75(06-08)	2007	0.3	0	0	0	0	0
Philadelphia	80	75(06-08)	2008	0	0	0	0	0	0
Philadelphia	80	75(08-10)	2008	0.3	0	0	0	0	0
Philadelphia	80	75(08-10)	2009	0	0	0	0	0	0
Philadelphia	80	75(08-10)	2010	0.4	0	0	0	0	0
Sacramento	60	55(06-08)	2006	0	0	0	0	0	0
Sacramento	60	55(06-08)	2007	0	0	0	0	0	0
Sacramento	60	55(06-08)	2008	0	0	0	0	0	0
Sacramento	60	55(08-10)	2008	0	0	0	0	0	0
Sacramento	60	55(08-10)	2009	0	0	0	0	0	0
Sacramento	60	55(08-10)	2010	0	0	0	0	0	0
Sacramento	70	55(06-08)	2006	0	0	0	0	0	0
Sacramento	70	55(06-08)	2007	0	0	0	0	0	0
Sacramento	70	55(06-08)	2008	0	0	0	0	0	0
Sacramento	70	55(08-10)	2008	0	0	0	0	0	0
Sacramento	70	55(08-10)	2009	0	0	0	0	0	0
Sacramento	70	55(08-10)	2010	0	0	0	0	0	0
Sacramento	80	55(06-08)	2006	0	0	0	0	0	0
Sacramento	80	55(06-08)	2007	0	0	0	0	0	0
Sacramento	80	55(06-08)	2008	0	0	0	0	0	0
Sacramento	80	55(08-10)	2008	0	0	0	0	0	0
Sacramento	80	55(08-10)	2009	0	0	0	0	0	0
Sacramento	80	55(08-10)	2010	0	0	0	0	0	0
Sacramento	60	60(06-08)	2006	0.3	0	0	0	0	0
Sacramento	60	60(06-08)	2007	0.4	0	0	0	0	0
Sacramento	60	60(06-08)	2008	0.8	0.1	0	0	0	0

Study Area	Exposure Benchmark (ppb-8hr)	Air Quality Scenario[1]	Year	% of all school-age children experiencing multiple exposures per O_3 season at or above benchmarks, adjusted air quality					
				≥1	≥2	≥3	≥4	≥5	≥6
Sacramento	60	60(08-10)	2008	0.8	0.1	0	0	0	0
Sacramento	60	60(08-10)	2009	0.1	0	0	0	0	0
Sacramento	60	60(08-10)	2010	0.1	0	0	0	0	0
Sacramento	70	60(06-08)	2006	0	0	0	0	0	0
Sacramento	70	60(06-08)	2007	0	0	0	0	0	0
Sacramento	70	60(06-08)	2008	0	0	0	0	0	0
Sacramento	70	60(08-10)	2008	0	0	0	0	0	0
Sacramento	70	60(08-10)	2009	0	0	0	0	0	0
Sacramento	70	60(08-10)	2010	0	0	0	0	0	0
Sacramento	80	60(06-08)	2006	0	0	0	0	0	0
Sacramento	80	60(06-08)	2007	0	0	0	0	0	0
Sacramento	80	60(06-08)	2008	0	0	0	0	0	0
Sacramento	80	60(08-10)	2008	0	0	0	0	0	0
Sacramento	80	60(08-10)	2009	0	0	0	0	0	0
Sacramento	80	60(08-10)	2010	0	0	0	0	0	0
Sacramento	60	65(06-08)	2006	4.7	0.9	0.2	0	0	0
Sacramento	60	65(06-08)	2007	1.6	0	0	0	0	0
Sacramento	60	65(06-08)	2008	4.3	0.9	0.2	0.1	0	0
Sacramento	60	65(08-10)	2008	3.8	0.7	0.2	0	0	0
Sacramento	60	65(08-10)	2009	2.2	0.2	0	0	0	0
Sacramento	60	65(08-10)	2010	0.9	0.1	0	0	0	0
Sacramento	70	65(06-08)	2006	0	0	0	0	0	0
Sacramento	70	65(06-08)	2007	0.2	0	0	0	0	0
Sacramento	70	65(06-08)	2008	0.2	0	0	0	0	0
Sacramento	70	65(08-10)	2008	0.2	0	0	0	0	0
Sacramento	70	65(08-10)	2009	0	0	0	0	0	0
Sacramento	70	65(08-10)	2010	0	0	0	0	0	0
Sacramento	80	65(06-08)	2006	0	0	0	0	0	0
Sacramento	80	65(06-08)	2007	0	0	0	0	0	0
Sacramento	80	65(06-08)	2008	0	0	0	0	0	0
Sacramento	80	65(08-10)	2008	0	0	0	0	0	0
Sacramento	80	65(08-10)	2009	0	0	0	0	0	0
Sacramento	80	65(08-10)	2010	0	0	0	0	0	0
Sacramento	60	70(06-08)	2006	10	3.4	1.3	0.5	0.2	0.1
Sacramento	60	70(06-08)	2007	3	0.3	0	0	0	0
Sacramento	60	70(06-08)	2008	8.3	2.4	0.9	0.4	0.1	0.1
Sacramento	60	70(08-10)	2008	7.6	2.1	0.8	0.3	0.1	0
Sacramento	60	70(08-10)	2009	5.9	1.3	0.3	0.1	0	0
Sacramento	60	70(08-10)	2010	2.2	0.3	0	0	0	0
Sacramento	70	70(06-08)	2006	0.4	0	0	0	0	0
Sacramento	70	70(06-08)	2007	0.5	0	0	0	0	0
Sacramento	70	70(06-08)	2008	0.9	0.1	0	0	0	0
Sacramento	70	70(08-10)	2008	0.8	0.1	0	0	0	0
Sacramento	70	70(08-10)	2009	0.1	0	0	0	0	0
Sacramento	70	70(08-10)	2010	0.1	0	0	0	0	0
Sacramento	80	70(06-08)	2006	0	0	0	0	0	0

Study Area	Exposure Benchmark (ppb-8hr)	Air Quality Scenario[1]	Year	% of all school-age children experiencing multiple exposures per O_3 season at or above benchmarks, adjusted air quality					
				≥1	≥2	≥3	≥4	≥5	≥6
Sacramento	80	70(06-08)	2007	0	0	0	0	0	0
Sacramento	80	70(06-08)	2008	0	0	0	0	0	0
Sacramento	80	70(08-10)	2008	0	0	0	0	0	0
Sacramento	80	70(08-10)	2009	0	0	0	0	0	0
Sacramento	80	70(08-10)	2010	0	0	0	0	0	0
Sacramento	60	75(06-08)	2006	16.5	7.4	3.7	2	1.1	0.6
Sacramento	60	75(06-08)	2007	6	1.2	0.3	0.1	0	0
Sacramento	60	75(06-08)	2008	13.5	5.4	2.5	1.2	0.7	0.4
Sacramento	60	75(08-10)	2008	12.8	4.9	2.2	1.1	0.6	0.3
Sacramento	60	75(08-10)	2009	11.2	3.9	1.5	0.6	0.3	0.1
Sacramento	60	75(08-10)	2010	4.5	0.9	0.2	0.1	0	0
Sacramento	70	75(06-08)	2006	2.6	0.4	0	0	0	0
Sacramento	70	75(06-08)	2007	1.1	0	0	0	0	0
Sacramento	70	75(06-08)	2008	2.7	0.5	0.1	0	0	0
Sacramento	70	75(08-10)	2008	2.3	0.4	0.1	0	0	0
Sacramento	70	75(08-10)	2009	1.1	0.1	0	0	0	0
Sacramento	70	75(08-10)	2010	0.5	0	0	0	0	0
Sacramento	80	75(06-08)	2006	0	0	0	0	0	0
Sacramento	80	75(06-08)	2007	0.2	0	0	0	0	0
Sacramento	80	75(06-08)	2008	0.2	0	0	0	0	0
Sacramento	80	75(08-10)	2008	0.2	0	0	0	0	0
Sacramento	80	75(08-10)	2009	0	0	0	0	0	0
Sacramento	80	75(08-10)	2010	0	0	0	0	0	0
St. Louis	60	55(06-08)	2006	0	0	0	0	0	0
St. Louis	60	55(06-08)	2007	0.1	0	0	0	0	0
St. Louis	60	55(06-08)	2008	0	0	0	0	0	0
St. Louis	60	55(08-10)	2008	0	0	0	0	0	0
St. Louis	60	55(08-10)	2009	0	0	0	0	0	0
St. Louis	60	55(08-10)	2010	0	0	0	0	0	0
St. Louis	70	55(06-08)	2006	0	0	0	0	0	0
St. Louis	70	55(06-08)	2007	0	0	0	0	0	0
St. Louis	70	55(06-08)	2008	0	0	0	0	0	0
St. Louis	70	55(08-10)	2008	0	0	0	0	0	0
St. Louis	70	55(08-10)	2009	0	0	0	0	0	0
St. Louis	70	55(08-10)	2010	0	0	0	0	0	0
St. Louis	80	55(06-08)	2006	0	0	0	0	0	0
St. Louis	80	55(06-08)	2007	0	0	0	0	0	0
St. Louis	80	55(06-08)	2008	0	0	0	0	0	0
St. Louis	80	55(08-10)	2008	0	0	0	0	0	0
St. Louis	80	55(08-10)	2009	0	0	0	0	0	0
St. Louis	80	55(08-10)	2010	0	0	0	0	0	0
St. Louis	60	60(06-08)	2006	0.1	0	0	0	0	0
St. Louis	60	60(06-08)	2007	1.5	0.1	0	0	0	0
St. Louis	60	60(06-08)	2008	0	0	0	0	0	0
St. Louis	60	60(08-10)	2008	0.4	0	0	0	0	0
St. Louis	60	60(08-10)	2009	0.4	0	0	0	0	0

Study Area	Exposure Benchmark (ppb-8hr)	Air Quality Scenario[1]	Year	% of all school-age children experiencing multiple exposures per O_3 season at or above benchmarks, adjusted air quality					
				≥1	≥2	≥3	≥4	≥5	≥6
St. Louis	60	60(08-10)	2010	0.5	0	0	0	0	0
St. Louis	70	60(06-08)	2006	0	0	0	0	0	0
St. Louis	70	60(06-08)	2007	0	0	0	0	0	0
St. Louis	70	60(06-08)	2008	0	0	0	0	0	0
St. Louis	70	60(08-10)	2008	0	0	0	0	0	0
St. Louis	70	60(08-10)	2009	0	0	0	0	0	0
St. Louis	70	60(08-10)	2010	0	0	0	0	0	0
St. Louis	80	60(06-08)	2006	0	0	0	0	0	0
St. Louis	80	60(06-08)	2007	0	0	0	0	0	0
St. Louis	80	60(06-08)	2008	0	0	0	0	0	0
St. Louis	80	60(08-10)	2008	0	0	0	0	0	0
St. Louis	80	60(08-10)	2009	0	0	0	0	0	0
St. Louis	80	60(08-10)	2010	0	0	0	0	0	0
St. Louis	60	65(06-08)	2006	2.5	0.2	0	0	0	0
St. Louis	60	65(06-08)	2007	7.3	2	0.6	0.2	0.1	0
St. Louis	60	65(06-08)	2008	0.2	0	0	0	0	0
St. Louis	60	65(08-10)	2008	2.4	0.2	0	0	0	0
St. Louis	60	65(08-10)	2009	2.4	0.1	0	0	0	0
St. Louis	60	65(08-10)	2010	6.3	1.1	0.2	0	0	0
St. Louis	70	65(06-08)	2006	0	0	0	0	0	0
St. Louis	70	65(06-08)	2007	0.4	0	0	0	0	0
St. Louis	70	65(06-08)	2008	0	0	0	0	0	0
St. Louis	70	65(08-10)	2008	0.1	0	0	0	0	0
St. Louis	70	65(08-10)	2009	0.2	0	0	0	0	0
St. Louis	70	65(08-10)	2010	0.1	0	0	0	0	0
St. Louis	80	65(06-08)	2006	0	0	0	0	0	0
St. Louis	80	65(06-08)	2007	0	0	0	0	0	0
St. Louis	80	65(06-08)	2008	0	0	0	0	0	0
St. Louis	80	65(08-10)	2008	0	0	0	0	0	0
St. Louis	80	65(08-10)	2009	0	0	0	0	0	0
St. Louis	80	65(08-10)	2010	0	0	0	0	0	0
St. Louis	60	70(06-08)	2006	9.3	2.3	0.7	0.2	0	0
St. Louis	60	70(06-08)	2007	16.9	7	3.4	1.7	0.8	0.4
St. Louis	60	70(06-08)	2008	1.2	0.1	0	0	0	0
St. Louis	60	70(08-10)	2008	6	1.1	0.2	0.1	0	0
St. Louis	60	70(08-10)	2009	6.2	0.9	0.1	0	0	0
St. Louis	60	70(08-10)	2010	14.9	5.2	1.9	0.7	0.2	0.1
St. Louis	70	70(06-08)	2006	0.5	0	0	0	0	0
St. Louis	70	70(06-08)	2007	2.7	0.3	0.1	0	0	0
St. Louis	70	70(06-08)	2008	0	0	0	0	0	0
St. Louis	70	70(08-10)	2008	0.5	0	0	0	0	0
St. Louis	70	70(08-10)	2009	0.7	0	0	0	0	0
St. Louis	70	70(08-10)	2010	0.9	0	0	0	0	0
St. Louis	80	70(06-08)	2006	0	0	0	0	0	0
St. Louis	80	70(06-08)	2007	0.2	0	0	0	0	0
St. Louis	80	70(06-08)	2008	0	0	0	0	0	0

Study Area	Exposure Benchmark (ppb-8hr)	Air Quality Scenario[1]	Year	% of all school-age children experiencing multiple exposures per O_3 season at or above benchmarks, adjusted air quality					
				≥1	≥2	≥3	≥4	≥5	≥6
St. Louis	80	70(08-10)	2008	0	0	0	0	0	0
St. Louis	80	70(08-10)	2009	0.1	0	0	0	0	0
St. Louis	80	70(08-10)	2010	0	0	0	0	0	0
St. Louis	60	75(06-08)	2006	18.1	7.8	3.7	1.6	0.8	0.4
St. Louis	60	75(06-08)	2007	25.8	13.8	8.2	5	3	1.8
St. Louis	60	75(06-08)	2008	3.5	0.5	0.1	0	0	0
St. Louis	60	75(08-10)	2008	9.5	2.4	0.6	0.2	0.1	0
St. Louis	60	75(08-10)	2009	10	2.2	0.5	0.1	0	0
St. Louis	60	75(08-10)	2010	21.1	9.6	4.8	2.3	1.2	0.6
St. Louis	70	75(06-08)	2006	2.9	0.2	0	0	0	0
St. Louis	70	75(06-08)	2007	8.1	2.2	0.7	0.2	0	0
St. Louis	70	75(06-08)	2008	0.2	0	0	0	0	0
St. Louis	70	75(08-10)	2008	1.1	0	0	0	0	0
St. Louis	70	75(08-10)	2009	1.4	0	0	0	0	0
St. Louis	70	75(08-10)	2010	3.6	0.4	0.1	0	0	0
St. Louis	80	75(06-08)	2006	0.1	0	0	0	0	0
St. Louis	80	75(06-08)	2007	1.1	0.1	0	0	0	0
St. Louis	80	75(06-08)	2008	0	0	0	0	0	0
St. Louis	80	75(08-10)	2008	0	0	0	0	0	0
St. Louis	80	75(08-10)	2009	0.2	0	0	0	0	0
St. Louis	80	75(08-10)	2010	0.1	0	0	0	0	0
Washington DC	60	55(06-08)	2006	0	0	0	0	0	0
Washington DC	60	55(06-08)	2007	0	0	0	0	0	0
Washington DC	60	55(06-08)	2008	0	0	0	0	0	0
Washington DC	60	55(08-10)	2008	0	0	0	0	0	0
Washington DC	60	55(08-10)	2009	0	0	0	0	0	0
Washington DC	60	55(08-10)	2010	0	0	0	0	0	0
Washington DC	70	55(06-08)	2006	0	0	0	0	0	0
Washington DC	70	55(06-08)	2007	0	0	0	0	0	0
Washington DC	70	55(06-08)	2008	0	0	0	0	0	0
Washington DC	70	55(08-10)	2008	0	0	0	0	0	0
Washington DC	70	55(08-10)	2009	0	0	0	0	0	0
Washington DC	70	55(08-10)	2010	0	0	0	0	0	0
Washington DC	80	55(06-08)	2006	0	0	0	0	0	0
Washington DC	80	55(06-08)	2007	0	0	0	0	0	0
Washington DC	80	55(06-08)	2008	0	0	0	0	0	0
Washington DC	80	55(08-10)	2008	0	0	0	0	0	0
Washington DC	80	55(08-10)	2009	0	0	0	0	0	0
Washington DC	80	55(08-10)	2010	0	0	0	0	0	0
Washington DC	60	60(06-08)	2006	0.6	0	0	0	0	0
Washington DC	60	60(06-08)	2007	0.1	0	0	0	0	0
Washington DC	60	60(06-08)	2008	0.1	0	0	0	0	0
Washington DC	60	60(08-10)	2008	0.3	0	0	0	0	0
Washington DC	60	60(08-10)	2009	0	0	0	0	0	0
Washington DC	60	60(08-10)	2010	0.6	0	0	0	0	0
Washington DC	70	60(06-08)	2006	0	0	0	0	0	0

Study Area	Exposure Benchmark (ppb-8hr)	Air Quality Scenario[1]	Year	% of all school-age children experiencing multiple exposures per O_3 season at or above benchmarks, adjusted air quality					
				≥1	≥2	≥3	≥4	≥5	≥6
Washington DC	70	60(06-08)	2007	0	0	0	0	0	0
Washington DC	70	60(06-08)	2008	0	0	0	0	0	0
Washington DC	70	60(08-10)	2008	0	0	0	0	0	0
Washington DC	70	60(08-10)	2009	0	0	0	0	0	0
Washington DC	70	60(08-10)	2010	0	0	0	0	0	0
Washington DC	80	60(06-08)	2006	0	0	0	0	0	0
Washington DC	80	60(06-08)	2007	0	0	0	0	0	0
Washington DC	80	60(06-08)	2008	0	0	0	0	0	0
Washington DC	80	60(08-10)	2008	0	0	0	0	0	0
Washington DC	80	60(08-10)	2009	0	0	0	0	0	0
Washington DC	80	60(08-10)	2010	0	0	0	0	0	0
Washington DC	60	65(06-08)	2006	2.4	0.3	0.1	0	0	0
Washington DC	60	65(06-08)	2007	1.8	0.2	0	0	0	0
Washington DC	60	65(06-08)	2008	0.9	0.1	0	0	0	0
Washington DC	60	65(08-10)	2008	3.4	0.6	0.1	0	0	0
Washington DC	60	65(08-10)	2009	0.1	0	0	0	0	0
Washington DC	60	65(08-10)	2010	5	1.2	0.4	0.2	0.1	0
Washington DC	70	65(06-08)	2006	0.1	0	0	0	0	0
Washington DC	70	65(06-08)	2007	0	0	0	0	0	0
Washington DC	70	65(06-08)	2008	0	0	0	0	0	0
Washington DC	70	65(08-10)	2008	0.2	0	0	0	0	0
Washington DC	70	65(08-10)	2009	0	0	0	0	0	0
Washington DC	70	65(08-10)	2010	0.2	0	0	0	0	0
Washington DC	80	65(06-08)	2006	0	0	0	0	0	0
Washington DC	80	65(06-08)	2007	0	0	0	0	0	0
Washington DC	80	65(06-08)	2008	0	0	0	0	0	0
Washington DC	80	65(08-10)	2008	0	0	0	0	0	0
Washington DC	80	65(08-10)	2009	0	0	0	0	0	0
Washington DC	80	65(08-10)	2010	0	0	0	0	0	0
Washington DC	60	70(06-08)	2006	6.7	1.8	0.5	0.2	0.1	0
Washington DC	60	70(06-08)	2007	6.5	1.7	0.6	0.2	0.1	0
Washington DC	60	70(06-08)	2008	3.4	0.6	0.1	0	0	0
Washington DC	60	70(08-10)	2008	9.2	2.7	0.8	0.2	0.1	0
Washington DC	60	70(08-10)	2009	1	0.1	0	0	0	0
Washington DC	60	70(08-10)	2010	12.5	5	2.4	1.3	0.7	0.4
Washington DC	70	70(06-08)	2006	0.7	0	0	0	0	0
Washington DC	70	70(06-08)	2007	0.1	0	0	0	0	0
Washington DC	70	70(06-08)	2008	0.2	0	0	0	0	0
Washington DC	70	70(08-10)	2008	1.3	0.1	0	0	0	0
Washington DC	70	70(08-10)	2009	0	0	0	0	0	0
Washington DC	70	70(08-10)	2010	1.4	0.1	0	0	0	0
Washington DC	80	70(06-08)	2006	0	0	0	0	0	0
Washington DC	80	70(06-08)	2007	0	0	0	0	0	0
Washington DC	80	70(06-08)	2008	0	0	0	0	0	0
Washington DC	80	70(08-10)	2008	0.1	0	0	0	0	0
Washington DC	80	70(08-10)	2009	0	0	0	0	0	0

Study Area	Exposure Benchmark (ppb-8hr)	Air Quality Scenario[1]	Year	% of all school-age children experiencing multiple exposures per O_3 season at or above benchmarks, adjusted air quality					
				≥1	≥2	≥3	≥4	≥5	≥6
Washington DC	80	70(08-10)	2010	0.1	0	0	0	0	0
Washington DC	60	75(06-08)	2006	12.5	4.6	2.1	1	0.4	0.2
Washington DC	60	75(06-08)	2007	12.9	4.9	2.2	1.1	0.5	0.3
Washington DC	60	75(06-08)	2008	7.3	1.8	0.5	0.1	0	0
Washington DC	60	75(08-10)	2008	18.5	8.2	4.1	2	0.9	0.4
Washington DC	60	75(08-10)	2009	4.1	0.6	0.1	0	0	0
Washington DC	60	75(08-10)	2010	23.4	12.5	7.7	4.9	3.3	2.1
Washington DC	70	75(06-08)	2006	2	0.2	0	0	0	0
Washington DC	70	75(06-08)	2007	1.1	0.1	0	0	0	0
Washington DC	70	75(06-08)	2008	0.8	0	0	0	0	0
Washington DC	70	75(08-10)	2008	4.8	0.8	0.1	0	0	0
Washington DC	70	75(08-10)	2009	0.2	0	0	0	0	0
Washington DC	70	75(08-10)	2010	6	1.4	0.4	0.1	0	0
Washington DC	80	75(06-08)	2006	0.1	0	0	0	0	0
Washington DC	80	75(06-08)	2007	0	0	0	0	0	0
Washington DC	80	75(06-08)	2008	0	0	0	0	0	0
Washington DC	80	75(08-10)	2008	0.8	0	0	0	0	0
Washington DC	80	75(08-10)	2009	0	0	0	0	0	0
Washington DC	80	75(08-10)	2010	0.6	0	0	0	0	0

[1] Abbreviation indicates 8 hour standard level and three year averaging period. For example, 75(08-10) represents simulated ambient concentrations just meeting the existing standard (75 ppb-8hr) using air quality years 2008-2010.

Table 5F-3. Mean and maximum number of **all school-age children** (and associated days per O_3 season) with at least one O_3 exposure at or above 60 ppb-8hr while at moderate or greater exertion.

Study area	Air quality scenario/ standard level (ppb)	Mean		Maximum	
		Number of persons	Number of days per year	Number of persons	Number of days per year
Atlanta	base	255487	827740	364916	1496000
Atlanta	75	127378	229661	166169	325800
Atlanta	70	64409	90926	93000	136700
Atlanta	65	24933	29770	41687	50720
Atlanta	60	5064	5359	10128	10840
Atlanta	55	818	841	2522	2599
Baltimore	base	130826	357038	186072	609600
Baltimore	75	61511	103377	95781	179700
Baltimore	70	35864	49008	59793	88270
Baltimore	65	15050	17414	27416	33050
Baltimore	60	3109	3218	5888	5998
Baltimore	55	346	346	817	817
Boston	base	177358	331700	287713	699100
Boston	75	124529	190220	198171	365900
Boston	70	81220	108757	141821	220300
Boston	65	30411	34767	60377	72810
Boston	60	6710	7064	15506	16620
Boston	55	538	538	1713	1713
Chicago	base	258923	429030	503935	1046000
Chicago	75	260946	449687	470011	922700
Chicago	70	174401	246130	303354	477300
Chicago	65	79122	99309	153346	198700
Chicago	60	22578	26014	41894	48110
Chicago	55	3831	4223	5842	6885
Cleveland	base	119740	234478	172100	423200
Cleveland	75	59189	88930	104388	183800
Cleveland	70	24462	31229	53736	75140
Cleveland	65	6525	7396	17557	20980
Cleveland	60	838	864	2195	2289
Cleveland	55	81	83	188	189
Dallas	base	310037	853320	452737	1894000
Dallas	75	141119	247643	251505	588800
Dallas	70	82161	122988	175109	327400
Dallas	65	32893	41095	83162	117000
Dallas	60	8561	9159	20508	23050
Dallas	55	846	860	2913	2913
Denver	base	129615	328720	178689	563800
Denver	75	95296	187450	143603	367300
Denver	70	57266	91933	106034	221600
Denver	65	21348	27215	53234	78130
Denver	60	1177	1192	2863	2928
Denver	55	21	21	39	39
Detroit	base	207174	402320	349520	922500

Study area	Air quality scenario/ standard level (ppb)	Mean		Maximum	
		Number of persons	Number of days per year	Number of persons	Number of days per year
Detroit	75	143352	230015	194330	374700
Detroit	70	74557	102732	104733	163400
Detroit	65	29881	35337	46936	61570
Detroit	60	3329	3457	6699	7225
Detroit	55	82	82	320	320
Houston	base	254991	574580	360732	1082000
Houston	75	110832	170062	173115	270900
Houston	70	64399	84835	115161	152500
Houston	65	25986	29950	55481	63640
Houston	60	2565	2610	6888	6947
Houston	55	41	41	236	236
Los Angeles	base	1244571	3978000	1423198	4981000
Los Angeles	75	342236	741990	368974	814100
Los Angeles	70	159498	290320	179329	334200
Los Angeles	65	39327	57289	54045	72580
Los Angeles	60	1548	1547	5831	5831
Los Angeles	55	15	15	75	75
New York	base	1148294	2740240	1368877	3664000
New York	75	418702	635750	729630	1311000
New York	70	125784	153098	253458	324700
New York	65	1602	1880	3241	3704
Philadelphia	base	388598	1094200	503583	1630000
Philadelphia	75	170184	279328	252907	459900
Philadelphia	70	87649	117594	145466	203900
Philadelphia	65	29333	33504	56832	65860
Philadelphia	60	7291	7693	20486	21890
Philadelphia	55	699	699	2891	2891
Sacramento	base	147074	469140	190752	764900
Sacramento	75	47859	80022	76891	148400
Sacramento	70	27070	37802	46556	71980
Sacramento	65	12503	14970	22069	27510
Sacramento	60	1744	1830	3892	4296
Sacramento	55	0	0	0	0
St. Louis	base	122408	315640	202543	689100
St. Louis	75	86067	162499	136172	316800
St. Louis	70	53629	82013	89003	160900
St. Louis	65	20760	25857	38381	54050
St. Louis	60	2915	3100	7840	8718
St. Louis	55	140	143	503	515
Washington	base	267667	771500	345115	1141000
Washington	75	127234	246782	226043	556700
Washington	70	63711	97776	121074	219700
Washington	65	22251	27721	48425	65920
Washington	60	2645	2776	5578	5803
Washington	55	0	0	0	0

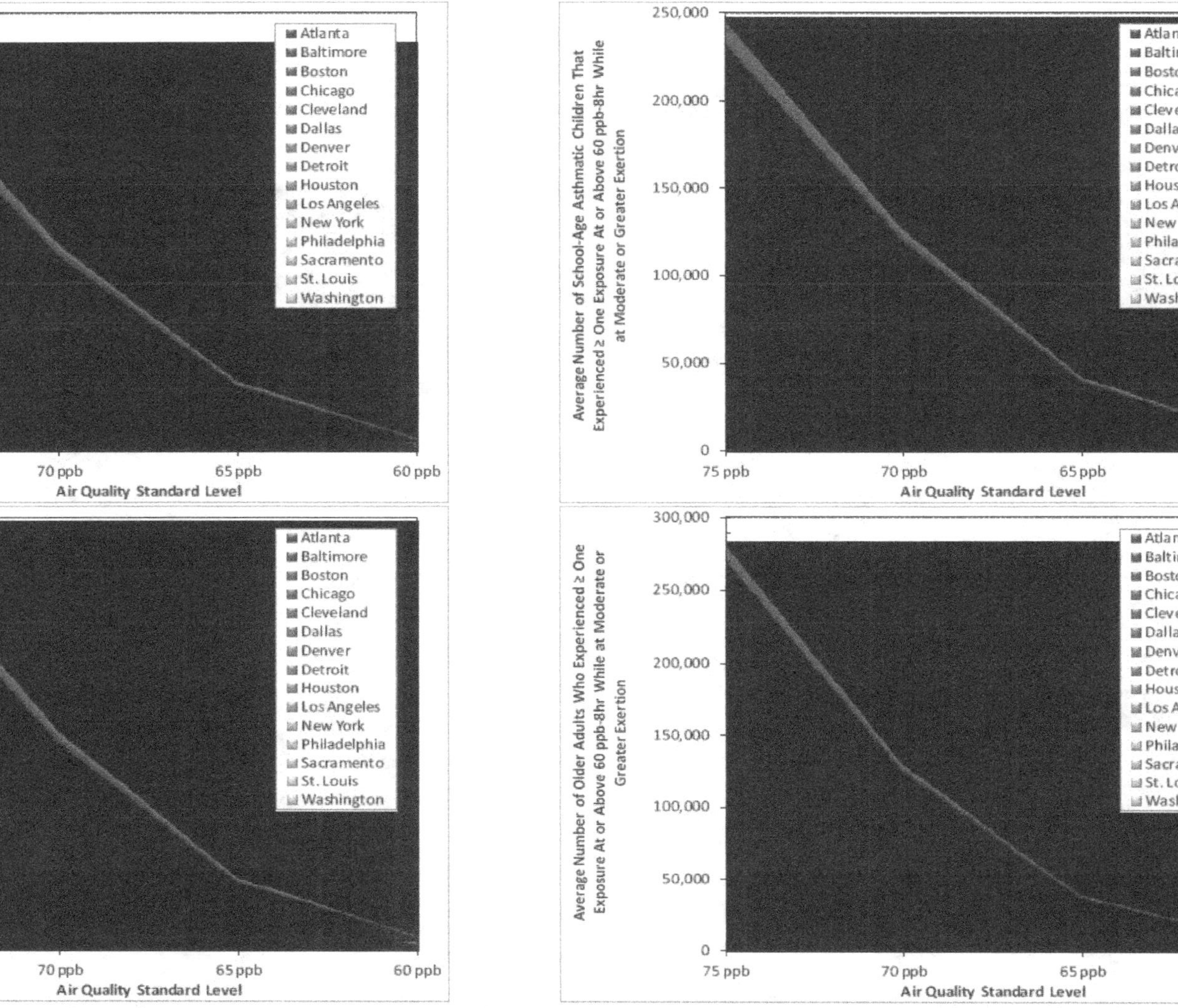

Figure 5F-14. Average number of persons with at least one 8-hour exposure at or above 60 ppb considering the existing and alternative standards, year 2006-2010 adjusted air quality. All school-age children (top left), asthmatic school-age children (top right), asthmatic adults (bottom left), older adults (bottom right).

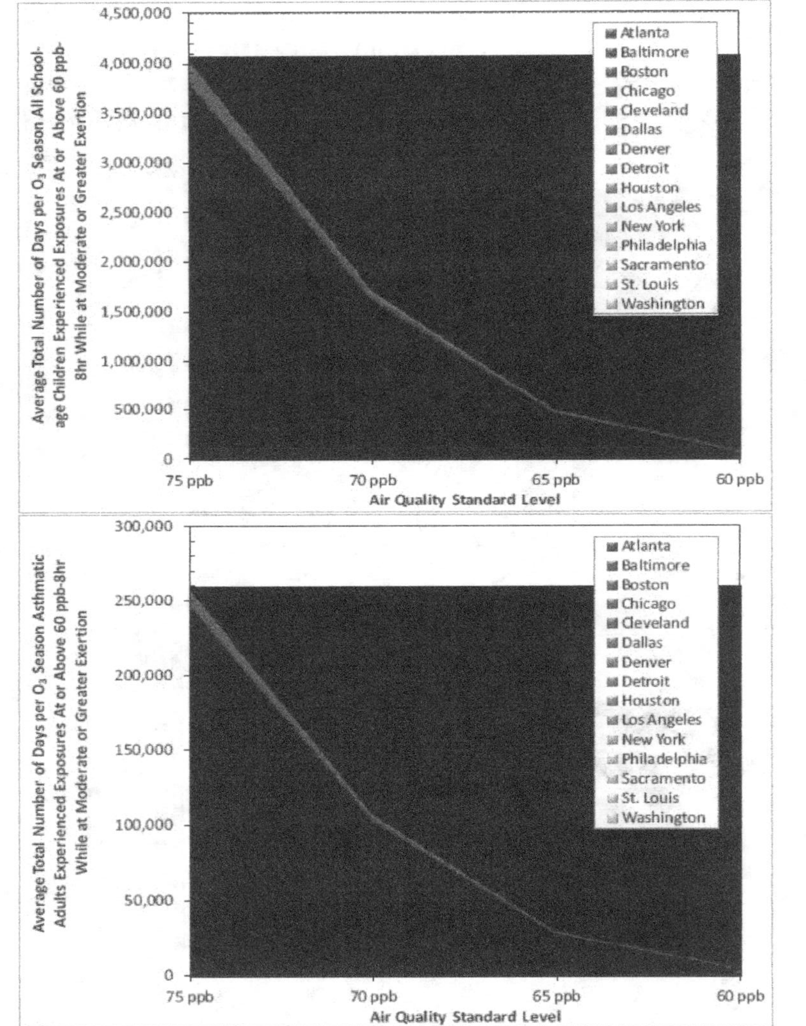

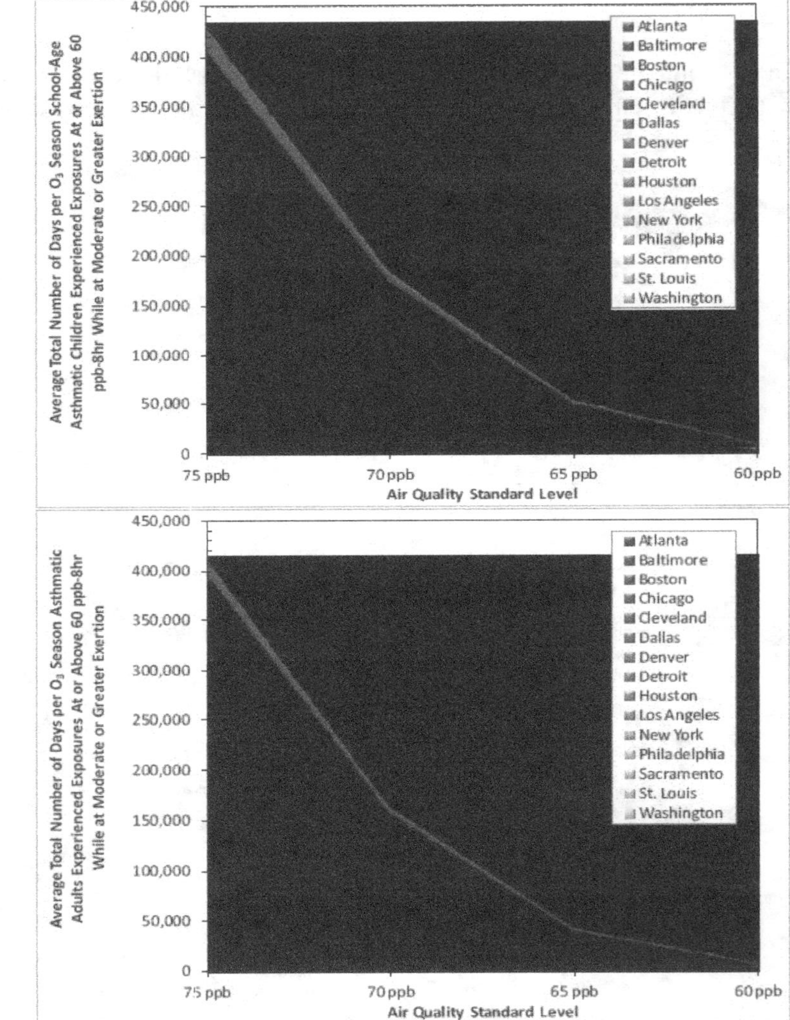

Figure 5F-15. Average total number of days in an O_3 season where simulated persons experienced 8-hour exposures at or above 60 ppb considering the existing and alternative standards, year 2006-2010 adjusted air quality. All school-age children (top left), asthmatic school-age children (top right), asthmatic adults (bottom left), older adults (bottom right).

Table 5F-4. Total number of persons experiencing at least one or two 8-hour exposures in all study areas by year, base air quality and air quality adjusted to just meeting the existing 75 ppb standard.

Exposure Study Group	Air Quality Scenario (3-year averaging period)	Year	8-hr Average Exposure Benchmark Level (ppb)	Number of Persons Exposed Per O_3 Season[1]	
				At least once	At least twice
Asthmatic School-age Children	75(2006-2008)	2006	60	254454	94876
			70	39422	4113
			80	2399	120
		2007	60	281062	113215
			70	51351	8788
			80	4236	133
		2008	60	155918	42881
			70	12603	713
			80	428	13
		2009	60	160535	49447
			70	19889	1532
			80	1159	39
		2010	60	265139	110871
			70	45688	6737
			80	3981	143
	base	2006	60	664077	381016
			70	322187	126882
			80	116538	26233
		2007	60	677867	397000
			70	311929	113719
			80	96028	16731
		2008	60	523387	280744
			70	209734	67291
			80	59813	11174
		2009	60	364614	157977
			70	115454	33741
			80	34482	5779
		2010	60	504479	263154
			70	165488	49314
			80	30418	3270
All School-age Children	75(2006-2008)	2006	60	2434809	900706
			70	350957	38834
			80	20489	708
		2007	60	2624485	1045539
			70	481637	74715
			80	45085	992

[1]Numbers of children are summed across urban case study areas in each year. Because Chicago does not have a simulation of the existing standard for the 2008-2010 three-year standard averaging period, year 2008 study area sums were based only the year 2008 simulations that used the 2006-2008 three-year standard averaging period.

Exposure Study Group	Air Quality Scenario (3-year averaging period)	Year	8-hr Average Exposure Benchmark Level (ppb)	Number of Persons Exposed Per O_3 Season[1]	
				At least once	At least twice
		2008	60	1492203	438479
			70	120656	10568
			80	3615	52
		2009	60	1529789	469352
			70	188071	16908
			80	12385	141
		2010	60	2439477	997079
			70	395217	50200
			80	27981	670
	base	2006	60	6294372	3653902
			70	3104290	1265681
			80	1140328	264098
		2007	60	6307074	3658057
			70	2926601	1067241
			80	892814	149031
		2008	60	4987568	2672573
			70	2021415	684051
			80	597040	132098
		2009	60	3553528	1574327
			70	1191875	372643
			80	370166	63075
		2010	60	4671276	2417136
			70	1528455	421992
			80	273651	27126

[1] Numbers of children are summed across urban case study areas in each year. Because Chicago does not have a simulation of the existing standard for the 2008-2010 three-year standard averaging period, year 2008 study area sums were based only the year 2008 simulations that used the 2006-2008 thee-year standard averaging period.

Appendix 5-G

Targeted Evaluation of Exposure Model Input and Output Data

Table of Contents

5G-1 ANALYSIS OF TIME-LOCATON-ACTIVITY DATA 4
5G-1.1 Personal Attributes of Survey Participants in CHAD and Used by APEX 4
5G-1.2 Afternoon Time Spent Outdoors for CHAD Survey Participants 7
5G-1.3 Afternoon Time Spent Outdoors For ATUS Survey Participants 10
5G-1.4 Outdoor Time and Exertion Level of Asthmatics and Non-Asthmatics In CHAD 12
5G-2 CHARACTERIZATION OF FACTORS INFLUENCING HIGH EXPOSURES 17
5G-3 ANALYSIS OF APEX SIMULATED LONGITUDINAL ACTIVITY PATTERNS..... 26
5G-4 EXPOSURE RESULTS FOR ADDITIONAL AT-RISK POPULATIONS AND LIFESTAGES, EXPOSURE SCENARIOS, AND AIR QUALITY INPUT DATA USED 32
5G-4.1 Exposure Estimated For All School-Age Children During Summer Months, Neither Attending School nor Performing Paid Work... 32
5G-4.2 Exposures Estimated For Adult Outdoor Workers During Summer Months....... 34
5G-4.3 Averting Behavior and Potential Impact to Exposure Estimates.......................... 42
5G-5 COMPARISON OF PERSONAL EXPOSURE MEASUREMENT AND APEX MODELED EXPOSURES 47
5G-6 REFERENCES 50

List of Tables

Table 5G-1. Personal attributes of survey participants within CHAD and used by APEX. 6
Table 5G-2. Comparison of outdoor time expenditure and exertion level among asthmatic and non-asthmatic diary days for CHAD diaries used by APEX. 14
Table 5G-3. Percent of waking hours spent outdoors at an elevated activity level. A comparison of CHAD with Shamoo et al. (1994) study asthmatics. 15
Table 5G-4. Percent of waking hours spent outdoors at an elevated activity level: a comparison of CHAD with EPRI (1992) study asthmatics. 16
Table 5G-5. Percent of waking hours spent outdoors at an elevated activity level: a comparison of CHAD with EPRI (1988) study asthmatics. 16
Table 5G-6. Range of R^2 fit statistics for ANOVA models used to evaluate daily maximum 8-hour O_3 exposure concentrations stratified by study area, air quality scenario, and exposure level. 19
Table 5G-7. Range of R^2 fit statistics for ANOVA models used to evaluate daily maximum 8-hour O_3 exposure concentrations in Los Angeles stratified by age group, air quality scenario, and exposure level. 19
Table 5G-8. Distribution of days per week spent performing outdoor work considering the BLS/O*NET data set and stratified by APEX/CHAD occupation groups. 36
Table 5G-9. Personal attributes and mean time spent working outdoors for CHAD diaries reporting at least two hours of outdoor work. 37
Table 5G-10. Distribution of days per week spent performing outdoor work considering the APEX simulated population and stratified by APEX/CHAD occupation and age groups. 41

List of Figures

Figure 5G-1. Participation rate in outdoor activities (top) and mean time spent outdoors (bottom) for CHAD diaries having at least one minute outdoors (left) and CHAD diaries having at least two hours outdoors (right) during the afternoon. 8
Figure 5G-2. Participation rate in outdoor activities (top) and mean time spent outdoors (bottom) for ATUS diaries having at least one minute outdoors (left) and ATUS diaries having at least two hours outdoors (right). 11
Figure 5G-3. Contribution of influential factors to daily maximum 8-hour ozone exposures using base air quality (left), air quality adjusted to just meet the existing standard (right), considering all person days (top) and those days where daily maximum 8-hour exposure exceeded 50 ppb (bottom) in Boston. 21
Figure 5G-4. Contribution of influential factors to daily maximum 8-hour ozone exposures using base air quality (left), air quality adjusted to just meet the existing standard (right), considering all person days (top) and those days where daily maximum 8-hour exposure exceeded 50 ppb (bottom) in Atlanta. 22
Figure 5G-5. Distributions of afternoon outdoor time expenditure and daily maximum 8-hour ambient O_3 concentrations for simulated Boston school-age children (ages 5 to 18) (left) and adults (ages 19 to 35) (right) using base air quality (top) and concentrations adjusted to just meet the existing standard (bottom) for person days having daily maximum 8-hour exposures either below or above 50 ppb..... 24
Figure 5G-6. Afternoon microenvironmental time (top) and activities performed during afternoon time outdoors (bottom) for school-age children (left) and adults (right)

experiencing 8-hour daily maximum O_3 exposures ≥ 50 ppb, Boston base air quality, 2006. 26
Figure 5G-7. Cumulative distribution of median time spent outdoors (top row), afternoon outdoor participation ≥ 1 minute/day (2nd row), and afternoon outdoor participation ≥ 2 hours/day (3rd row) for male and female school-age children in Atlanta (left column), Boston (middle column) and Denver (right column) study areas. Percent of school-age children with ≥ 2 hours outdoors during afternoon hours (4th row) and the number of particular CHAD study diary days used (bottom row) for each exposure simulation day. 30
Figure 5G-8. Cumulative distribution of median time spent outdoors (top row), afternoon outdoor participation ≥ 1 minute/day (2nd row), and afternoon outdoor participation ≥ 2 hours/day (3rd row) for male and female school-age children in Houston (left column), Philadelphia (middle column) and Sacramento (right column) study areas. Percent of school-age children with ≥ 2 hours outdoors during afternoon hours (4th row) and the number of particular CHAD study diary days used (bottom row) for each exposure simulation day 31
Figure 5G-9. Comparison of the percent of all school-age children having daily maximum O_3 concentration at or above 60 ppb-8hr (top), 70 ppb-8hr (middle), or 80 ppb-8hr (bottom) during June, July, and August in Detroit 2007: using any available CHAD diary ("All CHAD Diaries") or using CHAD diaries having no time spent in school or performing paid work ("No School/Work Diaries") 33
Figure 5G-10. Percent of persons between age 19-35 (left) and 36-55 (right) experiencing exposures at or above selected benchmark levels while at moderate or greater exertion using an outdoor worker scenario-based approach (top) and a general population-based approach (bottom) considering air quality adjusted to just meet the existing standard in Atlanta, GA, Jun-Aug, 2006. 42
Figure 5G-11. Distribution of daily personal O3 exposures (top row), outdoor time (2nd row from top), ambient O_3 concentrations (3rd row from top), and air exchange rate (bottom row) for DEARS study participants (left column) and APEX simulated individuals (right column) in Wayne County, MI, July-August 2006. 49

This appendix presents the complete results of several targeted evaluations and exposure simulations designed to provide additional insights to APEX input data or approaches used to estimate exposures, algorithm and model performance evaluations, and estimated exposures for additional exposure study groups and lifestages of interest.

5G-1 ANALYSIS OF TIME-LOCATON-ACTIVITY DATA

We first present an overview of the data currently available in the CHAD database used by APEX, including comparison with the version of CHAD used to estimate exposures in the 1st draft O_3 REA. This is followed by an analysis of time spent outdoors - one of the most important attributes influencing exposures at or above benchmark levels - using CHAD and recent time-location-activity pattern data from the American Time Use Survey (ATUS). And finally, CHAD diaries identified as coming from asthmatics are compared with that of non-asthmatics for afternoon outdoor time and activity level as well as compared with available independent studies of asthmatic activity patterns.

5G-1.1 Personal Attributes of Survey Participants in CHAD and Used by APEX

The survey participants whose diary data are within CHAD were asked a number of questions regarding their personal attributes. The number and type of attributes present for diaries in CHAD is driven largely by the original intent of the individual study. In our exposure assessment, we have strict requirements to simulate individuals using several personal attributes, namely age, sex, temperature (as a surrogate for seasonal variation in activity patterns), and day-of-week. These attributes are considered as important drivers influencing daily activity patterns (Graham & McCurdy, 2004) and when diaries do not have these particular attributes for a particular day, they will not be used by APEX.

This APEX modeling requirement serves as an initial screen to the number of available diaries in the complete CHAD master database (i.e., 54,373) and considering the age range of the simulated exposure study groups (persons between the ages of 5 and 95), the actual number of diary days having complete information and used by APEX in the 2nd draft O_3 REA is 41,474.[1] This represents an increase of about 8,700 diaries currently used by APEX compared with what was used by APEX in the 1st draft O_3 REA. Additionally, there have been eight new study data sets incorporated into CHAD and used in our current exposure assessment since the previous O_3 NAAQS review conducted in 2007, most of which were from recently conducted activity pattern

---

[1] Diaries from persons age 4 are included in this evaluation because they may be used in a simulation to represent a person aged 5 due to the probabilistic nature of APEX. Typically, a diary matching the attributes of the simulated individual has a greater probability of selection. Accommodations are allowed to increase the diary pool size (e.g., expand the age window of diaries available by value (one year) or percent (15%) of the simulated persons age.

studies (Appendix 5B, Section 5B-4). The diary data included from these new studies have more than doubled the total activity pattern data used for 2007 O_3 exposure modeling and has increased the number of children's diaries by about a factor of five.

Table 5G-1 presents a summary of the important personal attributes used by APEX in creating activity patterns for simulated persons, along with other attributes of potential interest (e.g., race/ethnicity). First, we compared the representation of several attributes in the current CHAD used by APEX versus that used in the 1st draft O_3 REA. Outside of increases in the number of persons, the general distribution of diaries within the APEX diary selection attributes (e.g., age, sex, temperature, day-of-week) is similar in both databases. Worth noting is the number and percent of diaries from each of the three decades analyzed. Currently, the majority of diaries (54%) from CHAD are taken from surveys conducted in the past decade, while the pre-1990s represent less than 15% of the total diaries available by APEX.

While there may be other personal or situational attributes that affect daily time expenditure, these are typically not included in our assessment to generate simulated individuals simply because the response to the attribute is missing for most persons. For example, income level is missing for just over 66% of the study participants and only about 30% of employed workers (persons ages 19 to 64) reported their occupation (Table 5G-1). Missing response data in CHAD results from either the study not having an income/occupation related survey question or perhaps the participant refused to answer the question. Note also, when any attribute is added to the development of a person's profile, the pool of diaries available for selection in simulating an individual is reduced. This could lead to an increased repetition of diaries used for simulated individuals, potentially artificially reducing variability in time expenditure. In addition, the desired study group to be simulated may have too few diaries within a diary pool if most diaries are missing the needed attribute, leading to a simulation failure. This is why personal attributes are carefully selected and prioritized according to both their prevalence in CHAD and whether attribute has a known significant influence on activity patterns.

Table 5G-1. Personal attributes of survey participants within CHAD and used by APEX.

		Current APEX CHAD (41,474 total days)		1st Draft APEX CHAD (32,788 total days)	
Personal Attribute	**Group Within Attribute**	**person days (n)**	**percent of attribute**	**person days (n)**	**percent of attribute**
Age (years)	4 - 18	17680	42.6	14111	43
	19 - 34	4490	10.8	4001	12.2
	35 - 50	7238	17.5	5957	18.2
	51 - 64	6181	14.9	5016	15.3
	65+	5885	14.2	3703	11.3
Sex	Female	21466	51.8	16840	51.4
	Male	20008	48.2	15948	48.6
Daily Maximum Temperature (°F)	84+	13817	33.3	12113	36.9
	55-83	17827	43	13078	39.9
	<55	9830	23.7	7597	23.2
Day of Week	Weekday	29031	70.0	23794	72.6
	Weekend	12443	30.0	8994	27.4
Decade	1980s	5999	14.5	6167	18.8
	1990s	12831	30.9	12390	37.8
	2000s	22644	54.6	14231	43.4
Employed (ages 19-64)	No	4651	26.0	3747	25.0
	Yes	12755	71.2	11227	75.0
	Missing/Unknown	503	2.8	n/a	n/a
Occupation (employed)	Known	3867	30.3	3012	26.8
	Missing/Unknown	8888	69.7	8215	73.2
Income Group	> 1.5 x Poverty	10347	24.9	10416	31.8
	≤ 1.5 x Poverty	3713	9.0	3730	11.4
	Missing/Unknown	27414	66.1	18642	56.9
Calendar Month	Calendar Months 5 - 9	19151	46.2	16812	51.3
	Calendar Months 1 - 4, 10 - 12	22323	53.8	15976	48.7
Race/Ethnicity	Asian	670	1.6	349	1.1
	Black	6993	16.9	4040	12.3
	Hispanic	2476	6.0	1339	4.1
	Native American	28	0.1	7	0
	White	25009	60.3	19569	59.7
	Other	770	1.9	609	1.9
	Missing/Unknown	5528	13.3	6875	21

5G-1.2 Afternoon Time Spent Outdoors for CHAD Survey Participants

There have been questions raised regarding the representativeness of the diaries from studies conducted in the 1980s and whether there are any recognizable patterns in time expenditure in the CHAD diaries across the time period when data were collected. Because time spent outdoors is a significant factor influencing daily maximum 8-hour O_3 exposures, we evaluated the current collection of CHAD diaries used by APEX for two metrics: outdoor participation rate and mean time spent outdoors. The participation rate is the percent of the person-days having at least one minute outdoors, and because high O_3 concentrations commonly occur during the afternoon hours of summer months, we restricted the analysis to those times of day (12 PM to 8 PM) and year (May through September). The same data set was used to calculate a mean outdoor time, though the calculation was further restricted to person days meeting an additional criterion: person-days having at least one minute outdoors and person-days having at least 2-hours outdoors. Separating the data into these sub-groups give us insight to the diaries most likely to be used in simulating a person that exceeds a selected benchmark level and protects (to a limited degree) from study sample design bias (15-minute time block diaries versus minute-by-minute event level diaries). Data were further stratified by five age groups (4-18, 19-34, 35-50, 51-64, 65+) and three decades (1980s, 1990s, and 2000s) using the year the particular activity pattern study was conducted. As a reminder, CHAD is composed of primarily cross-sectional data, thus the trend evaluated over the three decades is changes (if any) in participation rate and the time spent outdoors by the study group, not individuals.

Figure 5G-1 illustrates the trends in afternoon outdoor activity participation and mean time expended outdoors, considering three decades, five age groups, and whether the total afternoon time spent outdoors was at least one minute or two hours. Regardless of decade and duration of time spent outdoors, participation in outdoor activities follows an expected pattern considering age groups, that is, children tend to have the highest participation rate when compared with the other age groups, while the oldest persons (aged 65 or greater) tend to have the lowest participation rate (Figure 5G-1, top left panel). When considering decade and CHAD diaries having at least one minute spent outdoors, the participation rate appears to have a non-linear concave trend, whereas CHAD diaries collected during the 1990s exhibit the lowest outdoor participation rate (ranging from about 40-70%) while much greater participation is found with the CHAD diaries collected during the 1980s (80-90%) and 2000s (70-80%).

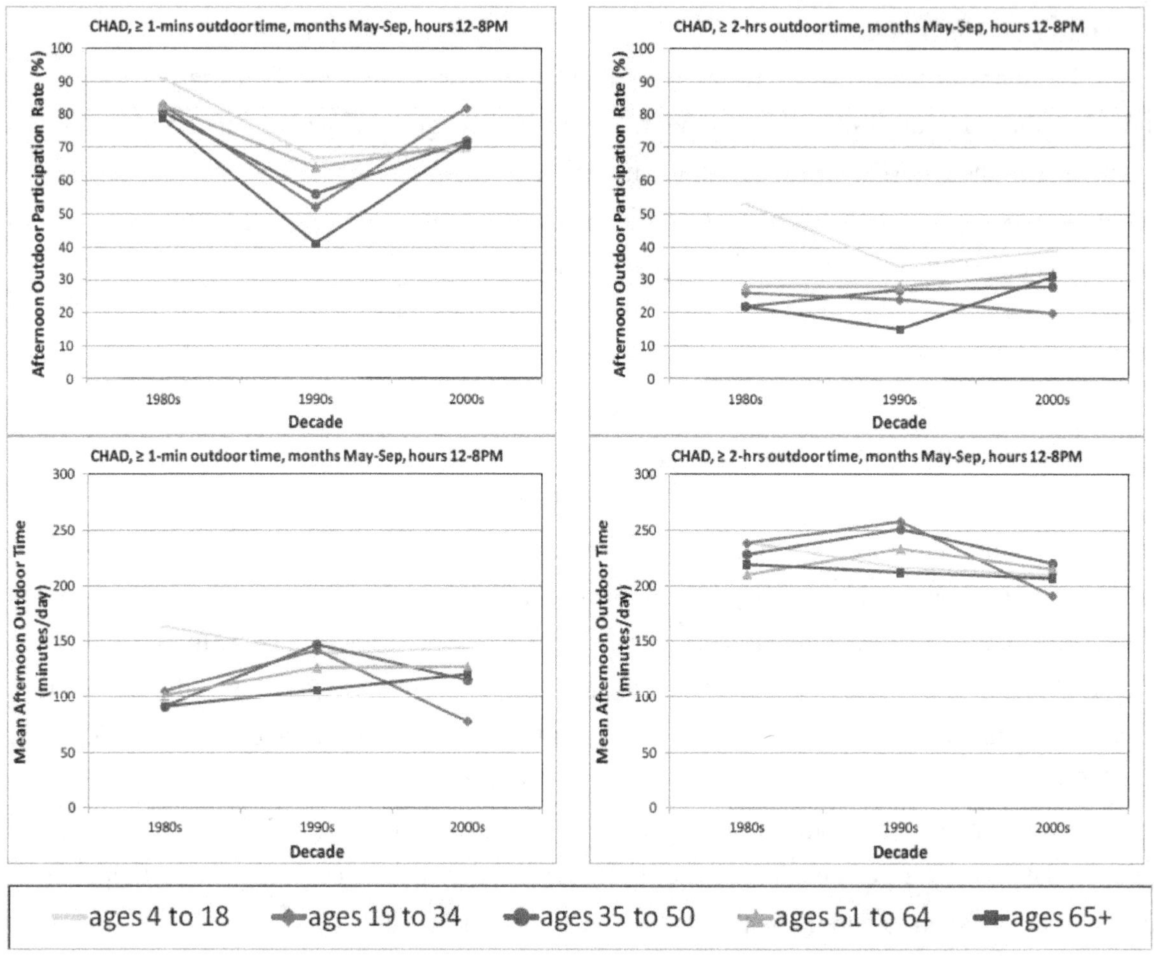

Figure 5G-1. Participation rate in outdoor activities (top) and mean time spent outdoors (bottom) for CHAD diaries having at least one minute outdoors (left) and CHAD diaries having at least two hours outdoors (right) during the afternoon.

It is possible that this observed pattern may be the result of the original study survey design. All of the CHAD diaries collected during the 1980s used an 'event' level approach, that is locations visited and activities performed were reported on a minute or longer basis and many of those same diaries also used a 'contemporaneous diary' approach, that is, real-time data reporting. The CHAD diaries collected during the 1990s used a mixture of event level and 15-minute time block data, though mostly using a 'recall' approach, that is, participants were asked about their activities performed the day before. It is likely that these diaries exhibit the lowest outdoor participation rate due to the participant missing or ignoring short duration events (< 15 minutes) that may have occurred outdoors (e.g., outdoors in a parking lot and walking to their vehicle). The CHAD diaries collected during the 2000s, while also a combination of studies that used the event level and 15-minute time block approach did have a few studies using the real-time diary approach, possibly responsible for the observed increase in outdoor participation rate during this decade.

When restricting the data set to person days having at least two hours of afternoon time spent outdoors, the above mentioned temporal pattern nearly disappears (Figure 5G-1, top right panel) as most age groups exhibit little to no variability in outdoor time participation rate across the three decades (~20-30% of person-days). However, the diaries from the 1980s for children ages 4-18 indicate the highest outdoor participation rate (i.e., 50%) compared to all other age groups and decade of collection. Most of these diaries (i.e., 96%) are from the California Children's Study and Cincinnati Activity Patterns Study. To date it remains unexplained why these two studies would have this unusual outdoor participation rate compared to the other studies and decades and could simply be a function of the two study selecting particularly active children by chance.[2] When considering the entire pool of all diaries available for this age group and used by APEX, these two studies contribute to about 19% of diaries having two or more hours of time spent outdoors during the afternoon. This translates to a small difference in the overall outdoor participation rate for diary pools that would include these earlier studies (39%) compared to the participation rate excluding these studies (36%), with both values similar to the findings reported recently by Marino et al. (2012) of 37.5% using a similar metric, though for pre-school age children. Thus, when considering participation in outdoor activities and the representativeness of the CHAD study data from the 1980s, it is unlikely that use of these older diaries would adversely influence exposure model estimates.

When considering the mean time spent outdoors during the afternoon hours for diaries having at least one minute recorded outdoors, the observed pattern is generally the inverse of participation. This pattern would be consistent with the above proposed reasoning: relatively more persons reporting shorter duration events leads to higher overall outdoor participation rate coinciding with a decrease in the mean time spent outdoors (Figure 5G-1, bottom left panel). In general, when restricting the data set to person-days that recorded at least two hours outdoors during the afternoon (Figure 5G-1, bottom right panel), there is variability in the amount of outdoor time over the three decades, with diaries from the 2000s exhibiting perhaps the lowest range of mean outdoor time (190-220 min/day) compared with the 1980s (210-240 min/day) and 1990s (212-258 min/day), a trend perhaps most notable trend when considering the children's diaries (a decrease in time spent outdoor of about 30 minutes over the period). However, the coefficient of variation (COV) for each of the age groups and across all decades for the cross-sectional data was consistently about 40% (data not shown), supporting a

[2] Restricting the data by any criterion will result in fewer diary days available that subset, thus any remaining or newly observed trends (e.g., an apparent linear decline in participation by the diaries from persons aged 19-34) should include an assessment of potential confounding factors. Given our current and other researcher's past evaluations of the CHAD data (Graham and McCurdy, 2003; Isaacs et al., 2012; McCurdy and Graham, 2004) often times, as is the case here, additional influential factors have not been measured and the observed phenomenon can only be identified as study related.

general conclusion of no large differences in the mean time spent outdoors for this set of diaries over the three decades of data collection. Thus, when considering all diaries having at least two hours of afternoon outdoors time and the representativeness of the CHAD study data from the 1980s, inclusion of these earlier diaries is also unlikely to have a significant adverse influence on exposure modeling outcomes.

5G-1.3 Afternoon Time Spent Outdoors For ATUS Survey Participants

We evaluated recent year (2002-2011) time expenditure data from the American Time Use Survey (ATUS) (US BLS, 2012a). As was done with the CHAD data set, the purpose of the evaluation was to evaluate trends in outdoor time over the period of time data were collected. A few strengths of the ATUS data are (1) its recent and ongoing data collection efforts, (2) large sample size (>120,000 diary days), (3) national representativeness, and (4) that varying diary approaches would not be an influential or confounding factor in evaluating trends over time.

ATUS does however have a few noteworthy limitations when compared with the CHAD data: (1) there are no survey participants under 15 years of age, (2) time spent at home locations is neither distinguished as indoors or outdoors, and (3) missing or unknown location data can comprise a significant portion of a persons' day (on average, about 40% (George and McCurdy, 2009)). To overcome the limitation afforded by the ambiguous home location, we identified particular activity codes most likely to occur outdoors (e.g., participation in a sport) to better approximate each ATUS individual's outdoor time expenditure. Missing several hours of location and activity information can be problematic when modeling exposures, an issue that renders the ATUS diaries generally unusable by APEX. However, the particular time of day the missing data occurs is more accommodating to the purpose of this analysis. While most diaries are missing location information for 6 or more hours per day, on average about 85% of the missing time information occurs outside of the hours of interest here (i.e., 12 PM-8 PM), with most missing time occurring between early morning (4 AM-9 AM) or late evening hours (10 PM-12 AM). Still though, we restricted the ATUS outdoor time analysis to diaries having no more than 1-hour of missing afternoon time while also only retaining diaries from ATUS identified metropolitan areas. Data were then stratified by the same five age groups as was done for the CHAD data, though here the time trends were assessed over individual survey years. Figure 5G-2 illustrates the results of the ATUS diary outdoor participation rates (top row) and the mean time spent outdoors during afternoon hours (bottom row) for persons having at least one minute of afternoon time spent outdoors (left column) or two hours or more outdoors (right column).

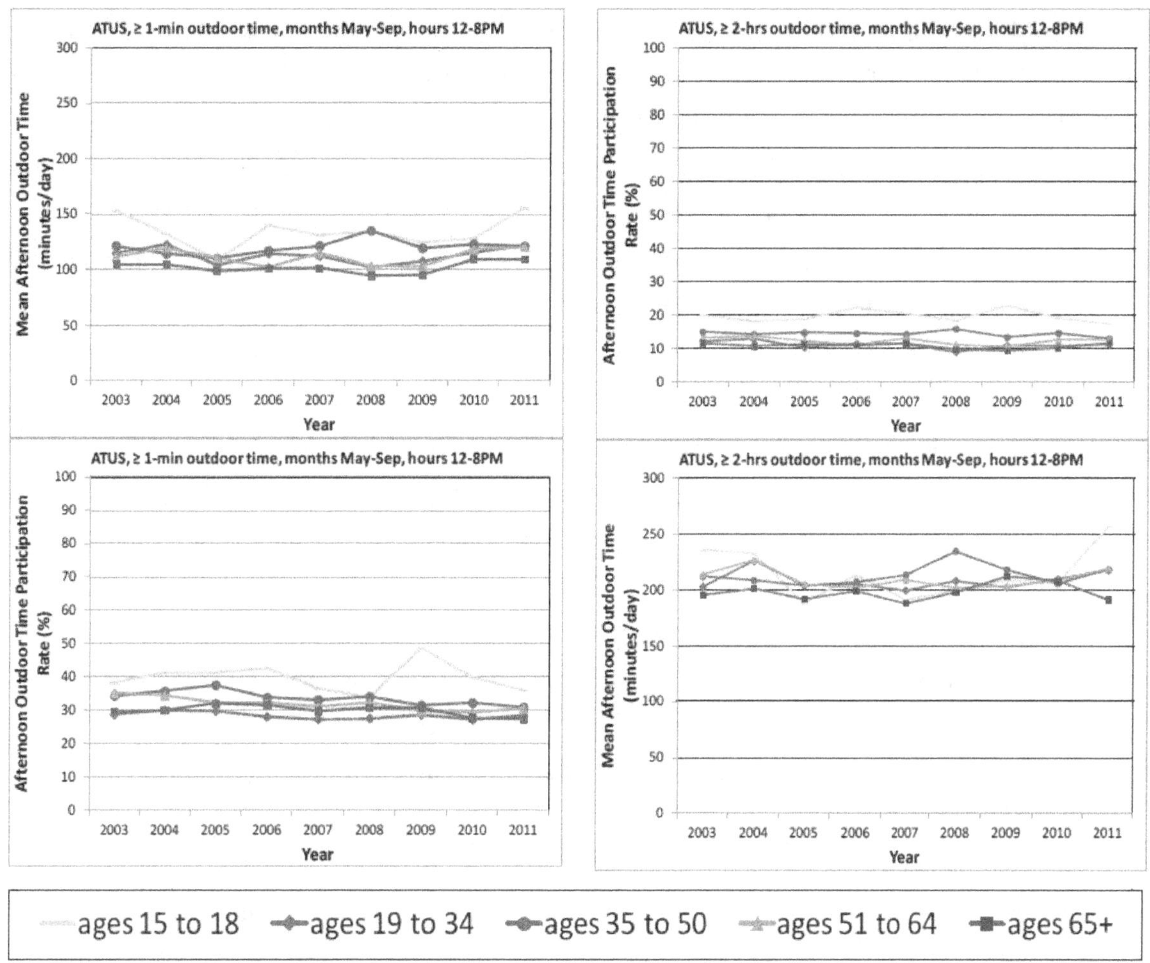

Figure 5G-2. Participation rate in outdoor activities (top) and mean time spent outdoors (bottom) for ATUS diaries having at least one minute outdoors (left) and ATUS diaries having at least two hours outdoors (right).

Not surprisingly given the lack of distinction regarding time indoors and outdoors while at home for ATUS participants as well as the diary approach used,[3] the outdoor activity participation rate for ATUS study subjects is lower than that of CHAD study subjects; about 30-40% of ATUS person-days have at least one minute of outdoor time (Figure 5G-2, top left panel). As was observed with the CHAD data, children (ATUS ages 15 to 18) are more likely to participate in outdoor activities. The mean time outdoors for persons that reported any amount of outdoor time is similar to the range indicated by CHAD diaries, generally between 100-150 minutes per day (Figure 5G-2, bottom left panel). When considering person-days having at least 2 hours of time spent outdoors, the range in ATUS diary outdoor participation rate (10-20%, Figure 5G-2, top right panel)) is lower than that observed for the CHAD data (generally between 20-40%), while the range in mean time spent outdoors (190-240 minutes per day, Figure 5G-2, top right panel) was similar to that of the CHAD data. Consistent also across the two studies is the participation rate of children being greater than that of the other age groups. There are no consistent trends over the nine year ATUS study period regarding either the participation rate or the mean time spent outdoors for any of the age groups.

5G-1.4 Outdoor Time and Exertion Level of Asthmatics and Non-Asthmatics In CHAD

Due to limited number of CHAD diaries with survey requested health information, all CHAD diaries are assumed appropriate for any simulated individual (i.e., whether asthmatic, non-asthmatic, or not indicated), provided they concur with age, sex, temperature, and day-of-week selection criteria. In general, the assumption of modeling asthmatics similarly to healthy individuals (i.e., using the same time-location-activity profiles) is supported by the findings of van Gent et al. (2007), at least when considering children 7 to 10 years in age. These researchers used three different activity-level measurement techniques; an accelerometer recording 1-minute time intervals, a written diary considering 15-minute time blocks, and a categorical scale of activity level. Based on analysis of 5-days of monitoring, van Gent et al. (2007) showed no difference in the activity data collection methods used as well as no difference between asthmatic children and healthy children when comparing their respective activity levels. Contrary to this, an analysis of 2000 BRFSS data by Ford et al. (2003) indicated a statistically significant difference between the percent of current asthmatics (30.9%) and non-asthmatics (27.8%) characterized as inactive. In addition, these researchers found small but statistically significant differences in the percent of asthmatic (26.6%) and non-asthmatic (28.1%) adults achieving recommended levels of physical activity (i.e., either moderate or greater activity levels).

[3] The ATUS time-use information was collected by subject recall of the prior day's activities and "conversational interviewing" rather than asking scripted questions.

Note though, the salient issue is not just outdoor time and activity levels, but the intersection of the two as well as recognizing the performance capabilities of persons with asthma. A person's overall physical activity level is strongly linked with their time spent outdoors and is considered an important correlate in encouraging increased physical activity among children and adults alike (e.g., Sallis et al., 1998). In addition, introducing regular exercise has been shown to improve physical fitness in asthmatic children, with statistically significant increases in ventilation measures such as maximum minute ventilation rate (VE_{max}) maximum oxygen uptake (VO_{2max}) (e.g., van Vledhoven et al., 2001). Further, in other related research, Santuz et al. (1997) indicated no statistically significant difference between asthmatic and non-asthmatic children when comparing maximum exercise performance levels, provided the individuals were conditioned through habitual exercise. Thus it appears that asthmatics perform activities at elevated levels and do so in outdoor microenvironments in similar fashion to non-asthmatics.

To provide further support to the assumption that any CHAD diary day can be used to represent the asthmatic study groups regardless of the study participants' characterization of having asthma or not, we first compared the amount of afternoon outdoor time and participation in elevated exertion levels among asthmatics and non-asthmatics. Because six of the 19 studies incorporated in CHAD reported whether the individual was asthmatic or non-asthmatic, we categorized the data and results using three categories (i.e., asthmatic, non-asthmatic, not classifiable). Afternoon hours were characterized as was done for above CHAD analyses, that is, the time between 12 PM and 8 PM and only those persons that did spend some time outdoors were retained. As is done by APEX in simulating individuals, level of exertion was estimated by sampling from the specific METS distributions assigned for each person's activity performed. Then, we selected for activities having a METS value of greater than 3 as times where a person was at moderate or greater exertion levels (US DHHS, 1999). Afternoon outdoor time was then stratified by exertion level, summed for two study groups of interest (children and adults), and presented in percent form within Table 5G-2.

When considering CHAD diaries used by APEX in our simulations, about 18% of the diaries are from either an asthmatic child or an asthmatic adult. Far fewer children's diaries are from persons whose asthmatic status is unknown (12%) when compared to adults (30%) though still, persons having unknown health status are a smaller proportion of the total available person-days. On average, about 43% of all children spent some afternoon time outdoors while asthmatic children have a higher participation rate (48.5%) when compared to non-asthmatic children (41.2%). About half of the adults whose asthmatic condition was known did spend afternoon time spent outdoors with participation rate generally similar for both asthmatic and non-asthmatic adults. Outdoor participation rate for persons having unknown asthma status varied from that of known persons; about 60% of the children's diaries and 31% of the adult diaries indicate some afternoon time was spent outdoors.

Table 5G-2. Comparison of outdoor time expenditure and exertion level among asthmatic and non-asthmatic diary days for CHAD diaries used by APEX.

	CHAD: Children (4 to 18)[1]			CHAD: Adults (19 to 95)[2]		
Asthmatic?	Yes	No	Unknown	Yes	No	Unknown
Persons (n)	3,206	12,346	2,128	1,254	15,465	7,075
Outdoor Habitué (n)	1,564	5,092	1,267	602	7,949	2,176
Outdoor Habitué (%)	48.8	41.2	59.5	48.0	51.4	30.8
Percent of Afternoon Hours Spent Outdoors (%)	28.5	27.5	28.9	26.2	27.2	22.2
Percent of Afternoon Outdoor Time at Moderate or Greater Exertion (%)	80.3	78.2	79.2	62.7	63.8	60.3

[1] CHAD studies for where a survey questionnaire response of whether or not child was asthmatic include CIN, ISR, NHA, NHW, OAB, and SEA (see REA Table 5-3 for study names).
[2] CHAD studies for where survey a questionnaire response of whether or not adult was asthmatic include CIN, EPA, ISR, NHA, NHW, NSA, and SEA.

The amount of time spent outdoors by the persons that did so varied little across the two study groups and three asthma categories. On average, diaries from children indicate approximately 2¼ hours of afternoon time is spent outdoors, 80% of which is at a moderate or greater exertion level, regardless of their asthma status. Slightly less afternoon time is spent outdoors by adults (about 125-130 minutes) when compared with children whose asthma status is known, though more notable is the lowered percent of afternoon time adults perform moderate or greater exertion level activities (about 63%). As noted above regarding the reduced participation rate for adults whose asthma status is unknown, diaries for these adults also have about 20 fewer minutes of afternoon time spent outdoors compared with those persons whose asthma status is known.

Outdoor time and activity levels of respective cohorts from three independent asthma activity pattern studies were compared to CHAD diary days using similar metrics. To make the CHAD data compatible with the independent asthma study data, the entire diary day was evaluated, not just the afternoon hours, and all persons (not just outdoor habitué) were considered. In addition and where possible, the demographics of the independent study participants was used to select for the most representative CHAD diaries (e.g., person's age, sex, month- or day-of-year, etc.). Table 5G-3, Table 5G-4, and Table 5G-5 summarize the data reported from the three asthma activity pattern studies and the compatible results generated using CHAD and the indicated asthma status of the study persons.

Table 5G-3. Percent of waking hours spent outdoors at an elevated activity level. A comparison of CHAD with Shamoo et al. (1994) study asthmatics.

Study	**Shamoo (1994)**[1]		**CHAD**[2]					
Location	**Los Angeles**		**Any**					
Time of Year	Summer Months (5-9)	Winter Months (11-3)	Summer Months (May-Sep)			Winter Months (Nov-Mar)		
Asthmatic?	Yes	Yes	Yes	No	Unknown	Yes	No	Unknown
Person days	336	314	375	4,812	1,049	211	2,512	1,998
Mean age (min-max)	33 (18 – 50)		37.3 (18 – 50)	37.9 (18 – 50)	35.2 (18 – 50)	30.4 (18 – 50)	32.3 (18 – 50)	34.3 (18 – 50)
Exertion Level	**Percent of Asthmatic Waking Hours Spent Outdoors at Given Exertion Level**							
Low	8.5	6.0	4.6	5.6	5.4	2.0	2.1	1.4
Moderate	1.9	1.7	6.2	6.9	4.6	2.0	2.7	2.0
Strenuous	0.2	0.2	1.1	1.2	1.0	0.6	0.7	0.5

[1] Based on number of minutes performing three self-rated activity levels for three locations per hour (indoor, outdoor, in-vehicles) over seven days. Non-random sample of 49 subjects selected from voluntary clinical studies.
[2] Combination of random and non random selection studies, national and city-specific, as well as varying diary protocol (see REA Appendix 5B). The APEX CHAD file (n=41,474) was additionally screened for persons having no sleep reported (n=141). Randomly sampled METS values from each activity-specific distribution were assigned to each person's activities. Moderate and vigorous activity levels were selected based on activities having a METS value of 3 to <6 and ≥6, respectively.

When considering the three independent asthma studies, the amount of time spent outdoors at moderate activity level ranges from a low of approximately 2% to a high of about 11% of waking hours. The estimates of outdoor time associated with moderate activity level using a similarly constructed cohort of CHAD diaries fall within that range (i.e., between 2.0 and 7.2%). A small but consistent trend of fewer minutes spent outdoors at moderate exertion was observed for the CHAD asthmatic cohort when compared with the CHAD non-asthmatic group. The general range in percent of outdoor time associated with strenuous activities using the CHAD person days (0.4% to 7.6%) was greater when compared to asthmatic persons from the three independent studies (0.2% to 3.3%).

We recognize that there are a number of differences that exist among the three asthmatic studies used for comparison along with the use of CHAD diary data from either asthmatics, non-asthmatics, or unclassified asthmatics that could contribute to variation in the time spent outdoors at elevated activity levels. This would include: the diary/survey collection methods used, the classification of activities performed and associated activity levels, the number of study subjects, and sample selection methods. The particulars regarding how each of these were addressed across the various studies is wide ranging and could potentially influence the results generated here. However, based on the mostly comparable results observed in time spent outdoors at moderate or greater exertion activity levels, we judge the use of a CHAD diary regardless of a persons' asthma condition is reasonably justified based on the available data analyzed.

Table 5G-4. Percent of waking hours spent outdoors at an elevated activity level: a comparison of CHAD with EPRI (1992) study asthmatics.

Study	**EPRI (1992)**[1]	**CHAD**		
Location	Cincinnati	Any		
Time of Year	August (Wed-Sun)	July and August (Thu-Sat)		
Asthmatic?	Yes	Yes	No	Unknown
Person days	408	711	3,085	1,209
Mean age (min-max)	26 (1 – 78)	17.9 (1 – 77)	34.9 (1 – 78)	32.3 (1 – 78)
Exertion Level	**Percent of Asthmatic Waking Hours Spent Outdoors at Given Exertion Level**			
Moderate	11	5.0	7.2	7.2
Strenuous	3.3	7.6	3.7	2.6

[1] Hour diary questionnaire form used up to three activities per hour. Random digit dialing and multiplicity sampling were used. Three consecutive diary days were collected from 136 asthmatics, mostly Thu-Sat though some Wednesday and Sunday data were included.

Table 5G-5. Percent of waking hours spent outdoors at an elevated activity level: a comparison of CHAD with EPRI (1988) study asthmatics.

Study	**EPRI (1988)**[1]	**CHAD**		
Location	Los Angeles	Any		
Time of Year	April (Fri-Mon)	April and May (Fri-Mon)		
Asthmatic?	Yes	Yes	No	Unknown
Person days	156			
Mean age (min-max)	- (18 – 37)[2]	26.2 (18 – 39)	29.7 (18 – 40)	30.9 (18 – 40)
Exertion Level	**Percent of Asthmatic Waking Hours Spent Outdoors at Given Exertion Level**			
Moderate	7	2.8	4.5	4.4
Strenuous	2.4	0.4	0.6	1.2

[1] Hour diary questionnaire form used up to three activities per hour. Non-random sample of 26 mild/moderate, 26 moderate/severe asthmatics selected from voluntary clinical studies. Three consecutive diary days were collected per person (either Fri-Sun or Sat-Mon).
[2] General age range approximated from twenty-nine subjects noted by EPRI (1988) as from Linn et al. (1987).

5G-2 CHARACTERIZATION OF FACTORS INFLUENCING HIGH EXPOSURES

We investigated the factors that influence estimated exposures, with a focus on persons experiencing the highest daily maximum 8-hour exposures six selected study areas – Atlanta, Boston, Denver, Houston, Philadelphia, and Sacramento.[4] This analysis required the generation of detailed APEX output files having varying time intervals, that is, the daily, hourly, and minute-by-minute (or *events*) files. Given that the size of these time-series files is dependent on the number of persons simulated, we simulated 5,000 persons and restricted the analysis to a single year (2006) to make this evaluation tractable.[5] Both the base case (unadjusted or 'as is' recent air quality conditions) and ambient O_3 adjusted to just meet the existing standard (75 ppb-8hr) air quality scenarios were evaluated in each of six study areas. All APEX conditions (e.g., ME descriptions, AERs, MET data) were consistent with the 200,000 person APEX simulations that generated all of summary output discussed in the main body of this chapter.

We were interested in identifying the specific microenvironments and activities most important to O_3 exposure and evaluating their duration and particular times of the day people were engaged in them. Because ambient O_3 concentrations peak mainly during the afternoon hours, we focused our microenvironmental time expenditure analysis on the hours between 12 PM and 8 PM. For every person and day of the exposure simulation, we aggregated the time spent outdoors, indoors, near-roadways, and inside vehicles during these afternoon hours (i.e., the time of interest summed to 480 minutes per person day). Data from several APEX output files were then combined to generate a single daily file for each person containing a variety of personal attributes (e.g., age, sex), their daily maximum 8-hour ambient and exposure concentrations, and the aforementioned time expenditure metrics.

We performed an analysis of variance (ANOVA) using SAS PROC GLM (SAS, 2012) to determine the factors contributing the greatest to the observed variability in the dependent variable, i.e., each person's daily maximum 8-hour O_3 exposure concentrations. This analysis was distinct for four age-groups of interest (i.e., 5-18, 19-35, 36-64, ≥65 years of age). The final

---

[4] For the 1st draft O_3 REA, this analysis was performed for four study areas: Atlanta, Denver, Los Angeles, and Philadelphia. One important difference between the exposure simulations at that time compared with this 2nd draft O_3 REA was the air quality data input to APEX: ambient monitoring data were used for the 1st draft O_3 REA along with a quadratic approach for adjusting air quality to just meet the existing standard.

[5] We recognize that there is year-to-year variability in ambient O_3 concentrations and it is possible that fewer persons simulated could result in differences in exposures compared to large-scale multi-year model simulations. Based on a similar detailed evaluation performed for the Carbon Monoxide REA (US EPA, 2010), it is expected any differences that exist between exposures estimated in a large simulation versus that using a smaller subset of persons would be small and of limited importance to this particular evaluation.

statistical models[6] included a total of seven explanatory variables: the main effects of (1) daily maximum 8-hour ambient O_3, (2 to 4) afternoon time spent outdoors, near-roads, and inside vehicles,[7] and (5) physical activity index (PAI), while also including interaction effects from (6) afternoon time outdoors by daily maximum 8-hour ambient concentration, and (7) PAI by afternoon time outdoors. Two conditions were considered: all person days of the simulation, and only those days where a person's 8-hour maximum exposure concentration was ≥50 ppb.[8] Selected output from this ANOVA included parameter estimates for each variable, model R-square statistic (R^2), and Type III model sums of squares (SS3).[9]

Model fits, as indicated by an R^2 value, were reasonable across each of the study areas (Table 5G-6). The selected factors explain about 40-80% of the total variability in 8-hour daily maximum exposures. Model fits were best when using all person days of the simulation though results were similar for both air quality scenarios. When considering only those days where persons had 8-hour daily maximum O_3 exposures ≥50 ppb, consistently less variability in maximum exposure concentrations was explained by the factors included in each model, though overall model fits were acceptable. Furthermore, the most robust models tended to be those developed using either school-age children aged 5 to 18 or adults 19 to 35 years old (e.g., see Table 5G-7 for Atlanta model R^2 results stratified by age groups).

We evaluated the relative contribution each variable had on the total explained variability using the SS3 in each respective model.[10] As with the R^2 statistics generated above, the percent contribution results were separated into four exposure scenarios for each study area, with estimates for Boston illustrated in Figure 5G-3. When considering all person days of the simulation (top row), the daily maximum 8-hour ambient O_3 concentration variable contributes the greatest to the explained model variance, consistently estimated to be about 85% across all age groups and for either the base or existing standard air quality scenarios. The interaction of this variable with afternoon outdoor time contributes an additional 7-10% to the explained variance, indicating that both ambient concentration and time spent outdoors collectively contribute to 90% or more of the explained model variance when evaluating all (e.g., high, mid-range, and low level) daily maximum 8-hour O_3 exposure concentrations. The main effect of

---

[6] In this investigation, we also evaluated the influence of sex, work and home districts, meteorological zones, each with varying statistical significance, though overall adding little to explaining variability beyond the variables selected for the final ANOVA model.

[7] Including indoor afternoon time creates a strict linear dependence among these four variables and generates biased estimates, thus it was neither included nor needed in this analysis.

[8] This exposure concentration was selected due to the reduced sample size needed for these simulations (i.e., 5,000 total persons), an issue of increasing importance when selecting for persons with the highest exposures.

[9] In each of the ANOVA models constructed, type II = type III = type IV sums of squares.

[10] Type III sums of squares (SS3) for a given effect are adjusted for all other effects evaluated in the model, regardless of whether they contain the given effect or not. Thus, the SS3 for each variable represents the individual effect sums of squares that sum to the total effect sums of squares (or the total model explained variance).

outdoor time contributed < 1% to the explained variance under these conditions as did contributions from the other included variables, except for time spent near-roads (about a 5% contribution). These results suggest that when considering the Boston exposure study groups broadly, the daily maximum 8-hour ambient O_3 concentration is the most important driver in estimating population-based O_3 exposures, nearly regardless of specific microenvironmental locations where exposure might occur.

Table 5G-6. Range of R^2 fit statistics for ANOVA models used to evaluate daily maximum 8-hour O_3 exposure concentrations stratified by study area, air quality scenario, and exposure level.

Study Area (O_3 season)	Base Case Model R^2 (sample sizes)		Existing Standard Model R^2 (sample sizes)	
	All Person Days	Person days with 8-hr daily max exposure ≥ 50 ppb	All Person Days	Person days with 8-hr daily max exposure ≥ 50 ppb
Atlanta (245 days)	0.71 – 0.78 (1,225,000)	0.62 – 0.70 (43,646)	0.67 – 0.74 (1,225,000)	0.61 – 0.70 (18,758)
Boston (183 days)	0.64 – 0.70 (915,000)	0.44 – 0.62 (23,496)	0.58 – 0.65 (915,000)	0.41 – 0.59 (16,184)
Denver (214 days)	0.61 – 0.68 (1,070,000)	0.49 – 0.62 (24,850)	0.55 – 0.62 (1,070,000)	0.47 – 0.61 (18,487)
Houston (365 days)	0.75 – 0.80 (1,825,000)	0.57 – 0.67 (17,779)	0.65 – 0.71 (1,825,000)	0.38 – 0.65 (7,322)
Philadelphia (214 days)	0.66 – 0.73 (1,070,000)	0.51 – 0.67 (39,561)	0.57 – 0.64 (1,070,000)	0.48 – 0.66 (16,841)
Sacramento (365 days)	0.75 – 0.81 (1,825,000)	0.50 – 0.71 (19,734)	0.68 – 0.74 (1,825,000)	0.48 – 0.73 (6,503)

Table 5G-7. Range of R^2 fit statistics for ANOVA models used to evaluate daily maximum 8-hour O_3 exposure concentrations in Los Angeles stratified by age group, air quality scenario, and exposure level.

Study Area	Age Group (years)	Base Case Model R^2		Existing Standard Model R^2	
		All Person Days	Person days with 8-hr daily max exposure ≥ 50 ppb	All Person Days	Person days with 8-hr daily max exposure ≥ 50 ppb
Atlanta	5-18	0.78	0.64	0.74	0.62
	19-35	0.71	0.70	0.67	0.70
	36-64	0.73	0.64	0.68	0.64
	≥65	0.74	0.62	0.70	0.61

When considering only person days having daily maximum 8-hour O_3 exposures ≥ 50 ppb and for either air quality scenario in Boston, collectively the main effects of ambient concentrations and outdoor time combined with their interaction similarly contribute to approximately 85% of the total explained variance (Figure 5G-3, bottom row). However, the main effect of the 8-hour daily maximum ambient O_3 concentration variable has a sharply lower contribution (generally about 5-15%) along with greater contribution from the main effects variable outdoor time (15-25% contribution) and its interaction with the ambient concentration variable (40-60%). These results suggest that for highly exposed persons in Boston, the most important influential factors are time spent outdoors corresponding with high daily maximum 8-hour ambient O_3 concentrations.

Results for Atlanta (Figure 5G-4), were generally similar to Boston with notable differences discussed here.[11] The contribution of the maximum 8-hour ambient O_3 concentration variable to the total explained variance (about 40-50%) was less than that observed in Boston when considering all person days (Figure 5G-3 and Figure 5G-4, top rows), while the contribution from the outdoor time/ambient O_3 interaction variable was greater in Atlanta (about 20-40% versus 10% in Boston).

This observed dissimilarity in the contribution by ambient concentrations and afternoon outdoor time may be driven by the A/C prevalence rates and AER distributions used for each study area.[12] Boston has lower A/C prevalence though overall higher AERs (even when considering mechanical ventilation), thus a greater contribution to exposure is expected from ambient concentrations by infiltrating to indoor microenvironments and hence, reflected in the strong main effects for the 8-hour daily maximum ambient O_3 concentration variable in Boston. Afternoon time spent near Atlanta roads was estimated to contribute to about 20-30% of the total explained variance when considering all person days and exposures, a value greater than that estimated for Boston (generally about 5%) again possibly reflecting an increased importance of this outdoor microenvironment in Atlanta (and Houston, Sacramento, not shown) relative to that in Boston (and Philadelphia, not shown).

---

[11] The discussion regarding the relative contribution of the variables to the total explained model variance also extends to the other four study areas, whereas results for Philadelphia were generally similar to Boston, Sacramento and Houston were similar to Atlanta, and Denver generally fell somewhere in between these extremes presented.

[12] A/C prevalence is highest in Houston (99%), Atlanta (98%), Philadelphia (95%), and Sacramento (93%) compared to Boston (86%) and Denver (67%). Boston and Philadelphia used the same (and highest) AER distributions; Sacramento, Houston, and Atlanta used separate but similar (and lower) AER distributions, while Denver AER distributions fall somewhere in between these two extremes (see Appendix B, Tables 5B-3 to 5B-5).

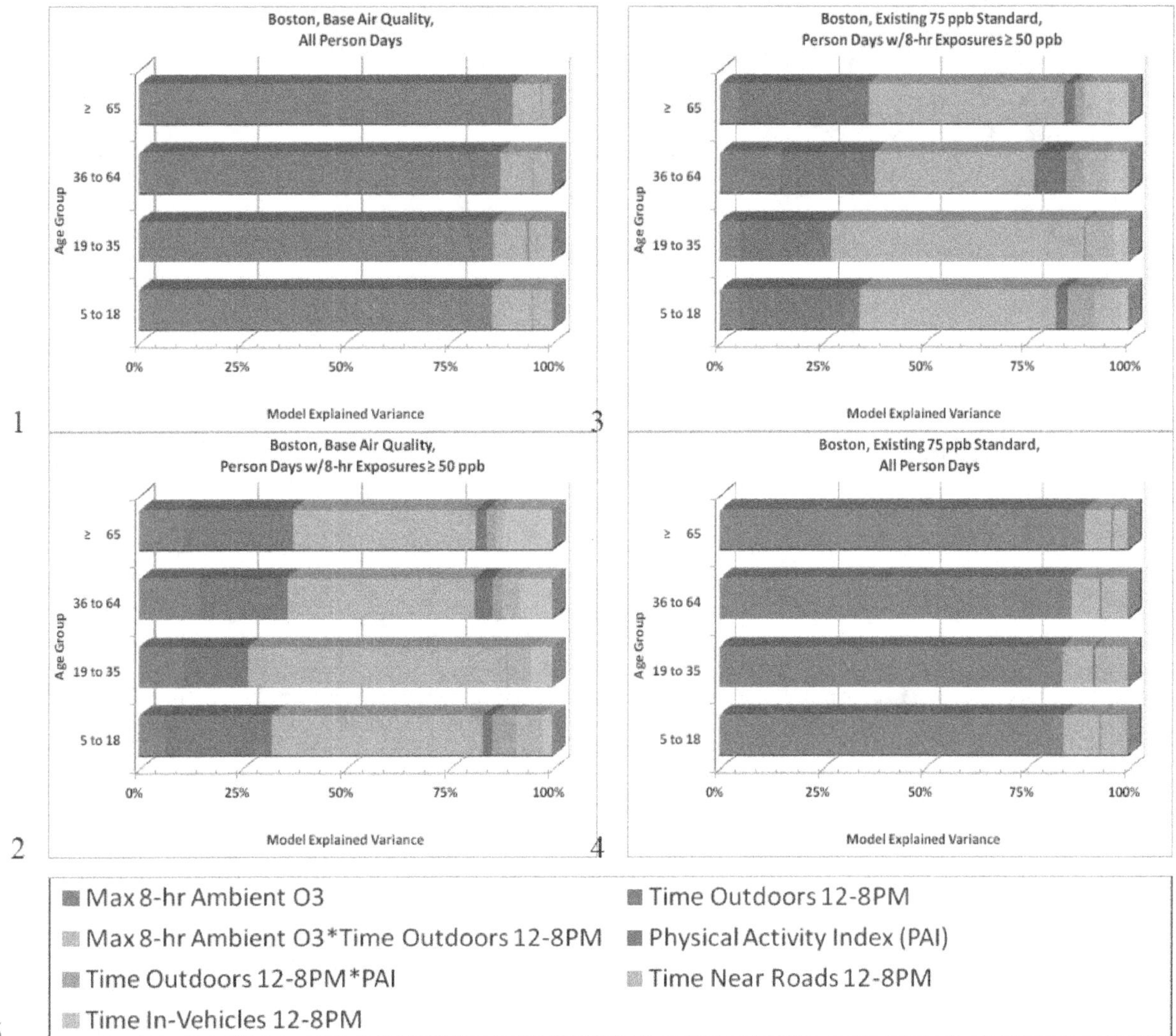

Figure 5G-3. Contribution of influential factors to daily maximum 8-hour ozone exposures using base air quality (left), air quality adjusted to just meet the existing standard (right), considering all person days (top) and those days where daily maximum 8-hour exposure exceeded 50 ppb (bottom) in Boston.

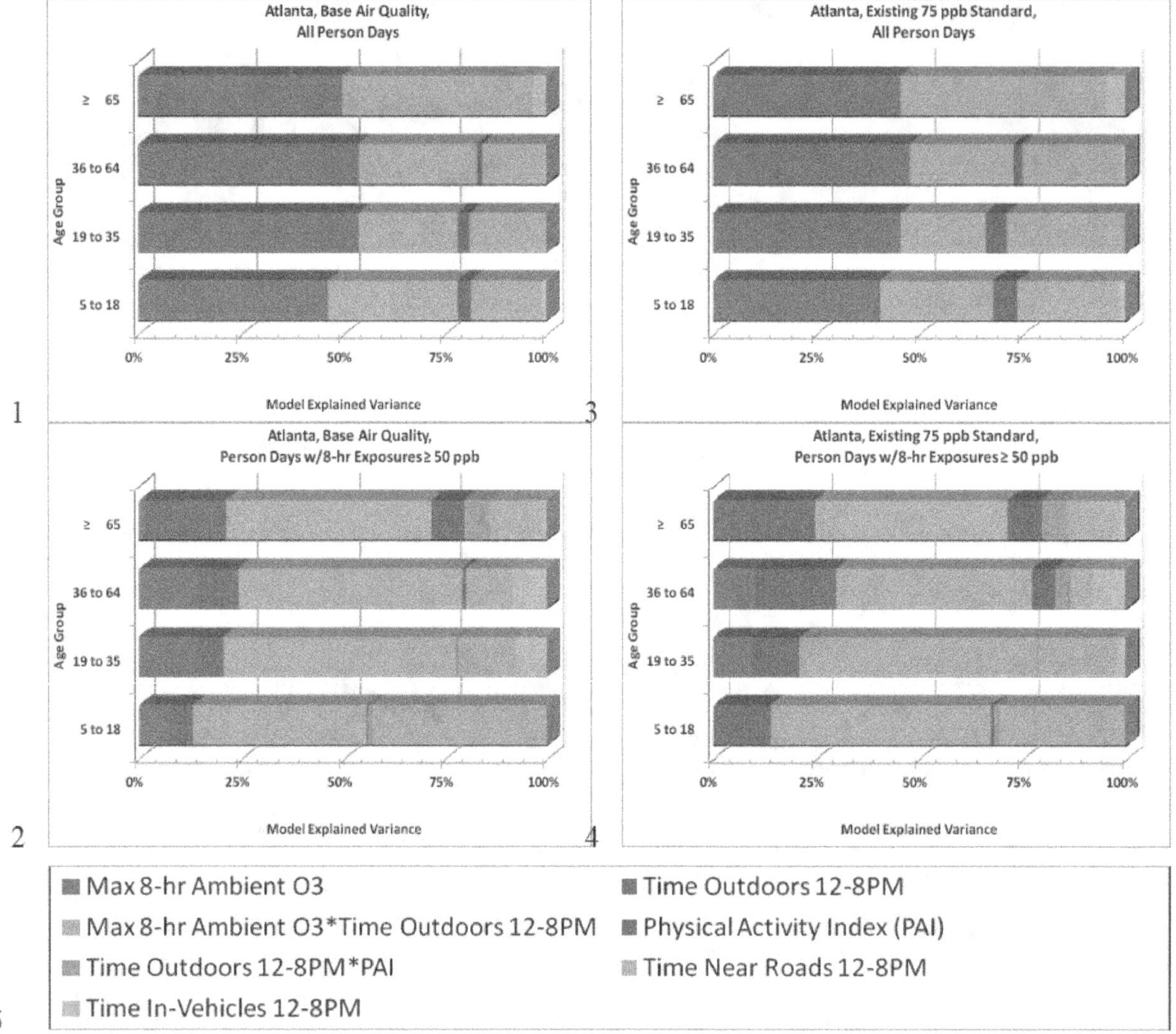

Figure 5G-4. Contribution of influential factors to daily maximum 8-hour ozone exposures using base air quality (left), air quality adjusted to just meet the existing standard (right), considering all person days (top) and those days where daily maximum 8-hour exposure exceeded 50 ppb (bottom) in Atlanta.

Because afternoon outdoor time expenditure and daily maximum 8-hour ambient O_3 concentrations are an important determinant for high O_3 exposures regardless of air quality scenario considered, we compared the distributions of these two variables using person days where daily maximum 8-hour O_3 exposures were either below or above 50 ppb. Figure 5G-5 presents this comparison for Boston[13] school-age children (ages 5 to 18) and adults (ages 19 to 35) and considering 2006 base air quality and air quality adjusted to just meet the existing O_3 8-hour standard. For school-age children that did not experience a daily maximum 8-hour

[13] The overall features of the outdoor time and ambient concentration distributions illustrated by simulated persons in Boston are similar to each of the other study areas (data not shown).

exposure at or above 50 ppb (e.g., top left panel, base air quality), over half of them did not spend afternoon time spent outdoors, while just under 20% of them spent at least two hours of their afternoon time spent outdoors, with fewer than 5% spending more than four hours of their afternoon time outdoors. In addition, nearly 70% would have their daily maximum 8-hour ambient concentrations below 50 ppb (please note, ambient is not exposure).

Not surprisingly, the distributions for both the outdoor time and ambient concentration variables are shifted to the right of the figure for school-age children's person days where daily maximum 8-hour exposures $\geq$ 50 ppb (e.g., Figure 5G-1, top left panel, base air quality), as for more than half of the days, highly exposed simulated individuals spend about 250 minutes outdoors during the afternoon hours along with experiencing daily maximum 8-hour ambient O_3 concentrations $\geq$ 75 ppb.

By design, when air quality is simulated to just meet the existing standard (e.g., Figure 5G-5, bottom left panel), upper percentile ambient concentrations are reduced compared to those comprising the base air quality such that the majority of ambient concentrations fall well below the existing standard level of 75 ppb. Given so few occurrences of very high 8-hour ambient O_3 concentrations for this air quality scenario, only those school-age children having a majority of their time spent outdoors experienced the highest daily maximum 8-hour O_3 exposure concentrations (Figure 5G-5, bottom left panel, right-most solid line). For additional completeness, we note the time and concentration distributions for adult person days (Figure 5G-5, right column) were similar with that estimated for school-age children.

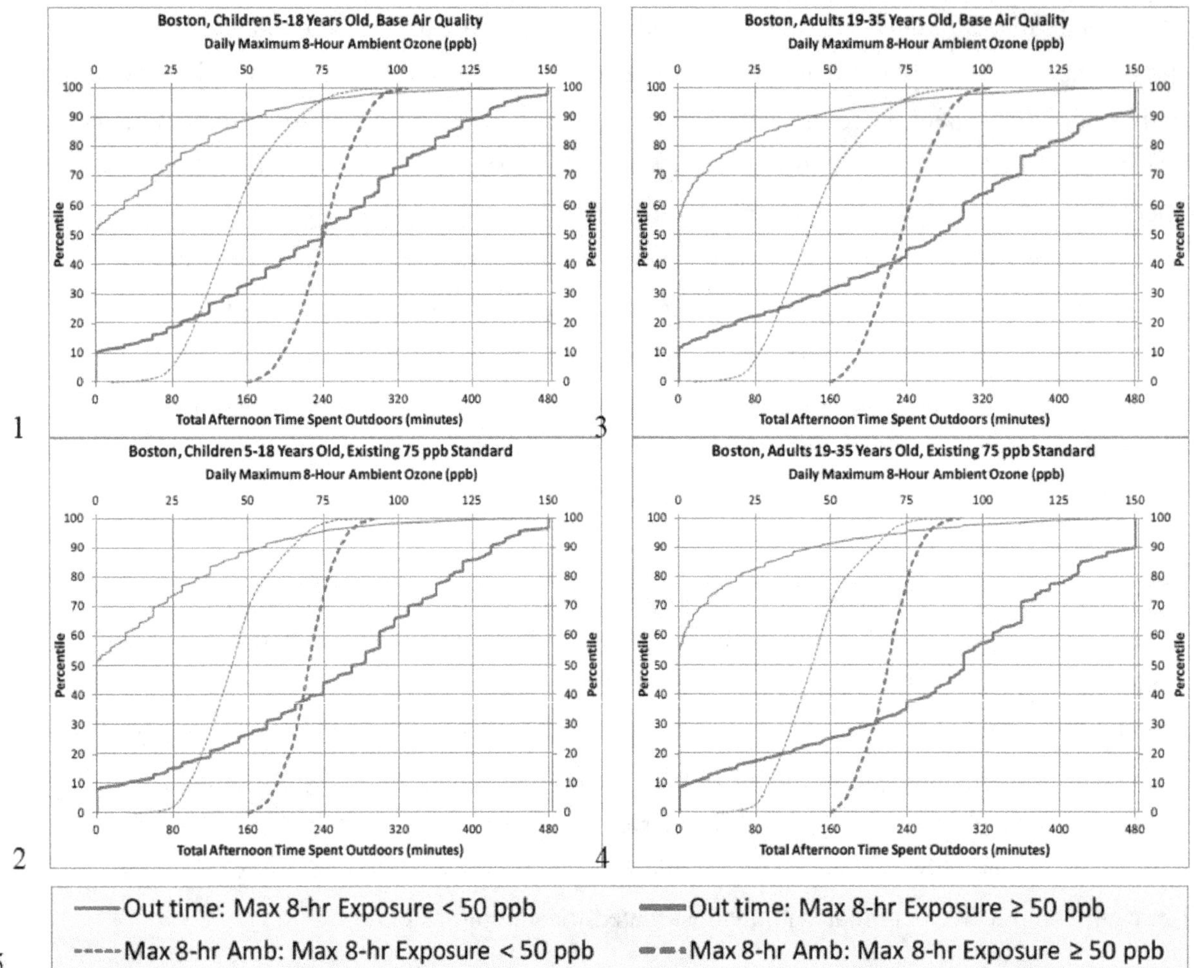

Figure 5G-5. Distributions of afternoon outdoor time expenditure and daily maximum 8-hour ambient O_3 concentrations for simulated Boston school-age children (ages 5 to 18) (left) and adults (ages 19 to 35) (right) using base air quality (top) and concentrations adjusted to just meet the existing standard (bottom) for person days having daily maximum 8-hour exposures either below or above 50 ppb.

By definition, any 8-hour average exposure is time-averaged across all microenvironmental concentrations; thus several different microenvironments may contribute to each person's daily maximum level. Understandably based on the above analyses, the outdoor microenvironment is most important for persons experiencing the highest O_3 exposures, but we are also interested in the percentage of time expenditure spent among detailed indoor, outdoor, and vehicular locations people may inhabit during the afternoon. We summed the afternoon time expended for highly exposed persons, considering a total of 12 microenvironments (i.e., 3 indoor, 5 outdoor, 2 near road, and 2 vehicular). As an example, Figure 5G-6 presents this microenvironmental information for Boston school-age children (Figure 5G-6, top left panel) and adults (Figure 5G-6, top right panel) for persons experiencing daily maximum 8-hour average O_3

exposures ≥ 50 ppb and considering base air quality conditions. On average, approximately 50% of school-age children's total afternoon time is spent outdoors, of which half of this portion is spent outdoors at home, with parks and other non-residential outdoor locations comprising the remaining portion. Approximately 45% of the school-age children's afternoon time on high exposure days is spent indoors, while just less than 10% of afternoon time is spent near-roads or inside motor vehicles. Afternoon microenvironmental time expenditure for highly exposed adults (ages 19-35) in Boston was generally similar with that estimated for school-age children (Figure 5G-6, top right panel).

A person's activity level plays an important role in estimating the risk of adverse health responses to inhaled ozone. As such, we evaluated the activities performed by highly exposed individuals while they spent time outdoors during the afternoon hours. Note there are over 100 specific activity codes used in CHAD/APEX, though not all of these will be used in an exposure modeling simulation depending on the particular diaries that are selected to represent the simulated study group. We summed the time spent in each specific activity across all highly exposed persons when spending afternoon time outdoors, ranked the activity sums, and identified the top eleven activities performed. An aggregate of any remaining less often performed activities was generated to complete this analysis of activity time expenditure.

Figure 5G-6 shows results for Boston school-age children (bottom left panel), indicating that greater than half of the time highly exposed children spent outdoors specifically involves performing a moderate or greater exertion level activity, such as a sporting activity. The same type of analysis was done for highly exposed adults in Boston (Figure 5G-6, bottom right panel), whereas about 30% of the outdoor time expenditure was spent engaged in a paid work related activity (though not necessarily a high exertion level activity), about 15% of the time was spent playing sports or other moderate or greater exertion level activity, with much of the remaining specific activities associated with low exertion level (e.g., eating, sitting, visiting) or other less frequently performed activities of variable exertion level. These results support our identification of school-age children as an important exposure group, largely a result of the combined outdoor time expenditure along with concomitantly performing moderate or high exertion level activities. It is worth noting, one important group not directly assessed in the general population-based exposure modeling and remaining as a limitation to the main body REA results is outdoor workers. This exposure study group was explicitly modeled using a scenario based approach and summarized in section 5G-4.2.

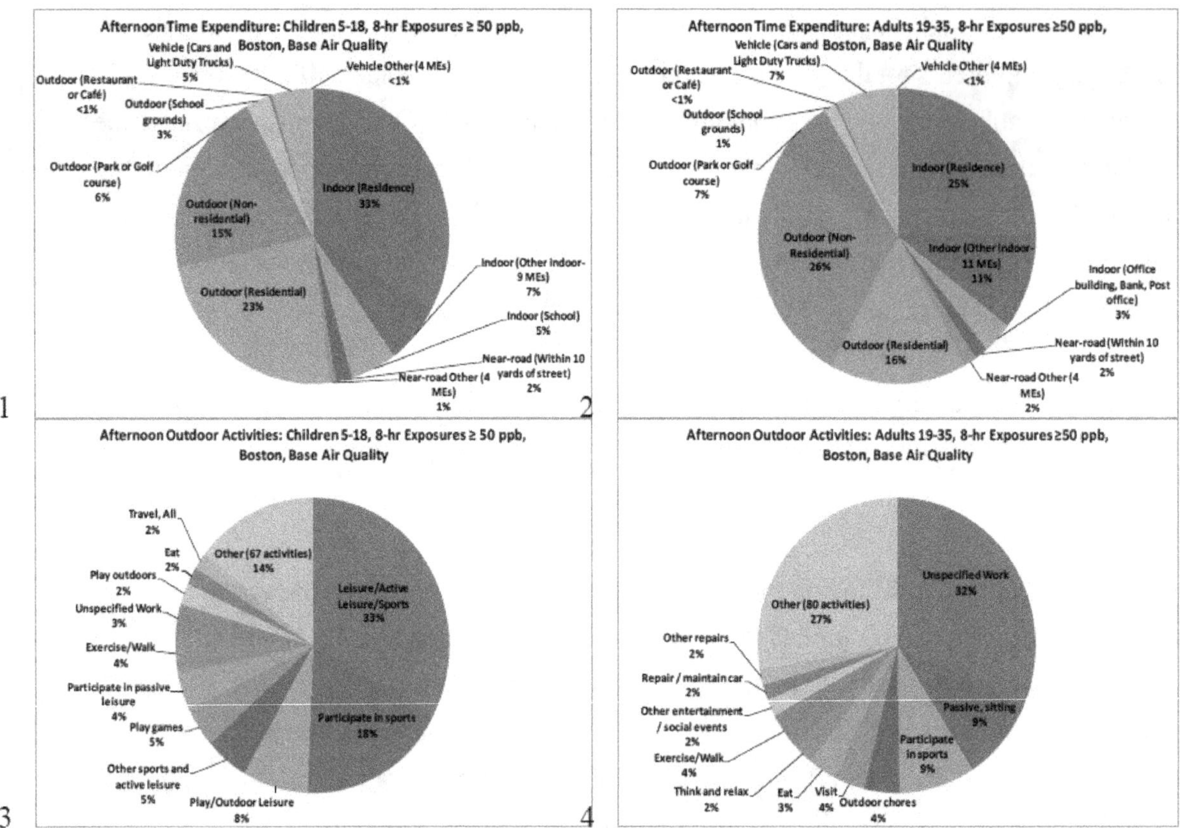

Figure 5G-6. Afternoon microenvironmental time (top) and activities performed during afternoon time outdoors (bottom) for school-age children (left) and adults (right) experiencing 8-hour daily maximum O_3 exposures ≥ 50 ppb, Boston base air quality, 2006.

5G-3 ANALYSIS OF APEX SIMULATED LONGITUDINAL ACTIVITY PATTERNS IN SCHOOL-AGE CHILDREN

We evaluated the APEX approach used for linking together cross-sectional activity pattern diaries to generate longitudinal profiles for our simulated individuals. Of particular interest were how well variability in outdoor participation rate and the amount of time expended were represented in our population-based exposure simulations. Our goal in developing the most reasonable longitudinal profiles is to capture expected, important features of population activity patterns, i.e., there is correlation within an individual's day-to-day activity patterns (though not exactly repeated nor entirely random) and variability across the modeled study group in day-to-day activity patterns (not every simulated individual in the study group does the same thing on the same day). As a reminder, the longitudinal approach is probabilistic, though guided by key variables influencing activity patterns (i.e., age, sex, day-of-week, commute time [employed person only], daily maximum temperature, and in our application, considers within and between variability in outdoor time expenditure). See REA Appendix 5B, section 5B-4.2.

We used the same event-level output data that was generated for the high exposure analysis above (section 5G-2), which includes the same six study areas – Atlanta, Boston, Denver, Houston, Philadelphia, and Sacramento – and focused the analysis on school-age children (ages 5-18). Total time spent outdoors during the afternoon hours (12 PM-8 PM) was calculated for each person-day of the simulation in each study area. Results of this analysis are presented in five individual plots for each study area, though combined in a multi-panel display, one per three study areas, designed to fit on a single page. The five individual plots generated for each study area are described as follows.

1) **Cumulative distribution summarizing each child's median time spent outdoors across an O_3 season:**[14] We first selected simulated individuals within the age group of interest (5-18) and then stratified these persons by sex. The median value (50th percentile) of afternoon time spent outdoors was determined for each simulated individual using all days in their study area's O_3 season. This data set, comprised of individual median values (in minutes) was ranked and plotted, stratified by sex.
2) **Cumulative distribution summarizing each child's afternoon outdoor participation (at least one minute/day) across an O_3 season:** We subset school-age children from the data set and stratified by sex. A categorical variable was developed by assigning a numeric value of 1 when an individual spent at least one minute during the afternoon hours outdoors on that given day. Then for each simulated individual, outdoor participation was determined by summing this variable across the simulation period (i.e., the number of days per O_3 season the persons spent at least one minute outdoor during the afternoon hours) and dividing by the total number of days in that study area's O_3 season. This data set, comprised of individual participation values (provided as a percent) was ranked and plotted, stratified by sex.
3) **Cumulative distribution summarizing each child's afternoon outdoor participation (at least two hours/day) across an O_3 season:** Calculated and presented in the same manner as #2 above, only that the categorical variable was assigned a numeric value of 1 if the simulated individual spent at least two hours outdoor during the afternoon hours on that given day.
4) **Daily time series of afternoon outdoor participation (at least two hours/day) by study group across O_3 season:** Using the categorical variable determined in #3, we calculated the study group's outdoor participation for every day of each study area's O_3 season. Data similar in fashion to our earlier analyses of outdoor time expenditure (e.g., section 5G-2), only differing in that presented are day-to-day variability in

[14] The number of days in an O_3 season varies across the six study areas. See Table 5G-6.

outdoor participation for the simulated study group across each study area's O_3 season.

5) **Daily time series of the number of study-specific CHAD diaries used across O_3 season:** The APEX daily file output can include the identity of the specific CHAD diary used to simulate every individual's daily activity. For every day simulated, we summed the number of CHAD diaries used to model the school age children's activity patterns for each day, though stratified by CHAD study identifier (e.g., see REA Appendix 5B, Table 5B-1). Plotted is the day-to-day variability in particular CHAD study diaries used across each study area's O_3 season.

The results of this longitudinal activity pattern analysis in given in Figure 5G-7 and Figure 5G-8. To begin, a few generalities regarding the features of each plot and where consistency is exhibited across study areas. In general, simulated female school-age children tend to spend less afternoon time outdoors than their male counterparts, consistent of course with expectations and the data used to develop these simulated profiles (Graham and McCurdy, 2004). About half of the simulated study group spends about half of their days with no afternoon time spent outdoors across their study area's ozone season, while about 10-20% spent just over 2 hours afternoon time spent outdoors for half of their days (top row of Figure 5G-7 and Figure 5G-8). Nearly every simulated individual participates in at least one afternoon outdoor activity across the O_3 season and exhibits a mostly monotonic relationship (2nd row of Figure 5G-7 and Figure 5G-8), though when considering durations of 2 hours or more, longitudinal outdoor participation drops dramatically (an non-linearly) for most persons comprising the study group (3rd row of Figure 5G-7 and Figure 5G-8). For more than half of simulated school age children, only approximately 1 out of every 5 days was spent outdoors during the afternoon hours for at least two hours, while a maximum value (around 3 of 5 days) was simulated for only about 10% or fewer children comprising the study group. Study group participation in at least two hours of afternoon time outdoors day-to-day ranges from about 5-40% across each study area's O_3 season (4th row of Figure 5G-7 and Figure 5G-8), though not surprisingly highest during typical summer months (June through September). And finally, the majority of CHAD diaries that are used come from the recently conducted ISR and OAB studies (bottom row of Figure 5G-7 and Figure 5G-8). This is also expected given that these two studies were designed to collect children's activity patterns and contribute to the bulk of the children's diaries in CHAD. Worthy of note is the shift in the source of diaries used across the calendar year; the contribution from the OAB study increases during the summer months while that of the ISR wanes. This is because of the days/seasons of the year the original study data were collected; most of the ISR data were collected during non-summer months while the OAB study was conducted during peak O_3

concentration days. While the use of these different studies in varying numbers over the simulation period likely drives some of the observed variability in the outdoor participation, at this time (and previously) staff treat the CHAD study data equally without bias, following our initial screen of the CHAD master data base that selected for the most complete data available.

Variability in the five longitudinal display plots across the six study areas is evident, though to a much smaller degree than that observed when considering the magnitude of the within study area variability. While different lengths of each study areas' O_3 season may negate direct comparability of the distributions presented, it is reasonable to conclude that simulated school-age children in the Atlanta, Houston, and Sacramento study areas had slightly overall greater participation in outdoor activities and spent more time outdoors than counterparts in Boston, Denver, and Philadelphia. That said, when considering the daily time series of participation rates, on many summer days Philadelphia and Sacramento school-age participation rates for males are as great or greater than participation rates observed most other study areas, including study areas likely having considerably warmer summer temperatures (Figure 5G-7 and Figure 5G-8, 4th row). It is possible that for study areas such as Atlanta, summertime maximum daily temperatures exceed the range affording outdoor comfort, yielding slightly lower rates of participation in outdoor activities.

Overall, the simulated longitudinal profiles indicate the method for linking together cross-sectional diaries generates a diverse mixture of persons having variable, though expected, activity patterns: a small fraction of the simulated population spend a limited amount of afternoon time outdoors and occurring at a low frequency across an O_3 season, a small fraction consistently spends a greater amount (> 2 hours) of time outdoors and occurring at greater frequency (e.g., 4/5 days per week), while the remaining simulated individuals fall somewhere in between these two lower and upper bounds regarding participation and total time. While we are not aware of a population database available to compare with these simulated results, we are comfortable with the method performance in representing the intended variability in longitudinal activity patterns

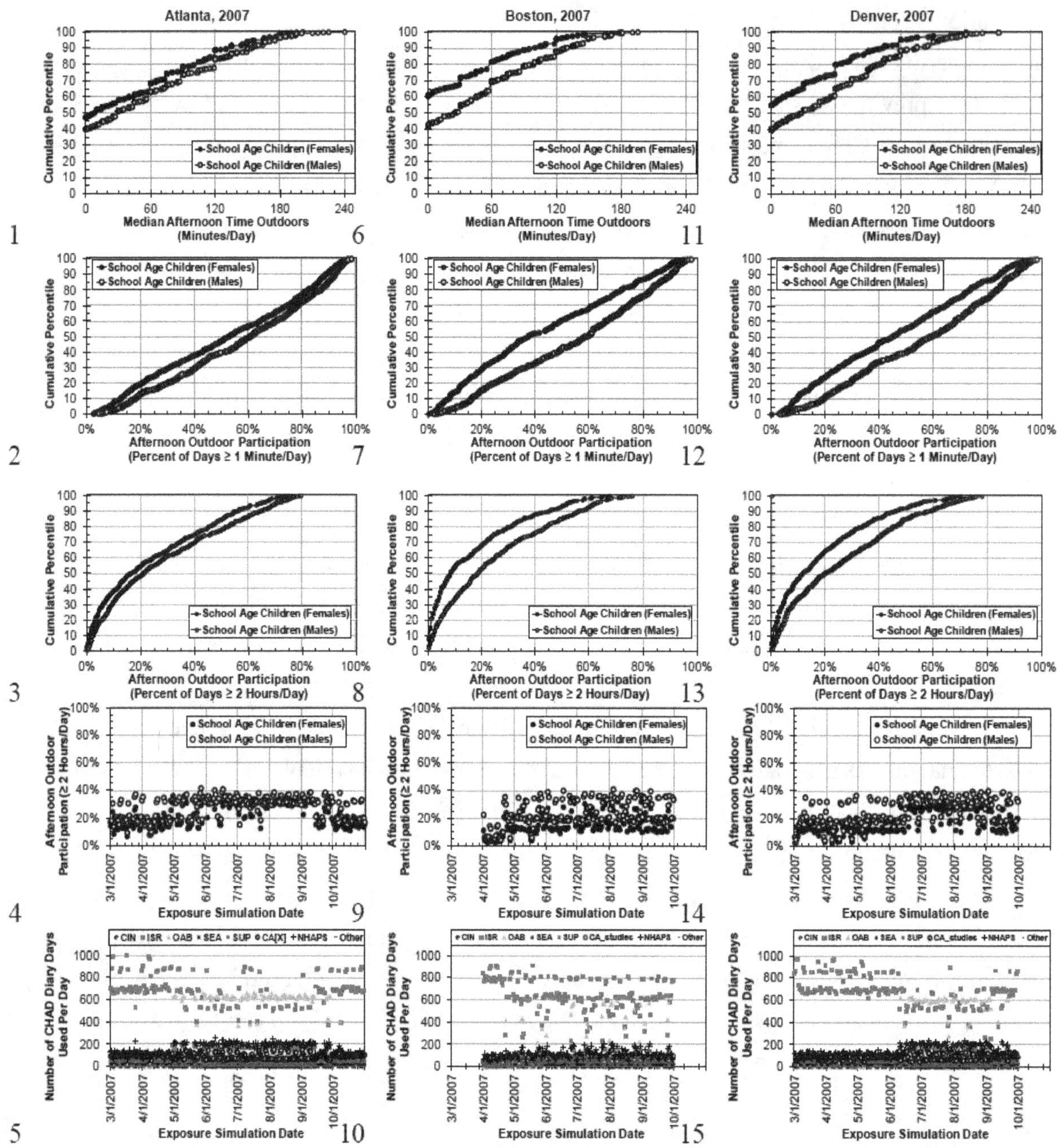

Figure 5G-7. Cumulative distribution of median time spent outdoors (top row), afternoon outdoor participation ≥ 1 minute/day (2nd row), and afternoon outdoor participation ≥ 2 hours/day (3rd row) for male and female school-age children in Atlanta (left column), Boston (middle column) and Denver (right column) study areas. Percent of school-age children with ≥ 2 hours outdoors during afternoon hours (4th row) and the number of particular CHAD study diary days used (bottom row) for each exposure simulation day.

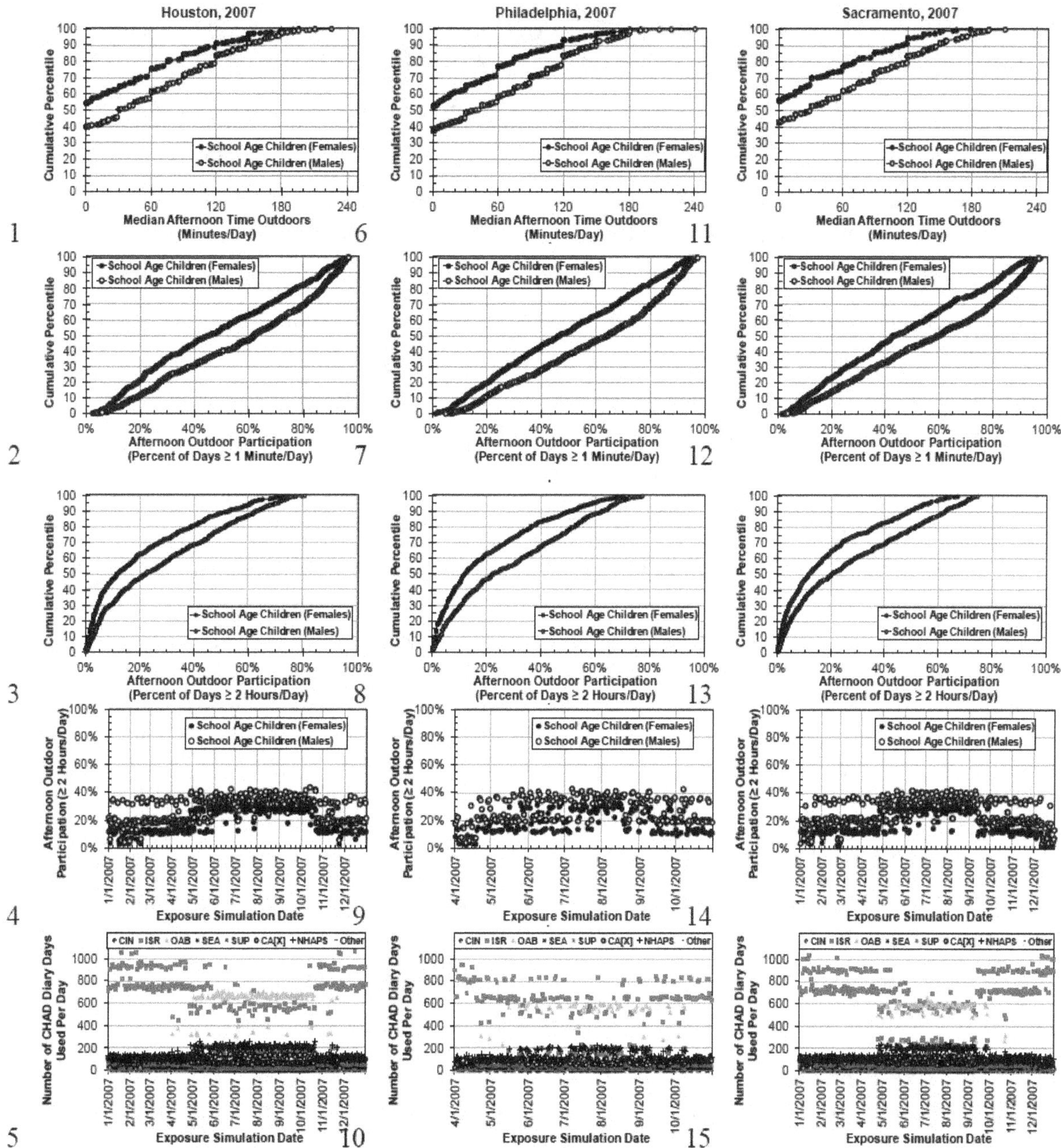

Figure 5G-8. Cumulative distribution of median time spent outdoors (top row), afternoon outdoor participation ≥ 1 minute/day (2nd row), and afternoon outdoor participation ≥ 2 hours/day (3rd row) for male and female school-age children in Houston (left column), Philadelphia (middle column) and Sacramento (right column) study areas. Percent of school-age children with ≥ 2 hours outdoors during afternoon hours (4th row) and the number of particular CHAD study diary days used (bottom row) for each exposure simulation day.

5G-4 EXPOSURE RESULTS FOR ADDITIONAL AT-RISK POPULATIONS AND LIFESTAGES, EXPOSURE SCENARIOS, AND AIR QUALITY INPUT DATA USED

This section includes results for three additional simulations designed to complement exposures estimated using our general population-based modeling approach presented in the main body of the REA. These simulations include (1) exposures estimated for school-aged children during summer months only (section 5G-4.1), (2) adult outdoor worker exposures (section 5G-4.2), and (3) exposures to school-age children and asthmatic school-age children assuming a portion of these study groups exhibit averting behavior in response to high O_3 concentration days (section 5G-4.3).

5G-4.1 Exposure Estimated For All School-Age Children During Summer Months, Neither Attending School nor Performing Paid Work

A targeted simulation was performed for the Detroit study area during the months of June through August 2007 to simulate summertime exposures by assuming all children were on a traditional calendar year summer vacation. To do this, a subset of the CHAD diaries used by APEX was created by including only those persons that did not have any time spent while at school or time performing paid work. Even though the school children age range in our exposure simulation is 5-18 years old, to maximize the number of diaries available for use by APEX we expanded the CHAD diary selection to include children from 4-19 years old. In considering these diary selection criteria, the resulting time location activity pattern data set input to APEX had a total of 10,226 diaries having 379,524 event entries. All simulation conditions were set identically to those set for the main REA exposure simulations in Detroit, though 75,000 children were explicitly simulated here for this targeted analysis. Four air quality scenarios were considered: just meeting the existing 8-hour standard of 75 ppb, and at alternative levels of 70, 65, and 60 ppb. The exposure results from these targeted simulations were compared with identical APEX simulations run using all available CHAD diaries during the same summer months (i.e., diary days that include locations visited and activities performed from persons reporting either school time or paid work).

Figure 5G-9 contains the exposure results for this simulation (*"No School/Work Diaries"*) and for a nearly identical simulation that differed only in that is used all CHAD diaries (*"all CHAD Diaries"*). When restricting the CHAD diary pool to include only those diaries having no time spent at school or performing paid work activities, there is about 1/3 or 33% increase in the number of children at or above each of the selected benchmark levels, a relationship also consistent when considering multiple exposures over the simulation period.

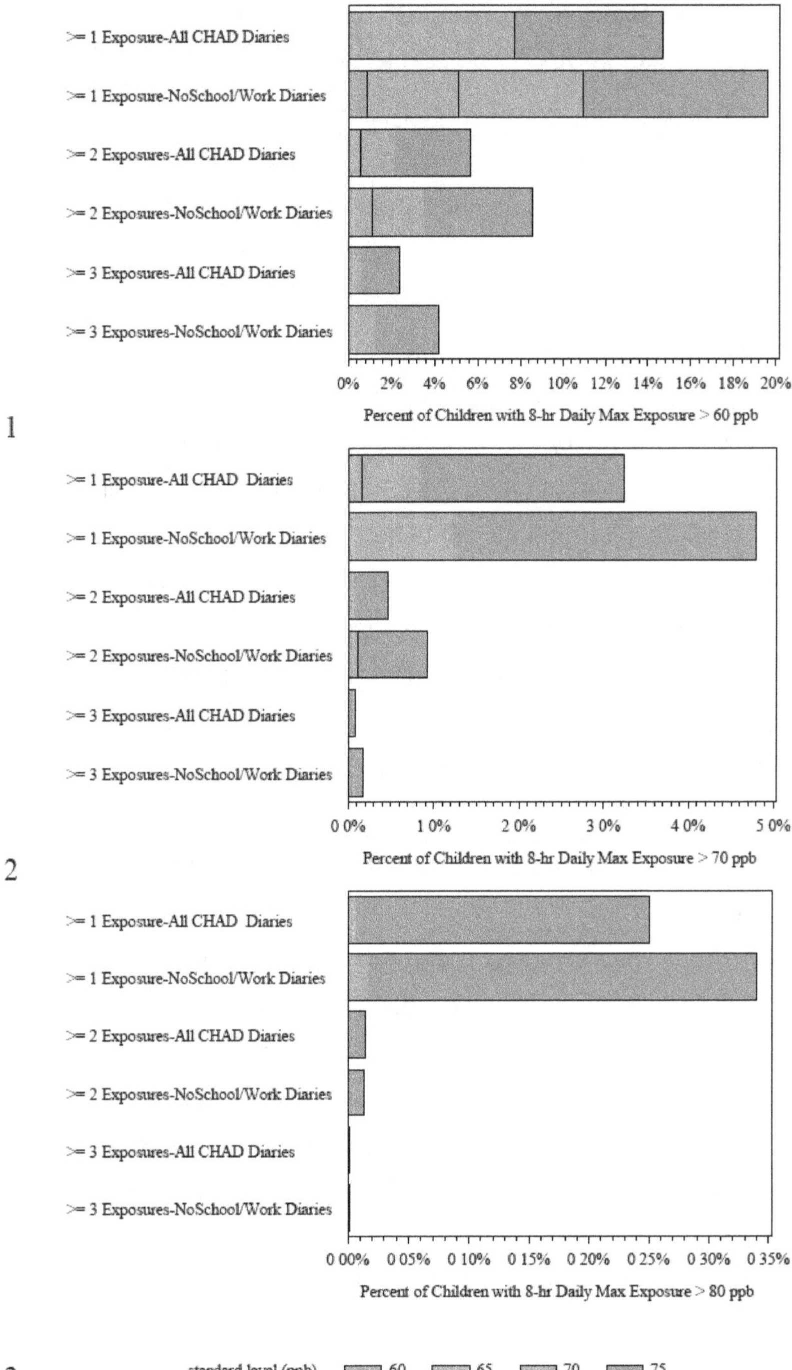

Figure 5G-9. Comparison of the percent of all school-age children having daily maximum O_3 concentration at or above 60 ppb-8hr (top), 70 ppb-8hr (middle), or 80 ppb-8hr (bottom) during June, July, and August in Detroit 2007: using any available CHAD diary ("All CHAD Diaries") or using CHAD diaries having no time spent in school or performing paid work ("No School/Work Diaries").

5G-4.2 Exposures Estimated For Adult Outdoor Workers During Summer Months

A targeted APEX simulation was performed for the Atlanta study area to simulate summertime exposures for two hypothetical adult outdoor worker study groups (ages 19-35 and ages 35-54), using 2006 air quality just meeting the existing O_3 standard. To do this, both the daily and longitudinal activity patterns used by APEX needed to best reflect patterns expected for adult outdoor workers (e.g., a standardized work schedule during weekdays) while also capturing variability in those patterns across various occupation types and the overall simulated adult outdoor worker study group. The development of reasonable time location activity pattern data to be input to APEX was a complex undertaking, attempting to account for a number of influential factors such as the distribution of adult outdoor workers, their varying occupation types, the probabilities associated with performing outdoor work, the linking of this information with the existing CHAD diaries and APEX METS distributions, all to be done within the existing APEX model framework and capabilities.

First, the complete distribution of all employed persons' occupations was estimated using data provided by the U.S. Bureau of Labor and Statistics (US BLS, 2012b).[15] The information of interest was obtained from the 2010 National Employment Matrix, data originally developed from the Occupational Employment Statistics (OES) survey and based on the 2010/2000 Standard Occupational Classification (SOC) system. The three variables retained for our purposes here included the SOC occupation titles and codes and the 2010 estimated number of persons employed, covering over 750 occupation titles.

Second, the identification of occupations where workers spend time outdoors for at least one or more days per week was determined using data from the Occupational Information Network (O*NET).[16] The O*NET was developed by a partnership of public and private organizations[17] via sponsorship by the US Department of Labor/Employment and Training Administration (USDOL/ETA). A wealth of information is provided by the O*NET regarding specific occupations including human interaction processes (e.g., amount of public speaking in a particular job, the likelihood of encountering angry people), physical work conditions (e.g., approximate time spent standing, whether exposed to radiation), and structural job characteristics (e.g., the degree of job automation, freedom to make decisions).

An advanced search of O*NET was performed using the web database. We first isolated the data of interest here by work context and selected physical work conditions. Data tables for two survey question responses were downloaded: the first consisted of persons responding to the

---

[15] US employment data by SOC codes were obtained from: http://www.bls.gov/emp/#tables: Table 1.2 Employment by occupation, 2010 and projected 2020.
[16] Additional information is available at http://www.onetonline.org.
[17] Current O*NET partners include Research Triangle Institute (RTI), the Human Resources Research Organization (HumRRO), North Carolina State University (NCSU), MCNC, and Maher & Maher.

question, "how often does this job require working outdoors, exposed to all weather conditions?" and the second "how often does this job require working outdoors, under cover (e.g., structure with roof but no walls)?". The tables contain the responses to these two questions, stratified by the occupation names/codes, and rated using a context score ranging from 0 to 100. According to the context scale provided by O*NET, occupations with a score of 75 were characterized as having at least one day per week outdoors, while a score of 100 indicated that every day work was performed outdoors by workers in that particular occupation, thus the greater the context score, the greater the likelihood of outdoor work participation.

To start, there were 862 unique occupation codes with context scores for question #1 (i.e., exposed entirely to weather), 30 of which also contained context scores for question #2 (there were no occupations with context scores for question #2 alone). Assuming ozone exposure would be similar for outdoor workers whether under cover or totally exposed to weather, we merged the data responses from the two questions by occupation code and assigned the highest context score of the two responses to each occupation. Given the context scaling information provided by O*NET, we then assumed the context scores 76-80, 81-85, 86-90, 91-95, and 96-100 characterized occupations as having 1, 2, 3, 4, or 5 days per week outdoors. Then, mapping of the O*NET occupations to the above described BLS SOC occupation data set was performed, and was generally agreeable, with a few exceptions.[18] Following additional processing, there were 144 unique occupation titles having one to five days per week where work was performed outdoors and the number of persons constituting each. See Attachment 1 for the final O*NET/BLS mapping and additional data processing assumptions.

These 144 specific occupations then needed to be mapped to the occupation-related activity codes used by APEX to generate METs in estimating energy expenditure. When CHAD was developed in the late 1990s, occupation codes from the 1990 US Census were mapped to twelve broad occupation categories[19] and were assigned METs distributions based on the most commonly performed activities associated with work tasks. In order to use the APEX/CHAD METs database in its current format and integrate the newly developed 2010 BLS/O*NET data set, the 1990 Census occupation codes were translated using two additional mapping files: a 1990 Census to 2000 Census code map file[20] and a 2000 Census to SOC code map file.[21] The

---

[18] There were seven O*NET codes that did not directly correlate with the BLS SOC codes and several O*NET occupations that were subcategories to broader BLS SOC codes.

[19] The list of 1990 Census codes used to map the occupation titles to the APEX METs distributions are found at https://usa.ipums.org/usa/volii/99occup.shtml. The twelve occupation groups are Executive Administrative Managerial (ADMIN), Administrative support (ADMSUP), Professional (PROF), Technical (TECH), Sales (SALE), Protective Services (PROTECT), Service (SERV), Farming Forestry Fishing (FARM), Precision Production (PREC), Machine Operators (MACH), Transportation (TRANS), and Laborers (LABOR). Household workers (HSHLD), while an APEX/CHAD occupation group, do not have any work days where time is spent outdoors.

[20] http://www.census.gov/people/io/files/techtab02.pdf.

[21] http://www.census.gov/people/eeotabulation/documentation/occcategories.pdf

mapping was generally agreeable among the various occupations and across the source files over the years however, in characterizing several of the O*NET occupation codes to the broad APEX/CHAD occupation groups, there were several instances where more than one broad occupation group could be assigned to a single O*NET occupation. Professional judgement was used to assign the most appropriate broad occupation category when there were multiple choices, with decisions made to complete the processing of the 34 occupation titles given in Attachment 2 of this appendix. The final distribution of days per week spent performing outdoor work considering the BLS/O*NET data set and stratified by APEX/CHAD occupation groups is provided in Table 5G-8. This data set provides the target for developing an activity pattern data set that reasonably reflects a distribution of outdoor workers properly weighted by occupation groups along with assignment of an approximate number of days per week they spend outdoors.

Table 5G-8. Distribution of days per week spent performing outdoor work considering the BLS/O*NET data set and stratified by APEX/CHAD occupation groups.

	BLS/O*NET Occupations						**Target for CHAD Diary Outdoor Work Days per Week (based on BLS/O*NET mean)**
		Number of Outdoor Work Days per Week			**Percent of Employed Persons Performing Outdoor Work**		
CHAD Group	**(n)**	**(mean)**	**(min)**	**(max)**	**Outdoor Workers only**	**All Workers**	
ADMIN	7	1.4	1	3	5.1	0.6	1
ADMSUP	7	3.4	1	5	4.6	0.5	3
FARM	19	4.0	1	5	18.3	2.1	4
LABOR	10	4.5	1	5	9.3	1.1	5
MACH	3	4.0	2	4	0.5	0.1	4
PREC	42	2.5	1	5	25.1	2.9	3
PROF	3	1.3	1	4	0.2	<0.1	1
PROTECT	15	3	1	5	7.7	0.9	3
SALE	1	1.0	1	1	2.4	0.3	1
SERV	4	3.2	1	5	1	0.1	3
TECH	6	3.4	3	4	3.1	0.4	3
TRANS	27	4.3	1	5	22.7	2.7	4
TOTAL	**144**	-	-	-	**100**	**11.7**	-

CHAD contains diary information from a number of persons who reported their time spent at work, and is indicated by CHAD activity codes beginning with either '100' or '101'. This, combined with outdoor location information (i.e., 60 location codes provided in Attachment 3 of this appendix) and selected for where the total time working outdoors for the day was ≥ 2 hours, yielded 1,510 CHAD daily activity patterns potentially usable in representing outdoor workers. We evaluated these diaries across a number of personal attributes and calculated the mean time spent outdoors for each diary day (Table 5G-9). As expected, very few diaries were from persons that identified their occupation, though where that information was present, these diaries retained that specific identifier and used to simulate persons having that particular occupation group. Because we were interested in generating a typical work week, that is, work was performed during weekday days, only weekday diaries were used to develop the target diary pool for the weekday schedule, giving a final pool of 1,403 usable diaries.

Table 5G-9. Personal attributes and mean time spent working outdoors for CHAD diaries reporting at least two hours of outdoor work.

Category	Attribute	Outdoor Worker Diary days (n)	Mean Outdoor Work Time (minutes)
CHAD Occupation Group	ADMIN	13	264
	ADMSUP	4	336
	FARM	11	478
	LABOR	14	469
	MACH	10	320
	PREC	52	416
	PROF	8	307
	PROTECT	4	402
	SALE	13	290
	SERV	5	291
	TECH	1	290
	TRANS	12	328
	X	1363	344
Age Range (years)	15-18	19	327
	19-24	78	420
	25-34	155	443
	35-44	364	359
	45-54	384	347
	55-64	360	306
	65-74	137	286
	75-84	11	243
	> 84	2	475
Sex	F	349	290
	M	1161	365
Day of Week	FRI	123	349
	MON	174	379
	SAT	66	399
	SUN	41	434
	THU	280	341
	TUE	357	342
	WED	469	328

We assumed outdoor worker diaries having missing occupation information could be used to represent outdoor worker diaries having an assigned occupation and, in the absence of any additional information to suggest stronger alternative, simply assigned CHAD occupation groups to these diaries equally, though weighted by the appropriate days per week outdoor work was performed by a particular occupation group. This is in part because, in performing an APEX simulation, the options for developing a multiday diary profile can be either controlled by a key variable (such as time spent outdoors), use the same diary to represent everyday of the simulated person's exposure period, or be constructed by an entirely random sequence of diaries. Outdoor time, a commonly used key variable to represent intra- and inter-personal variability activity patterns over multiple days by APEX, in general is an unknown for each occupation group and thus the key variable approach cannot be used to develop the longitudinal profiles. Rather than use the same diary for each person day, we elected to use a random selection of diaries, though having the selected diaries drawn from specific occupation groups developed from a large diary pool weighted by the target number of days per week where outdoor work was performed (Table 5G-8). To clarify how this was done, an example follows using a single occupation group of outdoor workers, i.e., those comprising the Transportation (TRANS) group.

According to information summarized in Table 5G-8, transportation-associated workers (TRANS) were estimated to, on average, spend four days per week working outdoors. We assigned the set of all available weekday outdoor worker diaries having 'missing' for their occupation (n=1,363) as now having an occupation of 'TRANS'. We then replicated this new set of outdoor worker diaries (including those few diaries having a known occupation, TRANS or otherwise, to total 1,403 outdoor worker diaries) to generate a data set now having four weekday days per person day. Thus from an APEX modeling perspective, all personal attributes for persons in that pool are identical from one day to the next and have an equal likelihood of selection. This four day by 1,403 person activity pattern data set, now principally comprised of outdoor workers having 'TRANS' as their occupation, was then combined with a single weekday by 1,403 person activity pattern data set, only that this particular one day data set, while still derived from the same set of outdoor worker diaries, differs in that all of the paid work time spent outdoors was changed to paid work time occurring within indoor locations for that one work day. Then, when APEX constructs a longitudinal diary for any person with the 'TRANS' occupation using completely random selection, all other personal attributes remaining the same, the probability of selecting an outdoor work diary for any day is 0.8 while that of an indoor diary is 0.2, appropriately reflecting, on average, the days per week that occupation group spends time working outdoors. This process of building up the activity pattern diary pool was repeated for each outdoor worker occupation group to reflect both the probability of performing either indoor or outdoor work during weekdays. This collection of weekday diaries was then combined with the pool of all remaining CHAD weekend diaries (n=13,953 person days), though where persons had missing occupation information (n=12,561 person days), any one of the twelve occupation groups was randomly assigned to a given weekend day. This complete outdoor worker activity pattern set now totaled 98,133 person days[22] having 3,656,560 activity events.

That said, it was soon apparent this input data set was too large for APEX to use when error messages were generated upon model execution. For these outdoor worker simulations to proceed, we determined the maximum size of the diary data set was approximately 42,000 diary days. Our current response to this limitation was to first restrict the exposure simulations using a tighter age range, thus permitting us to also limit the activity pattern input data set by similar age ranges. Two age groups of outdoor workers were of interest for our exposure simulations: 19-35 and 36-55 years old. To maximize the number of diaries used to model the 19-35 year olds,[23] we increased the range for APEX usable diaries from the large outdoor worker activity pattern data set initially developed to include ages from 16 to 42 years old (n=30,657 person days). As a reminder, only the *activity patterns* of 16-18 year olds characterized as outdoor workers (i.e., persons having ≥ 2 hours of paid work occurring outdoors) were available to simulate adults ages 19 or above. All of the anthropometric attributes of these simulated adults (e.g., body mass, resting metabolic rate, ventilation rate) were derived from their age appropriate data, distributions, and/or equations. There were adequate numbers of diaries available to model the 36-55 year olds such that the age range for inclusion in that data set was restricted to those between the ages of 36-54 (n=41,736 person days).

When modeling exposures using occupation groups, two additional input files were needed by APEX. The first was simply a file containing the CHAD ID and the specific occupation group identified for that person day. The second, a profile factors file, contains the probabilities a simulated person in the model domain will have a particular occupation, a profile variable that can be stratified by age, sex, and/or census tract (US EPA 2012a, b). Based on the information we developed in Table 5G-8, we only assigned specific probabilities for each particular occupation group, i.e., the percent of employed persons performing outdoor work using the outdoor worker proportions equally across ages, sexes and tracts. An additional modification to the APEX employment probabilities input file (*Employment2000_043003.txt*) was also needed to generate exposures and appropriate output summary tables only for employed persons. And finally, because of the generally limited number of diaries available in developing the diary pool for the 19-35 year olds and to use as many of the diaries available, we lengthened the age selection range (AgeCutPct = 30.0 and Age2Probab = 0.15) and only included two temperature diary pools (<84 or ≥84) to have an adequate number of diaries available in each diary pool for APEX to execute the desired simulation.

---

[22] In summary, the weekday data set was developed assuming 12 occupations for 5 days per work week (or 35 outdoor days+ 25 indoor days) for each of the 1,403 diary days and added to the 13,953 weekend days.

[23] Note from Table 5G-9, there are just over 250 diaries from persons aged 15-34.

Finally, a 10,000 person exposure simulation was performed for each age group of outdoor workers in Atlanta for Jun 1- Aug 30, 2006 using air quality just meeting the existing standard. In addition, as a point of comparison for the longitudinal approaches developed here and used for estimating general population-based exposures, identical simulations were performed in Atlanta during the same time of year, air quality scenario, and age groups only differing by using the approach described in our primary REA exposure simulations, i.e., using outdoor time as the key variable in developing longitudinal profiles, sampling from all available CHAD diaries, and not explicitly addressing simulated worker occupations and work performed (structured schedules and associated METs values).[24]

We first summarized the outdoor work time performed by each simulated outdoor worker during weekdays, stratified by particular occupation, to ascertain whether or not the exposure simulation met our defined goal. In comparing the results of Table 5G-10 with those provided in Table 5G-8 we see that the goal was met for both age groups of outdoor workers, i.e., the longitudinal approach was structured correctly to reproduce the distribution of the outdoor worker occupation groups and the number of days persons in a particular group spent working outdoors. Estimated exposures are presented in (Figure 5G-10) for each of the two outdoor worker study groups and considering either a longitudinal scenario-based approach designed specifically to reflect an outdoor worker weekday schedule or using our general population-based modeling approach. It is clear that when accounting for a structured schedule that includes repeated occurrences of time spent outdoors for a specified study group, all while more consistently performing work tasks that may be at or above moderate or greater exertion levels, a greater percent of the study group experiences exposures at or above the selected health effect benchmark levels than that estimated using our general population-based approach. The differences between exposures estimated for the two longitudinal approaches become much greater when considering the percent of persons experiencing multiple exposure days at or above benchmark levels. For example, ≤ 2% of the general population-based exposure group was estimated to have two or more exposures at or above 60 ppb-8hr, while >17% of specifically simulated outdoor workers were estimated to experience exposures at or above that same level. In general, there was little difference in exposures estimated for the two age groups of outdoor workers.

---

[24] Because most CHAD IDs have unknown occupations, METs are sampled from a 'composite' triangular distribution min, peak, max [1.2, 1.9, 5.6] developed from the METs distributions used for all occupation groups. As such, exertion levels achieved by laborers, service, and transportation industries are not well represented using a general population approach given their respective METs distributions triangular[3.6, 8.1, 13.8], triangular[1.6, 5.6, 8.4], and lognormal geometric mean, standard deviation [3.0, 1.5] truncated to min, max [1.3, 8.4].

Table 5G-10. Distribution of days per week spent performing outdoor work considering the APEX simulated population and stratified by APEX/CHAD occupation and age groups.

Age Group	CHAD Occupation Group	Percent of Simulated Outdoor Workers	Number of Outdoor Work Days per Week		
			(mean)	(min)	(max)
19-35	ADMIN	5.0	1.2	0	2
	ADMSUP	4.6	3.0	2	4
	FARM	19.1	3.9	3	5
	LABOR	9.5	4.9	5	5
	MACH	0.5	4.0	3	5
	PREC	24.3	3.0	2	4
	PROF	0.2	1.4	1	3
	PROTECT	7.8	3.0	2	4
	SALE	2.4	1.2	0	2
	SERV	1.1	3.0	2	4
	TECH	3.2	3.0	2	4
	TRANS	22.4	4.0	3	5
36-55	ADMIN	5.4	1.1	0	2
	ADMSUP	4.7	3	2	4
	FARM	17.5	3.9	3	5
	LABOR	9.6	4.9	5	5
	MACH	0.6	3.9	3	4
	PREC	24.8	3	2	4
	PROF	0.3	1	0	2
	PROTECT	7.5	3	2	4
	SALE	2.5	1.1	0	2
	SERV	0.9	3	2	4
	TECH	3.1	3	2	4
	TRANS	23.2	3.9	3	5

Additional context regarding the estimate of the *number of persons* exposed between the two approaches can added with the following. Approximately 30% of our outdoor worker study group ages 19-55 was estimated to experience at least one exposure at or above 60 ppb-8hr while at moderate or greater exertion. Assuming a 92% employment rate and that outdoor workers constitute approximately 12% of the workforce (Table 5G-8), outdoor workers experiencing at least one exposure at or above 60 ppb-8hr would comprise about 3.3% of a total simulated population in that study area. For the same air quality scenario we estimated using the general population-based approach, about 5-8% of the study group would experience exposures at or above the same benchmark. To some extent, the general population-based approach will simulate exposure profiles of outdoor workers (persons with frequent and above average time spent working outdoors experiencing instances of high exposure concentrations at elevated exertion levels) by applying our standard longitudinal diary selection method. Intuitively, outdoor workers probably constitute a significant portion of the overall population-based study group that could exceed benchmark levels though at this time it is unknown whether the portion estimated here of 40-60% is accurate. It is however reasonable to conclude given the comparative exposure results that the general population-based approach would tend to underestimate the multiday exposures at or above selected benchmark levels experienced by persons adhering to a more rigid outdoor work schedule.

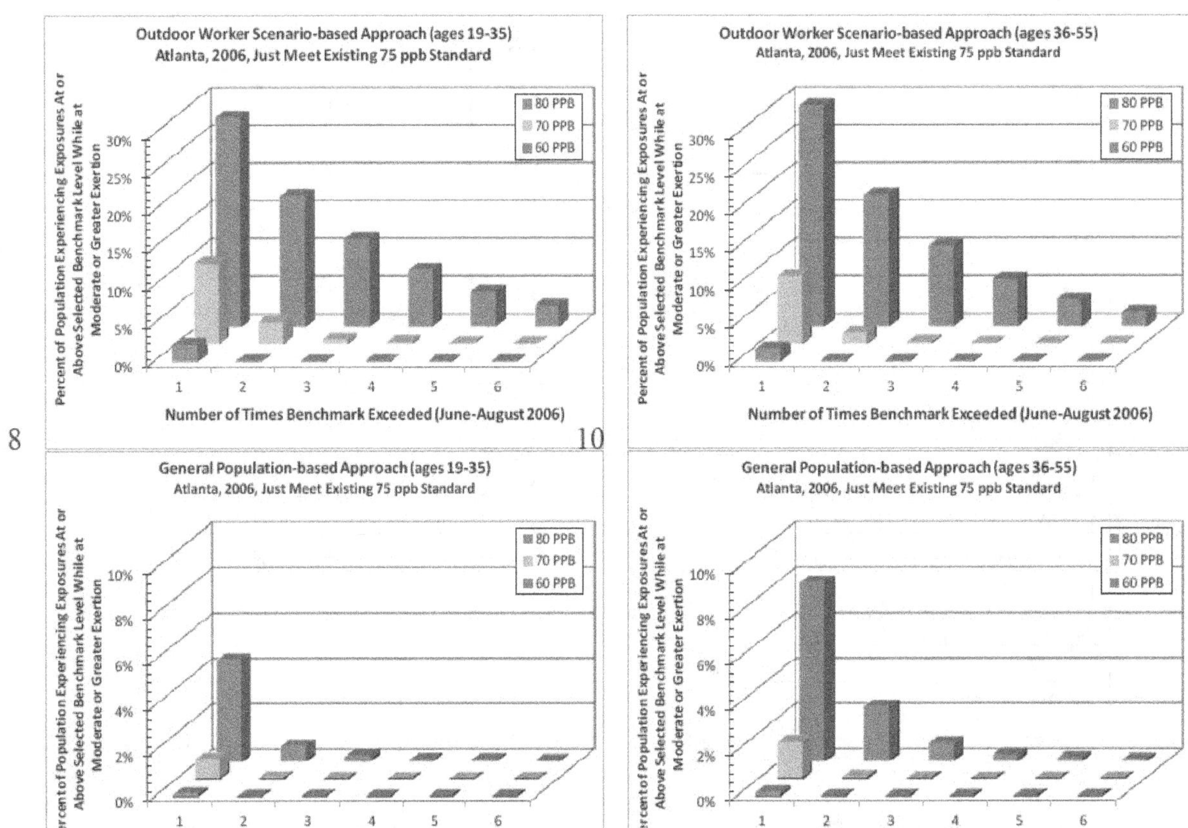

Figure 5G-10. Percent of persons between age 19-35 (left) and 36-55 (right) experiencing exposures at or above selected benchmark levels while at moderate or greater exertion using an outdoor worker scenario-based approach (top) and a general population-based approach (bottom) considering air quality adjusted to just meet the existing standard in Atlanta, GA, Jun-Aug, 2006.

5G-4.3 Averting Behavior and Potential Impact to Exposure Estimates

A growing area of air pollution research involves evaluating the actions persons might perform in response to high O_3 concentration days (ISA, section 4.1.1). Most commonly termed *averting behaviors*, they can be broadly characterized as personal activities that either reduce pollutant emissions or limit personal exposure levels. The latter topic is of particular interest in this REA due to the potential negative impact it could have on O_3 concentration-response (C-R) functions used to estimate health risk and on time expenditure and activity exertion levels recorded in the CHAD diaries used by APEX to estimate O_3 exposures. To this end, we have performed an additional review of the available literature here beyond that summarized in the ISA to include several recent technical reports that collected and/or evaluated averting behavior data. Our purpose is to generate a few reasonable quantitative approximations that allow us to better understand how averting behavior might affect our current population-based exposure and risk estimates. We expect that the continued development and communication of air quality information via all levels of environmental, health, and meteorological organizations will only further increase awareness of air pollution, its associated health effects, and the recommended actions to take to avoid exposure, thus making averting behaviors and participation rates an even more important consideration in future O_3 exposure and risk assessments. The following is a summary of our literature review, with details provided by Graham (2012). Later in this section, preliminary results of an exposure simulation designed to account for averting behavior are provided.

The first element considered in our evaluation is peoples' general perception of air pollution and whether they were aware of alert notification systems. The prevalence of awareness was variable; about 50% to 90% of survey study participants acknowledged or were familiar with air quality systems (e.g., Blanken et al., 1991; KS DOH, 2006; Mansfield et al., 2006; Semenza et al., 2008) and was dependent on several factors. In studies that considered a persons' health status, e.g., asthmatics or parents of asthmatic children, there was a consistently greater degree of awareness (approximately a few to 15 percentage points) when compared to that of non-asthmatics. Residing in an urban area was also an important influential factor raising awareness, as both the number of high air pollution events and their associated alerts are greater when compared to rural areas. Of lesser importance, though remaining a statistically significant influential variable, were several commonly correlated demographic attributes such as age, education-level, and income-level, with each factor positively associated with awareness.

The second element considered in our evaluation was the type of averting behaviors performed. For our purposes in this O_3 REA, the most relevant studies were those evaluating outdoor time expenditure, more specifically, the duration of outdoor events and the associated exertion level of activities performed while outdoors. This is because both of these variables are necessary to understanding O_3 exposure and associated adverse effects, and hence, in accurately estimating human health risk.

As stated above regarding air quality awareness, asthmatics consistently indicated a greater likelihood of performing averting behaviors compared to non-asthmatics – estimated to differ by about a factor of two. This difference could be the combined effect of those persons having been advised by health professional to avoid high air pollution events and them being aware of alert notification systems. Based on the survey studies reviewed, we estimate that 30% of asthmatics may reduce their outdoor activity level on alert days (e.g., KS DOH, 2006; McDermott et al., 2006; Wen et al., 2009).[25] An estimate of 15%, derived from reductions in public attendance at outdoor events (Zivin and Neidell, 2009) would be consistent with our estimate above when considering that the Zivin and Neidell (2009) study group is likely comprised mainly of non-asthmatics. That said, both attenuation and the re-establishment of averting behavior was apparent when considering a few to several days above high pollution alert levels (either occurring over consecutive days or across an entire year) (McDermott et al., 2006; Zivin and Neidell, 2009), suggesting that participation in averting behavior over a multiday period for an individual is complex and likely best represented by a time and activity-dependent function rather than a simple point estimate.

There were only a few studies offering quantitative estimates of durations of averting behavior, either considering outdoor exertion level or outdoor time (Bresnahan et al., 1997; Mansfield et.al, 2006, Neidell, 2010; Sexton, 2011). Each of these studies considered outdoor time expenditure during the afternoon hours. Based on the studies reviewed, we estimate that outdoor time/exertion during afternoon hours may be reduced by about 20-40 minutes in response to an air quality alert notification. Generally requisite factors include: a high alert level for the day (e.g., red or greater on the AQI), high O_3 concentrations (above the NAAQS), and persons having a compromised health condition (e.g., asthmatic or elderly).

The third element considered in our preliminary evaluation was how to further define the impact of averting behavior on modeled exposure estimates.[26] As described in section REA 5.3.2, APEX uses time location activity data (diaries) from CHAD to estimate population-based exposures. These diaries originate from a number of differing studies; some were generated as part of an air pollution research study, some were collected during a summer/ozone season, while some diary days may have corresponded with high O_3 concentration and air quality alert days. At this time, none of the diary days used by APEX have been specifically identified as representing days where a person did or did not adjust their activity pattern reduce their exposure. In considering the above discussion regarding the potential rate of participation and averting actions performed, it is possible that some of the CHAD diary days express instances where that selected individual may have reduced their time spent outdoors or reduced their exertion level while outdoors. Currently, without having a personal identifier for averting behavior in CHAD, the diaries are assigned to a simulated persons' day without directly

---

[25] Many of these studies do not account for one important factor when using a recall questionnaire design: whether the participant's stated response to air pollution is the same as the action they performed.

[26] The discussion of another important effect of averting behavior is on concentration-response functions (more relevant to the risk assessment in chapter 8). This is presented in the ISA (section 4.1.2).

considering ambient O_3 concentration levels[27]. Therefore, it is possible that there are instances where, on a given APEX simulation day, the simulated person may use a diary day from a person that did engage in one or more types of averting behavior (e.g., a diary having less time than usual spent outdoors in the afternoon), while for most other persons simulated on the same day (or the same person on a different high concentration day) the diaries used are from persons that did not actively engage in averting behavior. As a result, the effect of averting behavior may already be incorporated into our exposure modeling, albeit to an unknown though likely small degree,[28] though definitely generating low-biased estimates of exposures (and reduced number of persons at or above selected 8-h O_3 benchmark concentrations) that would occur in the complete absence of averting behavior.

With this in mind, we performed an APEX simulation to reflect the instance that a fraction of a selected study group spends less time outdoors on high concentration ozone days. First, a general APEX simulation was performed during June-August 2007 in Detroit to identify a short time period where a high number of children/asthmatic children were estimated to be exposed at or above the 8-hour O_3 benchmark levels of interest. To maintain a degree of tractability in the simulations, the development of the new CHAD input data, and the analysis of the exposure results, we restricted the exposure simulations to 5,000 total persons. One such high exposure event occurred over a two-day period considering base year air quality – August 1-2, 2007. Because conditions in APEX simulations can be controlled by using an identical random number seed, APEX daily files were output to explicitly identify all of the CHAD diaries used for this two-day simulation.

The activity pattern data from the identified CHAD diaries used to simulate the two-day exposure period were then used to generate a new activity pattern input data set, one adjusted for the above estimated parameters used to reflect averting performed by the two exposure study groups of interest. We did this after determining the following:

1) There were a total of 1,988 diaries used to simulate the maximum exposure day for each person, obviously some CHAD diaries were used more than once to simulate different people on their maximum exposure day. Note also, 48 diaries were used to simulate both days for the same person, 37 of these occasions were for unique individuals while for the remaining 11 instances the same diary was also used in either two or three persons two-day simulation.
2) We calculated the total time spent outdoors by hour of day for the CHAD diaries used by APEX for each persons' maximum exposure day, estimated the mean outdoor time

[27] APEX uses maximum temperature in assigning diaries for a select day in an area, capturing some relevant variability in O_3 concentrations.
[28] Neither the participation rate nor the duration of averting for simulated persons is being strictly controlled for by APEX when simulating exposures.

and number of persons by hour of day, and then stratified these results by the number of times a diary was used in the simulation. Ideally, the most effective adjustment to outdoor time considering averting would be applied to the afternoon hours. We observed that at hour 15 (3 PM-4 PM), there were both a reasonable number of CHAD diaries and diaries having an appropriate total time spent outdoors to achieve our desired adjustment of 15% and 30% participation for children and asthmatics, respectively, for on average 40 minutes in a day. This was determined using the results in the following table. To simulate averting for children, we selected the collection of diaries from the '2' and '3' times a diary was used category (n=766 total diary days or 15.3% of all diaries used for each persons maximum exposure day) and reduced all outdoor time events to 0 minutes spent outdoors during 3 PM-4 PM. To simulate averting for asthmatic children, we selected the collection of diaries from the '1' through '5' times a diary was used category (n=1,518 total diary days or 30.4% of all diaries used for each persons' maximum exposure day) and reduced all outdoor time events to 0 minutes spent outdoors during 3 PM-4 PM.

Number of times a select diary was used	Number of persons	Mean Outdoor Time	Total Diary Days
1	181	43.2	181
2	182	43.4	364
3	134	44.6	402
4	84	45.6	336
5	47	40.6	235
6	20	48.4	120
7	13	44.1	91
8	7	45.4	56
9	4	39.5	36
10	5	48.0	50
11	5	40.6	55
12	2	35.0	24
13	1	5.0	13
14	1	45.0	14
15	3	60.0	45
17	1	60.0	17

3) Following the selection of the hour where outdoor time would be adjusted to reflect averting, we evaluated whether diaries used for both simulation days per person and when used for multiple persons for both simulation days would affect the targeted reduction in outdoor time. Even considering the small number of instances identified in #1) above (48 single diaries used for both days of the two day simulation per person out of 5,000 possible cases), only two diaries had time spent outdoors during

the hours of 3 PM-4 PM where both days would be adjusted for averting, thus likely having a negligible effect the meeting of our approximate averting goals.

Both the outdoor time adjusted CHAD and the standard CHAD input files were separately used to simulate exposures to children and asthmatic children. Exposure results for the four simulations (simulated averting vs. no averting for both children and asthmatic children) are found in the main body of the REA.

5G-5 COMPARISON OF PERSONAL EXPOSURE MEASUREMENT AND APEX MODELED EXPOSURES

A new evaluation of APEX was performed using a subset of personal O_3 exposure measurements obtained from the Detroit Exposure and Aerosol Research Study (DEARS) (Meng et. al, 2012). For five consecutive days, personal O_3 outdoor concentrations along with daily time-location activity diaries were collected from 36 study participants in Wayne County Michigan during July and August 2006. The majority of participants were female (80%) having a mean age of 40.6 (min 20, max 72); mean age for males was 41.4 (min 22, max 65). Rather than using daily personal exposures estimated below the reported detection limit of 3 ppb (i.e., 0, 1 and 2 ppb), we approximated those falling below this level using a random assignment of concentrations of 1 and 2 ppb.

An APEX simulation was performed considering these same geographic and temporal features, followed with the sub-setting of APEX output data according to important personal attributes of the DEARS study participants (specific 5-day collection study periods, age/sex distributions, outdoor time, ambient concentrations, and air exchange rate). For both data sets and considering the output variables independently, the median daily values for each study participant attribute was generated using each individual's 5 person-days of data, then ranked median values were plotted along with each individual's corresponding minimum and maximum value. Distributions for four of these variables of (personal O_3 exposure, outdoor time, ambient O_3 concentrations, and air exchange rate from each of the two data sets are presented in Figure 5G-11.

Distributions of time spent outdoors and ambient concentrations were similar by design of the APEX population-based sample selection method. The upper percentiles the DEARS participant AER distribution was greater than that of AER of APEX simulated persons. For example, 40% of DEARS participants had a median value of two air exchanges per hour, while the same rate was only observed for 5% of APEX simulated individuals. In contrast, over 50% of APEX simulated individuals had median daily O_3 exposure concentrations above 10 ppb, while only 3% of DEARS participants' median values exceeded 10 ppb. This difference in exposure is surprising given the sharply higher residential indoor air exchange rate for the DEARS participants (i.e., indoor microenvironmental exposures would be expected to have a greater influence on total DEARS exposure compared with the APEX simulated exposures) all while holding all other potential influential variables the same between the two data sets and is subject to further investigation.

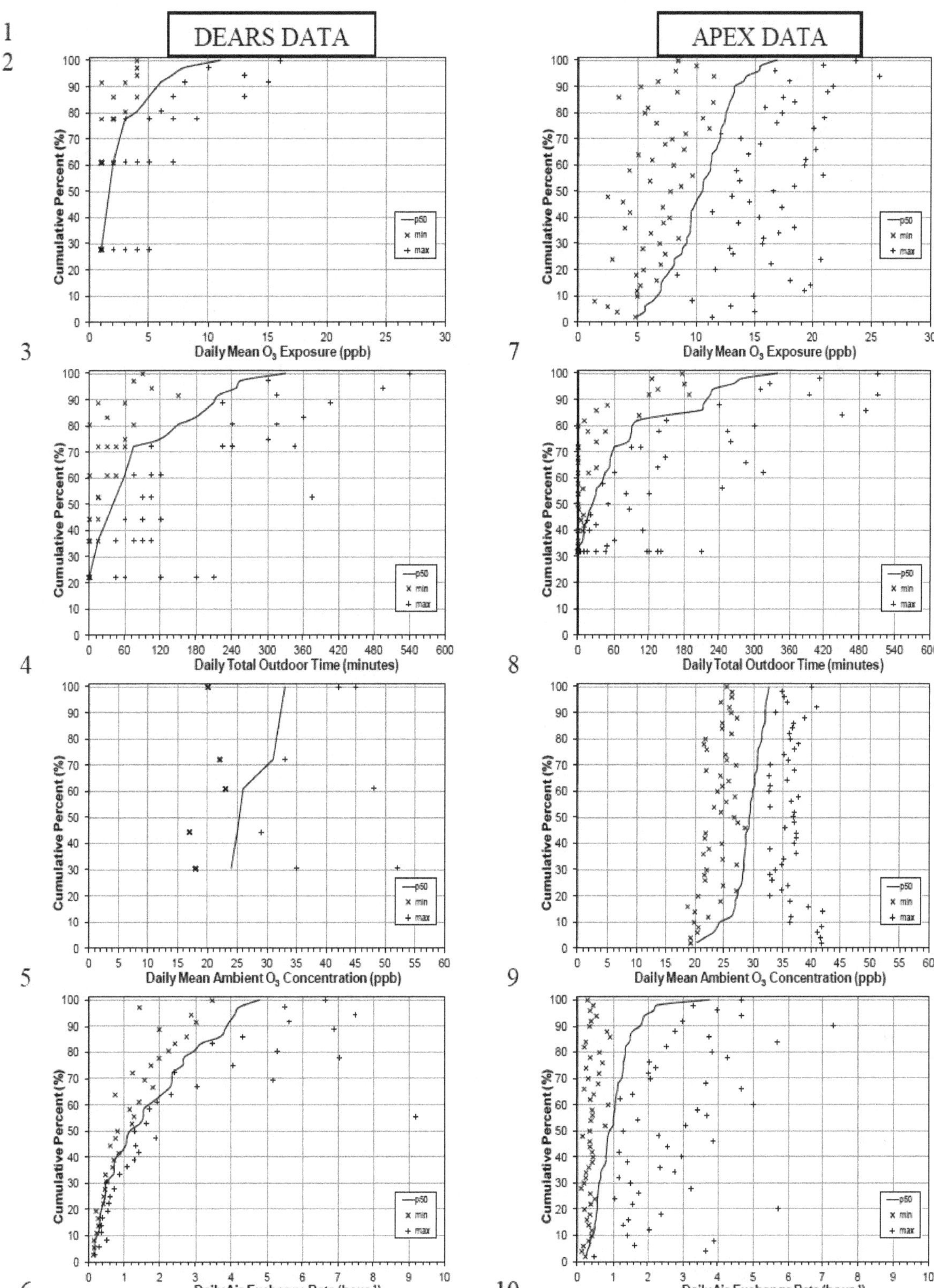

Figure 5G-11. Distribution of daily personal O3 exposures (top row), outdoor time (2nd row from top), ambient O_3 concentrations (3rd row from top), and air exchange rate (bottom row) for DEARS study participants (left column) and APEX simulated individuals (right column) in Wayne County, MI, July-August 2006.

5G-6 REFERENCES

Ainsworth, B. E., Haskell, W. L., Leon, A. S., Jacobs, D. R., Jr., Montoye, H. J., Sallis, J. F., Paffenbarger, R. S., Jr. (1993). Compendium of physical activities: classification of energy costs of human physical activities. *Med Sci Sports Exer.* 25: 71-80.

Arcus-Arth, A. and Blaisdell, J. (2007). Statistical distributions of daily breathing rates for narrow age groups of infants and children. Risk Analysis. 27(1):97-110.

Akland, G. G., Hartwell, T. D., Johnson, T. R., Whitmore, R. W. (1985). Measuring human exposure to carbon monoxide in Washington, D. C. and Denver, Colorado during the winter of 1982-83. *Environ Sci Technol.* 19: 911-918.

Blanken, P. D., Dillon, J., Wismann, G. (2001). The impact of an air quality advisory program on voluntary mobile source air pollution reduction. *Atmos Environ.* 35:2417-2421.

Bresnahan, B. W., Dickie, M., Gerking, S. (1997). Averting behavior and air pollution. *Land Econ.* 73:340-357.

Brochu, P., Ducre-Robitaille, J.-F., Brodeur, J. (2006a). Physiological daily inhalation rates for free-living individuals aged 1 month to 96 years, using data from doubly labeled water measurements: a proposal for air quality criteria, standard calculations and health risk assessment. Human and Ecological Risk Assessment. 12:675-701.

Brochu, P., Ducre-Robitaille, J.-F., Brodeur, J. (2006b). Supplemental Material for Physiological daily inhalation rates for free-living individuals aged 1 month to 96 years, using data from doubly labeled water measurements: a proposal for air quality criteria, standard calculations and health risk assessment. Human and Ecological Risk Assessment. 12:1-12.

EPRI. (1988). A Study of Activity Patterns Among a Group of Los Angeles Asthmatics. Research Project 940-5. Electric Power Research Institute. Prepared by Roth Associates. November 1988.

EPRI. (1992). A Survey of Daily Asthmatic Activity Patterns in Cincinnati. TR-101396. Research Project 940-05. Electric Power Research Institute. Prepared by Roth Associates. November 1992.

Ford, E. S., Heath, G. W., Mannino, D. M., Redd, S. C. (2003). Leisure-time physical activity patterns among US adults with asthma. *Chest.* 124:432-437.

Freedson, P. S. (1989). Field monitoring of physical activity in children. *Pediatr Exerc Sci.* 1:8-18.

George, B. J., McCurdy, T. (2009). Investigating the American Time Use Survey from an exposure modeling perspective. *JESEE.* 21:92-105.

Geyh, A. S., Xue, J., Özkaynak, H., and Spengler, J. D. (2000). The Harvard Southern California chronic ozone exposure study: assessing ozone exposure of grade-school-age children in two Southern California communities. *Environ Health Persp.* 108:265-270.

Glen, G., Smith, L., Isaacs, K., McCurdy, T., Langstaff, J. (2008). A new method of longitudinal diary assembly for human exposure modeling. *J Expos Sci Environ Epidem.* 18:299-311.

Graham, S. E., McCurdy, T. (2004). Developing meaningful cohorts for human exposure models. *J Expo Anal Environ Epidemiol.* 14: 23-43.

Graham, S. E., McCurdy, T. (2005). Revised ventilation rate (VE) equations for use in inhalation-oriented exposure models. Report no. EPA/600/X-05/008 is Appendix A of US EPA (2009b). Available at: http://cfpub.epa.gov/ncea/cfm/recordisplay.cfm?deid=202543.

Graham, S. (2009). Response to peer-review comments on Appendix A, prepared by S. Graham (US EPA). Response is Appendix D of US EPA (2009b). Available at: http://cfpub.epa.gov/ncea/cfm/recordisplay.cfm?deid=202543.

Graham, S. E. (2012). Comprehensive review of published averting behavior studies and available technical documents. Memo to Bryan Hubbell, Group Leader, Risk and Benefits Group, Office of Air Quality Planning and Standards. Docket # EPA-HQ-OAR-2008-0699-0085.

Hartwell, T. D., Clayton, C. A., Ritchie, R. M., Whitmore, R. W., Zelon, H. S., Jones, S. M., Whitehurst, D. A. (1984). Study of Carbon Monoxide Exposure of Residents of Washington, DC and Denver, Colorado. Research Triangle Park, NC: U.S. Environmental Protection Agency, Office of Research and Development, Environmental Monitoring Systems Laboratory. EPA-600/4-84-031.

Houtven, G. V., Johnson, F. R., Mansfiled, C., Yang, J.-C., Pyles, A. (2003). Parental Averting Behavior With Respect to Ozone Alerts. For presentation at AEA/ASSA Meeting, San Diego CA, January 2004.

Isaacs, K., Glen, G., McCurdy, T., Smith, L. (2008). Modeling energy expenditure and oxygen consumption in human exposure models: accounting for fatigue and EPOC. *J Expos Sci Environ Epidemiol.* 18: 289–298.

Johnson, T. (1984). A Study of Personal Exposure to Carbon Monoxide in Denver, Colorado. Research Triangle Park, NC: U.S. Environmental Protection Agency, Environmental Monitoring Systems Laboratory. EPA-600/4-84-014.

Johnson, T. (1989). Human Activity Patterns in Cincinnati, Ohio. Palo Alto, CA: Electric Power Research Institute. EPRI EN-6204.

Johnson, T., Capel, J., McCoy, M. (1996a). Estimation of Ozone Exposures Experienced by Urban Residents Using a Probabilistic Version of NEM and 1990 Population Data. Prepared by International Technology Air Quality Services for Office of Air Quality Planning and Standards, U.S. Environmental Protection Agency, Research Triangle Park, NC. Contract no. 63-D-30094. April 1996. Available at: http://www.epa.gov/ttn/naaqs/standards/ozone/s_o3_pr_td.html.

Johnson, T., Capel, J., McCoy, M., Warnasch, J. (1996b). Estimation of Ozone Exposures Experienced by Outdoor Children in Nine Urban Areas Using a Probabilistic Version of NEM. Prepared by International Technology Air Quality Services for Office of Air Quality Planning and Standards, U.S. Environmental Protection Agency, Research Triangle

Park, NC. Contract no. 63-D-30094. April 1996. Available at: http://www.epa.gov/ttn/naaqs/standards/ozone/s_o3_pr_td.html

Johnson, T., Capel, J., McCoy, M., Warnasch, J. (1996c). Estimation of Ozone Exposures Experienced by Outdoor Workers in Nine Urban Areas Using a Probabilistic Version of NEM. 1996. Prepared by International Technology Air Quality Services for Office of Air Quality Planning and Standards, U.S. Environmental Protection Agency, Research Triangle Park, NC. Contract no. 63-D-30094. April, 1996. Available at: http://www.epa.gov/ttn/naaqs/standards/ozone/s_o3_pr_td.html

Klepeis, N. E., Tsang, A. M., Behar, J. V. (1996). Analysis of the National Human Activity Pattern Survey (NHAPS) Respondents from a Standpoint of Exposure Assessment. Washington, DC: U.S. Environmental Protection Agency, Office of Research and Development. EPA/600/R-96/074.

Knowledge Networks. (2009). Field Report: National Scale Activity Survey (NSAS). Conducted for Research Triangle Institute. Submitted to Carol Mansfield November 13, 2009.

KS DOH. (2006). Environmental Factors, Outdoor Air Quality, and Activity Level. Results from 2005 Kansas Behavioral Risk Factor Surveillance System. Office of Health Promotion, Kansas Department of Health and Environment. Available at: http://www.kdheks.gov/brfss/PDF/cste_report_final.pdf

Langstaff, J. E. (2007). OAQPS Staff Memorandum to Ozone NAAQS Review Docket (OAR-2005-0172). Subject: Analysis of Uncertainty in Ozone Population Exposure Modeling. [January 31, 2007]. Available at: http://www.epa.gov/ttn/naaqs/standards/ozone/s_o3_cr_td.html

Linn, W. S., Avol, E. L., Peng, R. C., Shamoo, D. A., Hackney, J. D. (1987). Replicated dose-response study of sulfur dioxide effects in normal, atopic, and asthmatic volunteers. *Am Rev Respir Dis*. 136:1127-1134.

Lioy, P.J. (1990). The analysis of total human exposure for exposure assessment: A multi-discipline science for examining human contact with contaminants. *Environ Sci Technol*. 24: 938-945.

Liu, L.-J. S., Box, M., Kalman, D., Kaufman, J., Koenig, J., Larson, T., Lumley, T., Sheppard, L., Wallace, L. (2003). Exposure assessment of particulate matter for susceptible populations in Seattle. *Environ Health Persp*. 111: 909-918.

Mansfield, C., Houtven, G. V., Johnson, F. R., Yang, J.-C. (2009). Environmental Risks and Behavior: Do children spend less time outdoors when ozone pollution is high? ASSA annual meeting, January 5, 2009. Update of Houtven et al. (2003), using the OAB CHAD data set, and related to Mansfield et al. (2006).

Mansfield, C., Johnson, F. R., Van Houtven, G. (2006). The missing piece: averting behavior for children's ozone exposures. *Resource Energy Econ*. 28:215-228.

Marino, A. J., Fletcher, E. N., Whitaker, R. C., Anderson, S. E. (2012). Amount and environmental predictors of outdoor playtime at home and school: Across-sectional analysis of a national sample of preschool-aged children attending Head Start. *Health & Place*. 18: 1224-1230.

McCurdy, T. (2000). Conceptual basis for multi-route intake dose modeling using an energy expenditure approach. *J Expo Anal Environ Epidemiol*. 10: 1-12.

McCurdy, T., Glen, G., Smith, L., Lakkadi, Y. (2000). The National Exposure Research Laboratory's Consolidated Human Activity Database. *J Expo Anal Environ Epidemiol*. 10: 566-578.

McDermott, M., Srivastava, R., Croskell, S. (2006). Awareness of and compliance with air pollution advisories: A comparison of parents of asthmatics with other parents. *J Asthma*. 43:235-239.

Meng, Q., Williams, R., Pinto, J. P. (2012). Determinants of the associations between ambient concentrations and personal exposures to $PM_{2.5}$, NO_2, and O_3 during DEARS. *Atmos Environ*. 63:109-116.

Montoye, H. J., Kemper, H. C. G., Saris, W.H.N., Washburn, R.A. (1996). Measuring Physical Activity and Energy Expenditure. Champaign IL: Human Kinetics.

National Research Council (1991). Human Exposure Assessment for Airborne Pollutants: Advances and Opportunities. Washington, DC: National Academy of Sciences.

Neidell, M. (2010). Air quality warnings and outdoor activities: evidence from Southern California using a regression discontinuity approach design. *J Epidemiol Community Health*. 64:921-926.

Richmond H., Palma, T, Langstaff, J., McCurdy, T., Glenn, G., Smith, L. (2002). Further Refinements and Testing of APEX (3.0): EPA's population exposure model for criteria and air toxic inhalation exposures. Joint meeting of the International Society of Exposure Analysis and International Society of Environmental Epidemiology, Vancouver, CAN. August 11-15, 2002.

Robinson, J. P., Wiley, J. A., Piazza, T., Garrett, K., and Cirksena, K. (1989). Activity Patterns of California Residents and their Implications for Potential Exposure to Pollution. California Air Resources Board, Sacramento, CA. CARB-A6-177-33.

Sallis, J. F., Bauman, A., Pratt, M. (1998). Environmental and policy interventions to promote physical activity. *American Journal of Preventive Medicine*. 15(4):379-397.

Santuz, P., Baraldi, E., Filippone, M., Zacchello, F. (1997). Exercise performance in children with asthma: Is it different from that of healthy controls? *Eur Respir J*. 10:1254-1260.

SAS (2012). SAS/STAT 9.2 User's Guide, Second Edition. Available at: http://support.sas.com/documentation/cdl/en/statug/63033/PDF/default/statug.pdf

Semenza J. C., Wilson, D. J., Parra J., Bontempo, B. D., Hart, M., Sailor, D. J., George, L. A. (2008). Public perception and behavior change in relationship to hot weather and air pollution. *Environ Res*. 107:401-411.

Sexton, A. L. (2011). Responses to Air Quality Alerts: Do Americans Spend Less Time Outdoors? Available at: http://www.apec.umn.edu/prod/groups/cfans/@pub/@cfans/@apec/documents/asset/cfans_asset_365645.pdf

Shamoo, D. A., Linn, W. S., Peng, R.-C., Solomon, J. C., Webb, T. L., Hackney, J. D., Hong, H. (1994). Time-activity patterns and diurnal variation of respiratory status in a panel of asthmatics: implications for short-term air pollution effects. *J Expo Anal Environ Epidemiol.* 4(2):133-148.

Spier, C. E., Little, D. E., Trim, S. C., Johnson, T. R., Linn, W. S., Hackney, J. D. (1992). Activity patterns in elementary and high school students exposed to oxidant pollution. *J Expo Anal Environ Epidemiol.* 2:277-293.

Suh, H. (2008). Personal Communication, August 27, 2008, regarding personal exposure data for Atlanta.

Tsang, A. M., Klepeis, N. E. (1996). Descriptive Statistics Tables from a Detailed Analysis of the National Human Activity Pattern Survey (NHAPS) Data. U.S. Environmental Protection Agency. EPA/600/R-96/148.

US BLS. (2012a). American Time Use Survey User's Guide. Understanding ATUS 2003 to 2011. United States Bureau of Labor Statistics (BLS). August 2012. Data and documentation available at: http://www.bls.gov/tus/

US BLS. (2012b). Employment outlook: 2010–2020. Overview of projections to 2020. Available at: http://www.bls.gov/opub/mlr/2012/01/art1full.pdf.

US Census Bureau. (2007). 2000 Census of Population and Housing. Summary File 3 (SF3) Technical Documentation, available at: http://www.census.gov/prod/cen2000/doc/sf3.pdf. Individual SF3 files '30' (for income/poverty variables pct50) for each state were downloaded from: http://www2.census.gov/census_2000/datasets/Summary_File_3/.

USDA. (2000). Continuing Survey of Food Intake by Individuals (CSFII) 1994-96, 1998. Agricultural Research Service, CD-ROM. Available at: http://www.ars.usda.gov/Services/docs.htm?docid=14531

US DHHS. (1999). Promoting physical activity: a guide for community action. Champaign, IL: Human Kinetics. Department of Health and Human Services, Public Health Service, Centers for Disease Control and Prevention, National Center for Chronic Disease Prevention and Health Promotion, Division of Nutrition and Physical Activity. See Table 2-1 available at http://www.cdc.gov/nccdphp/dnpa/physical/pdf/PA_Intensity_table_2_1.pdf.

US EPA. (1986). Air Quality Criteria for Ozone and Other Photochemical Oxidants. Office of Health and Environmental Assessment, Environmental Criteria and Assessment Office, U.S. Environmental Protection Agency, Research Triangle Park, NC. EPA-600/8-84-020aF-eF. Available from: NTIS, Springfield, VA., PB87-142949.

US EPA. (1996a). Review of National Ambient Air Quality Standards for Ozone: Assessment of Scientific and Technical Information - OAQPS Staff Paper. Office of Air Quality Planning and Standards, U.S. Environmental Protection Agency, Research Triangle Park,

NC. EPA/452/R-96-007. Available at: http://www.epa.gov/ttn/naaqs/standards/ozone/s_o3_pr_sp.html

US EPA . (1996b). Air Quality Criteria for Ozone and Related Photochemical Oxidants. Office of Research and Development, National Center for Environmental Assessment, U.S. Environmental Protection Agency, Research Triangle Park, NC. EPA/600/P-93/004aF-cF. Available at: http://cfpub.epa.gov/ncea/cfm/recordisplay.cfm?deid=2831

US EPA. (2002). Consolidated Human Activities Database (CHAD) Users Guide. Database and documentation available at: http://www.epa.gov/chadnet1/

US EPA . (2006). Air Quality Criteria for Ozone and Related Photochemical Oxidants (Final). National Center for Environmental Assessment, U.S. Environmental Protection Agency, Research Triangle Park, NC. EPA/600/R-05/004aF-cF. Available at: http://cfpub.epa.gov/ncea/cfm/recordisplay.cfm?deid=149923

US EPA. (2007a). Review of National Ambient Air Quality Standards for Ozone: Policy Assessment of Scientific and Technical Information - OAQPS Staff Paper. Office of Air Quality Planning and Standards, U.S. Environmental Protection Agency, Research Triangle Park, NC. EPA-452/R-07-007. Available at: http://www.epa.gov/ttn/naaqs/standards/ozone/data/2007_07_ozone_staff_paper.pdf

US EPA. (2007b). Ozone Population Exposure Analysis for Selected Urban Areas. Office of Air Quality Planning and Standards, U.S. Environmental Protection Agency, Research Triangle Park, NC. Available at: http://www.epa.gov/ttn/naaqs/standards/ozone/s_o3_cr_td.html

US EPA. (2008). Risk and Exposure Assessment to Support the Review of the NO_2 Primary National Ambient Air Quality Standard. EPA-452/R-08-008a. November 2008. Available at: http://www.epa.gov/ttn/naaqs/standards/nox/data/20081121_NO2_REA_final.pdf.

US EPA . (2009a). Risk and Exposure Assessment to Support the Review of the SO_2 Primary National Ambient Air Quality Standard. EPA-452/R-09-007. August 2009. Available at http://www.epa.gov/ttn/naaqs/standards/so2/data/200908SO2REAFinalReport.pdf.

US EPA. (2009b). Metabolically derived human ventilation rates: a revised approach based upon oxygen consumption rates. National Center for Environmental Assessment, Washington, DC; EPA/600/R-06/129F. Available at: http://cfpub.epa.gov/ncea/cfm/recordisplay.cfm?deid=202543

US EPA. (2010). Quantitative Risk and Exposure Assessment for Carbon Monoxide – Amended. EPA Office of Air Quality Planning and Standards. EPA-452/R-10-009. July 2010. Available at: http://www.epa.gov/ttn/naaqs/standards/co/data/CO-REA-Amended-July2010.pdf

US EPA. (2011). Exposure Factors Handbook: 2011 Edition. National Center for Environmental Assessment, Washington, DC; EPA/600/R-09/052F. Available at: http://www.epa.gov/ncea/efh/pdfs/efh-complete.pdf

US EPA. (2012a). Total Risk Integrated Methodology (TRIM) - Air Pollutants Exposure Model Documentation (TRIM.Expo / APEX, Version 4.4) Volume I: User's Guide. Office of Air

Quality Planning and Standards, U.S. Environmental Protection Agency, Research Triangle Park, NC. EPA-452/B-12-001a. Available at: http://www.epa.gov/ttn/fera/human_apex.html

US EPA. (2012b). Total Risk Integrated Methodology (TRIM) - Air Pollutants Exposure Model Documentation (TRIM.Expo / APEX, Version 4.4) Volume II: Technical Support Document. Office of Air Quality Planning and Standards, U.S. Environmental Protection Agency, Research Triangle Park, NC. EPA-452/B-12-001b. Available at: http://www.epa.gov/ttn/fera/human_apex.html

US EPA. (2013). Integrated Science Assessment of Ozone and Related Photochemical Oxidants. U.S. Environmental Protection Agency, National Center for Environmental Assessment, U.S. Environmental Protection Agency, Research Triangle Park, N. EPA 600/R-10/076F. Available at: http://cfpub.epa.gov/ncea/isa/recordisplay.cfm?deid=247492#Download

University of Michigan. (2012). The Panel Study of Income Dynamics (PSID). Data and documentation available at http://psidonline.isr.umich.edu.

van Gent, R., van der Ent, K., van Essen-Zandvliet, L. E. M., Rovers, M. M., Kimpen, J. L. L., de Meer, G., Klijn, P. H. C. (2007). No Difference in physical activity in (un)diagnosed asthma and healthy controls. *Pediatric Pulmonology*. 42:1018-1023.

van Veldhoven, N. H. M. J., Vermeer, A., Bogaard, J. M., Hessels, M. G. P., Wijnroks, L., Colland, V. T., van Essen-Zandvliet, E. E. M. (2001). Children with asthma and physical exercise: effects of an exercise programme. *Clinical Rehabilitation*. 15:360-370.

Wen, X.-J., Balluz, L., Mokdad, A. (2009). Association between media alerts of Air Quality Index and change of outdoor activity among adult asthma in six states, BRFSS, 2005. *J Comm Health*. 34:40-46.

Wheeler, A., Zanobetti, A., Gold, D. R., Schwartz, J., Stone, P., Suh, H. H. (2006). The relationship between ambient air pollution and heart rate variability differs for individuals with heart and pulmonary disease. *Environ Health Perspect*. 114(4):560-6.

Whitfield, R., Biller, W., Jusko, M., Keisler, J. (1996). A Probabilistic Assessment of Health Risks Associated with Short- and Long-Term Exposure to Tropospheric Ozone. Argonne National Laboratory, Argonne, IL.

Wiley, J. A., Robinson, J. P., Piazza, T., Garrett, K., Cirksena, K., Cheng, Y.-T., Martin, G. (1991a). Activity Patterns of California Residents: Final Report. California Air Resources Board, Sacramento, CA. ARB/R93/487. Available from: NTIS, Springfield, VA., PB94-108719.

Wiley, J. A., Robinson, J. P., Cheng, Y.-T., Piazza, T., Stork, L., Pladsen, K. (1991b). Study of Children's Activity Patterns: Final Report. California Air Resources Board, Sacramento, CA. ARB-R-93/489.

Williams, R., Suggs, J., Creason, J., Rodes, C., Lawless, P., Kwok, R., Zweidinger, R., Sheldon, L. (2000). The 1998 Baltimore particulate matter epidemiology-exposure study: Part 2. Personal exposure associated with an elderly population. *J Expo Anal Environ Epidemiol*. 10(6): 533-543.

Williams, R., Suggs J., Rea, A., Leovic, K., Vette, A., Croghan, C., Sheldon, L., Rodes, C., Thornburg, J., Ejire, A., Herbst, M., Sanders Jr., W. (2003a). The Research Triangle particulate panel study: PM mass concentrations relationships. *Atmos Environ.* 37:5349-5363.

Williams, R., Suggs, J., Rea, A., Sheldon, L., Rodes, C., Thornburg, J. (2003b). The Research Triangle particulate panel study: modeling ambient source contributions to personal and residential PM mass concentrations. *Atmos Environ.* 37:5365-5378.

Xue, J., McCurdy, T., Spengler, J., Özkaynak, H. (2004). Understanding variability in time spent in selected locations for 7-12-year old children. *J Expo Anal Environ Epidemiol.* 14(3):222-33.

Zivin, J. G., Neidell M. (2009). Days of haze: Environmental information disclosure and intertemporal avoidance behavior. *J Environment Econ Manag.* 58(2):119-128.

Attachment 1. Occupations estimated to have at least one day per week where work is performed outdoors.

Major BLS SOC code	Major BLS SOC name	BLS SOC name	BLS SOC code	Employed (1000's)	O*NET Code	O*NET Name	Context	Out days/wk	Comment
11	Management Occupations	Industrial Production Managers	11-0000	150.3	11-3051.02	Geothermal Production Managers	77	1	
11	Management Occupations	Farmers, Ranchers, and O her Agricultural Managers	11-9013	400.8	11-9013.01	Nursery and Greenhouse Managers	96	5	divided bls code 11-9013 (Farm/Ranch/O her Ag Managers- 1202.5) by 3, possibly an overes imate
11	Management Occupations	Farmers, Ranchers, and O her Agricultural Managers	11-9013	400.8	11-9013.03	Aquacultural Managers	82	2	divided bls code 11-9013 (Farm/Ranch/O her Ag Managers- 1202.5) by 3, possibly an overes imate
11	Management Occupations	Construction Managers	11-9021	523.1	11-9021.00	Construction Managers	78	1	
13	Business and Financial Opera ions Occupations	Insurance Appraisers, Auto Damage	13-0000	10.6	13-1032.00	Insurance Appraisers, Auto Damage	89	3	
13	Business and Financial Opera ions Occupations	Compliance Officers	13-1041	216.6	13-1041.04	Government Property Inspectors and Investigators	82	2	
13	Business and Financial Opera ions Occupations	Farm Labor Contractors	13-1074	0.3	13-1074.00	Farm Labor Contractors	79	1	
13	Business and Financial Opera ions Occupations	Appraisers and Assessors of Real Estate	13-2021	77.8	13-2021.02	Appraisers, Real Estate	77	1	
17	Architecture and Engineering Occupations	Surveyors	17-0000	25.6	17-1022.00	Surveyors	90	3	divided bls code 17-1022 (surveyors- 51.2) by 2, estimate should be ok
17	Architecture and Engineering Occupations	Surveyors	17-1022	25.6	17-1022.01	Geodetic Surveyors	78	1	divided bls code 17-1022 (surveyors- 51.2) by 2, estimate should be ok
17	Architecture and Engineering Occupations	Surveying and Mapping Technicians	17-3031	56.9	17-3031.01	Surveying Technicians	92	4	
19	Life, Physical, and Social Science Occupations	Conservation Scientists	19-0000	7.8	19-1031.01	Soil and Water Conservationists	79	1	divided bls code 19-1031 (Conservation Scientists- 23.4) by 3, possibly an overestimate
19	Life, Physical, and Social Science Occupations	Conservation Scientists	19-1031	7.8	19-1031.03	Park Naturalists	78	1	divided bls code 19-1031 (Conservation Scientists- 23.4) by 3, possibly an overestimate
19	Life, Physical, and Social Science Occupations	Forest and Conservation Technicians	19-4093	36.5	19-4093.00	Forest and Conservation Technicians	90	3	
29	Healthcare Practitioners and Technical Occupations	Emergency Medical Technicians and Paramedics	29-0000	226.5	29-2041.00	Emergency Medical Technicians and Paramedics	87	3	
33	Protective Service Occupations	First-Line Supervisors of Police and Detec ives	33-0000	106.1	33-1012.00	First-Line Supervisors of Police and Detectives	82	2	
33	Protective Service Occupations	First-Line Supervisors of Fire Fighting and Prevention Workers	33-1021	30	33-1021.01	Municipal Fire Fighting and Preven ion Supervisors	88	3	divided bls code 33-1021 (First-Line Supervisors of Fire Fighting and Prevention Workers- 60.1) by 2, estimate should be ok
33	Protective Service Occupations	First-Line Supervisors of Fire Fighting and Prevention Workers	33-1021	30	33-1021.02	Forest Fire Fighting and Preven ion Supervisors	86	3	divided bls code 33-1021 (First-Line Supervisors of Fire Fighting and Prevention Workers- 60.1) by 2, estimate should be ok
33	Protective Service Occupations	Firefighters	33-2011	155.2	33-2011.01	Municipal Firefighters	80	1	divided bls code 33-2011 (Firefighters- 310.4) by 2, estimate should be ok
33	Protective Service Occupations	Firefighters	33-2011	155.2	33-2011.02	Forest Firefighters	87	3	divided bls code 33-2011 (Firefighters- 310.4) by 2, estimate should be ok
33	Protective Service Occupations	Fire Inspectors and Investigators	33-2021	6.8	33-2021.01	Fire Inspectors	86	3	divided bls code 33-2021 (Fire Inspectors and Investigators- 13.6) by 2, estimate should be ok
33	Protective Service Occupations	Fire Inspectors and Investigators	33-2021	6.8	33-2021.02	Fire Investigators	84	2	divided bls code 33-2021 (Fire Inspectors and Investigators- 13.6) by 2, estimate should be ok
33	Protective Service Occupations	Forest Fire Inspectors and Prevention Specialists	33-2022	1.6	33-2022.00	Forest Fire Inspectors and Preven ion Specialists	77	1	

Major BLS SOC code	Major BLS SOC name	BLS SOC name	BLS SOC code	Employed (1000's)	O*NET Code	O*NET Name	Context	Out days/wk	Comment
33	Protec ive Service Occupations	Detectives and Criminal Investigators	33-3021	23.9	33-3021.01	Police Detec ives	87	3	divided bls code 33-3021 (Detectives and Criminal Investigators- 119.4) by 5, possibly an overestimate
33	Protec ive Service Occupations	Detectives and Criminal Investigators	33-3021	23.9	33-3021.03	Criminal Investigators and Special Agents	76	1	divided bls code 33-3021 (Detectives and Criminal Investigators- 119.4) by 5, possibly an overestimate
33	Protec ive Service Occupations	Detectives and Criminal Investigators	33-3021	23.9	33-3021.05	Immigration and Customs Inspectors	85	2	divided bls code 33-3021 (Detectives and Criminal Investigators- 119.4) by 5, possibly an overestimate
33	Protec ive Service Occupations	Fish and Game Wardens	33-3031	7.6	33-3031.00	Fish and Game Wardens	96	5	
33	Protec ive Service Occupations	Parking Enforcement Workers	33-3041	9.8	33-3041.00	Parking Enforcement Workers	95	4	
33	Protec ive Service Occupations	Police and Sheriff's Patrol Officers	33-3051	332	33-3051.01	Police Patrol Officers	91	4	divided bls code 33-3051 (Police and Sheriff Patrol Officers- 663.9) by 2, estimate should be ok
33	Protec ive Service Occupations	Police and Sheriff's Patrol Officers	33-3051	332	33-3051.03	Sheriffs and Deputy Sheriffs	88	3	divided bls code 33-3051 (Police and Sheriff Patrol Officers- 663.9) by 2, estimate should be ok
33	Protec ive Service Occupations	Animal Control Workers	33-9011	15.5	33-9011.00	Animal Control Workers	87	3	
33	Protec ive Service Occupations	Private Detectives and Investigators	33-9021	34.7	33-9021.00	Private Detectives and Investigators	76	1	
33	Protec ive Service Occupations	Crossing Guards	33-9091	69.3	33-9091.00	Crossing Guards	100	5	
37	Building and Grounds Cleaning and Maintenance Occupa ions	First-Line Supervisors of Landscaping, Lawn Service, and Groundskeeping Workers	37-0000	202.9	37-1012.00	First-Line Supervisors of Landscaping, Lawn Service, and Groundskeeping Workers	94	4	
37	Building and Grounds Cleaning and Maintenance Occupa ions	Pest Control Workers	37-2021	68.4	37-2021.00	Pest Control Workers	98	5	
37	Building and Grounds Cleaning and Maintenance Occupa ions	Landscaping and Groundskeeping Workers	37-3011	1151.5	37-3011.00	Landscaping and Groundskeeping Workers	99	5	
37	Building and Grounds Cleaning and Maintenance Occupa ions	Pesticide Handlers, Sprayers, and Applicators, Vegetation	37-3012	29.5	37-3012.00	Pesticide Handlers, Sprayers, and Applicators, Vegetation	89	3	
37	Building and Grounds Cleaning and Maintenance Occupa ions	Tree Trimmers and Pruners	37-3013	50.6	37-3013.00	Tree Trimmers and Pruners	99	5	
39	Personal Care and Service Occupations		39-0000	29.3	39-4031.00	Morticians, Undertakers, and Funeral Directors	77	1	used employment data from bls code 39-4831 (funeral dir., etc), estimate should be ok
39	Personal Care and Service Occupations	Baggage Porters and Bellhops	39-6011	46	39-6011.00	Baggage Porters and Bellhops	82	2	
43	Office and Administrative Support Occupations	Couriers and Messengers	43-0000	116.2	43-5021.00	Couriers and Messengers	89	3	
43	Office and Administrative Support Occupations	Meter Readers, Utilities	43-5041	40.5	43-5041.00	Meter Readers, Utilities	99	5	
43	Office and Administrative Support Occupations	Postal Service Mail Carriers	43-5052	316.7	43-5052.00	Postal Service Mail Carriers	97	5	
45	Farming, Fishing, and Forestry Occupations	First-Line Supervisors of Farming, Fishing, and Forestry Workers	45-0000	11.8	45-1011.05	First-Line Supervisors of Logging Workers	95	4	divided bls code 45-1011 (First-Line Supervisors of Farming, Fishing, and Forestry Workers- 47) by 4, estimate should be ok

Major BLS SOC code	Major BLS SOC name	BLS SOC name	BLS SOC code	Employed (1000's)	O*NET Code	O*NET Name	Context	Out days/wk	Comment
45	Farming, Fishing, and Forestry Occupations	First-Line Supervisors of Farming, Fishing, and Forestry Workers	45-1011	11.8	45-1011.06	First-Line Supervisors of Aquacultural Workers	87	3	divided bls code 45-1011 (First-Line Supervisors of Farming, Fishing, and Forestry Workers- 47) by 4, estimate should be ok
45	Farming, Fishing, and Forestry Occupations	First-Line Supervisors of Farming, Fishing, and Forestry Workers	45-1011	11.8	45-1011.07	First-Line Supervisors of Agricultural Crop and Horticultural Workers	87	3	divided bls code 45-1011 (First-Line Supervisors of Farming, Fishing, and Forestry Workers- 47) by 4, estimate should be ok
45	Farming, Fishing, and Forestry Occupations	First-Line Supervisors of Farming, Fishing, and Forestry Workers	45-1011	11.8	45-1011.08	First-Line Supervisors of Animal Husbandry and Animal Care Workers	84	2	divided bls code 45-1011 (First-Line Supervisors of Farming, Fishing, and Forestry Workers- 47) by 4, estimate should be ok
45	Farming, Fishing, and Forestry Occupations	Animal Breeders	45-2021	11.5	45-2021.00	Animal Breeders	91	4	
45	Farming, Fishing, and Forestry Occupations		45-2091	186.6	45-2091.00	Agricultural Equipment Operators	95	4	used employment data from bls code 45-2090 (misc ag- 746.4) divided by 4, possibly an overestimate
45	Farming, Fishing, and Forestry Occupations		45-2092	186.6	45-2092.01	Nursery Workers	76	1	used employment data from bls code 45-2090 (misc ag- 746.4) divided by 4, possibly an overestimate
45	Farming, Fishing, and Forestry Occupations		45-2092	186.6	45-2092.02	Farmworkers and Laborers, Crop	98	5	used employment data from bls code 45-2090 (misc ag- 746.4) divided by 4, possibly an overestimate
45	Farming, Fishing, and Forestry Occupations		45-2093	186.6	45-2093.00	Farmworkers, Farm, Ranch, and Aquacultural Animals	83	2	used employment data from bls code 45-2090 (misc ag- 746.4) divided by 4, possibly an overestimate
45	Farming, Fishing, and Forestry Occupations	Fishers and Related Fishing Workers	45-3011	32	45-3011.00	Fishers and Related Fishing Workers	93	4	
45	Farming, Fishing, and Forestry Occupations		45-3021	0.6	45-3021.00	Hunters and Trappers	99	5	diff of parent bls 45-3000 (fish hunt- 32.6) and sub 45-3011 (fishing- 32), estimate should be ok
45	Farming, Fishing, and Forestry Occupations	Forest and Conservation Workers	45-4011	13.7	45-4011.00	Forest and Conserva ion Workers	83	2	
45	Farming, Fishing, and Forestry Occupations	Fallers	45-4021	9.6	45-4021.00	Fallers	99	5	
45	Farming, Fishing, and Forestry Occupations	Logging Equipment Operators	45-4022	35.1	45-4022.00	Logging Equipment Operators	95	4	
45	Farming, Fishing, and Forestry Occupations	Log Graders and Scalers	45-4023	3.8	45-4023.00	Log Graders and Scalers	81	2	
47	Construc ion and Extrac ion Occupations	First-Line Supervisors of Construction Trades and Extraction Workers	47-0000	558.5	47-1011.00	First-Line Supervisors of Construction Trades and Extraction Workers	83	2	
47	Construc ion and Extrac ion Occupations	Brickmasons and Blockmasons	47-2021	89.2	47-2021.00	Brickmasons and Blockmasons	99	5	
47	Construc ion and Extrac ion Occupations	Stonemasons	47-2022	15.6	47-2022.00	Stonemasons	92	4	
47	Construc ion and Extrac ion Occupations	Carpenters	47-2031	500.8	47-2031.01	Construction Carpenters	78	1	divided bls code 47-2031 (Carpenters-1001.7) by 2, es imate should be ok
47	Construc ion and Extrac ion Occupations	Carpenters	47-2031	500.8	47-2031.02	Rough Carpenters	83	2	divided bls code 47-2031 (Carpenters- 1001.7) by 2, es imate should be ok
47	Construc ion and Extrac ion Occupations	Cement Masons and Concrete Finishers	47-2051	144.7	47-2051.00	Cement Masons and Concrete Finishers	100	5	
47	Construc ion and Extrac ion Occupations	Construction Laborers	47-2061	998.8	47-2061.00	Construction Laborers	97	5	
47	Construc ion and Extrac ion Occupations	Paving, Surfacing, and Tamping Equipment Operators	47-2071	51.6	47-2071.00	Paving, Surfacing, and Tamping Equipment Operators	94	4	
47	Construc ion and	Pile-Driver Operators	47-2072	4.1	47-2072.00	Pile-Driver Operators	100	5	

Major BLS SOC code	Major BLS SOC name	BLS SOC name	BLS SOC code	Employed (1000's)	O*NET Code	O*NET Name	Context	Out days/wk	Comment
	Extraction Occupa ions								
47	Construction and Extraction Occupa ions	Operating Engineers and Other Construction Equipment Operators	47-2073	349.1	47-2073.00	Operating Engineers and Other Construction Equipment Operators	99	5	
47	Construction and Extraction Occupa ions	Glaziers	47-2121	41.9	47-2121.00	Glaziers	91	4	
47	Construction and Extraction Occupa ions	Insulation Workers, Floor, Ceiling, and Wall	47-2131	23.2	47-2131.00	Insulation Workers, Floor, Ceiling, and Wall	87	3	
47	Construction and Extraction Occupa ions	Pipelayers	47-2151	53.1	47-2151.00	Pipelayers	97	5	
47	Construction and Extraction Occupa ions	Plumbers, Pipefitters, and Steamfitters	47-2152	419.9	47-2152.02	Plumbers	79	1	
47	Construction and Extraction Occupa ions	Reinforcing Iron and Rebar Workers	47-2171	19.1	47-2171.00	Reinforcing Iron and Rebar Workers	94	4	
47	Construction and Extraction Occupa ions	Roofers	47-2181	136.7	47-2181.00	Roofers	100	5	
47	Construction and Extraction Occupa ions	Structural Iron and Steel Workers	47-2221	59.8	47-2221.00	Structural Iron and Steel Workers	97	5	
47	Construction and Extraction Occupa ions	Helpers--Brickmasons, Blockmasons, Stonemasons, and Tile and Marble Setters	47-3011	29.4	47-3011.00	Helpers--Brickmasons, Blockmasons, Stonemasons, and Tile and Marble Setters	86	3	
47	Construction and Extraction Occupa ions	Helpers--Pipelayers, Plumbers, Pipefitters, and Steamfitters	47-3015	57.9	47-3015.00	Helpers--Pipelayers, Plumbers, Pipefitters, and Steamfitters	87	3	
47	Construction and Extraction Occupa ions	Helpers--Roofers	47-3016	12.7	47-3016.00	Helpers--Roofers	88	3	
47	Construction and Extraction Occupa ions	Construction and Building Inspectors	47-4011	102.4	47-4011.00	Construction and Building Inspectors	88	3	
47	Construction and Extraction Occupa ions	Fence Erectors	47-4031	32.1	47-4031.00	Fence Erectors	99	5	
47	Construction and Extraction Occupa ions	Highway Maintenance Workers	47-4051	148.5	47-4051.00	Highway Maintenance Workers	84	2	
47	Construction and Extraction Occupa ions	Rail-Track Laying and Maintenance Equipment Operators	47-4061	15	47-4061.00	Rail-Track Laying and Maintenance Equipment Operators	93	4	
47	Construction and Extraction Occupa ions	Sep ic Tank Servicers and Sewer Pipe Cleaners	47-4071	25.3	47-4071.00	Septic Tank Servicers and Sewer Pipe Cleaners	92	4	
47	Construction and Extraction Occupa ions	Segmental Pavers	47-4091	1.3	47-4091.00	Segmental Pavers	84	2	
47	Construction and Extraction Occupa ions	Derrick Operators, Oil and Gas	47-5011	18.9	47-5011.00	Derrick Operators, Oil and Gas	100	5	
47	Construction and Extraction Occupa ions	Rotary Drill Operators, Oil and Gas	47-5012	22.5	47-5012.00	Rotary Drill Operators, Oil and Gas	99	5	
47	Construction and Extraction Occupa ions	Service Unit Operators, Oil, Gas, and Mining	47-5013	40.7	47-5013.00	Service Unit Operators, Oil, Gas, and Mining	99	5	
47	Construction and Extraction Occupa ions	Earth Drillers, Except Oil and Gas	47-5021	17.8	47-5021.00	Ear h Drillers, Except Oil and Gas	95	4	
47	Construction and Extraction Occupa ions	Explosives Workers, Ordnance Handling Experts, and Blasters	47-5031	6.8	47-5031.00	Explosives Workers, Ordnance Handling Experts, and Blasters	98	5	
47	Construction and Extraction Occupa ions	Rock Splitters, Quarry	47-5051	3.5	47-5051.00	Rock Splitters, Quarry	94	4	
47	Construction and	Roustabouts, Oil and Gas	47-5071	52.7	47-5071.00	Roustabouts, Oil and Gas	100	5	

Major BLS SOC code	Major BLS SOC name	BLS SOC name	BLS SOC code	Employed (1000's)	O*NET Code	O*NET Name	Context	Out days/wk	Comment
	Extraction Occupations								
49	Installation, Maintenance, and Repair Occupations	Telecommunications Equipment Installers and Repairers, Except Line Installers	49-0000	194.9	49-2022.00	Telecommunications Equipment Installers and Repairers, Except Line Installers	79	1	
49	Installation, Maintenance, and Repair Occupations	Electrical and Electronics Repairers, Powerhouse, Substation, and Relay	49-2095	23.4	49-2095.00	Electrical and Electronics Repairers, Powerhouse, Substation, and Relay	92	4	
49	Installation, Maintenance, and Repair Occupations	Automotive Glass Installers and Repairers	49-3022	18.1	49-3022.00	Automo ive Glass Installers and Repairers	88	3	
49	Installation, Maintenance, and Repair Occupations	Bus and Truck Mechanics and Diesel Engine Specialists	49-3031	242.2	49-3031.00	Bus and Truck Mechanics and Diesel Engine Specialists	83	2	
49	Installation, Maintenance, and Repair Occupations	Farm Equipment Mechanics and Service Technicians	49-3041	32.9	49-3041.00	Farm Equipment Mechanics and Service Technicians	83	2	
49	Installation, Maintenance, and Repair Occupations	Mobile Heavy Equipment Mechanics, Except Engines	49-3042	124.6	49-3042.00	Mobile Heavy Equipment Mechanics, Except Engines	78	1	
49	Installation, Maintenance, and Repair Occupations	Rail Car Repairers	49-3043	21.7	49-3043.00	Rail Car Repairers	81	2	
49	Installation, Maintenance, and Repair Occupations	Motorboat Mechanics and Service Technicians	49-3051	20.8	49-3051.00	Motorboat Mechanics and Service Technicians	86	3	
49	Installation, Maintenance, and Repair Occupations	Mechanical Door Repairers	49-9011	12.8	49-9011.00	Mechanical Door Repairers	93	4	
49	Installation, Maintenance, and Repair Occupations	Control and Valve Installers and Repairers, Except Mechanical Door	49-9012	43.8	49-9012.00	Control and Valve Installers and Repairers, Except Mechanical Door	88	3	
49	Installation, Maintenance, and Repair Occupations	Heating, Air Conditioning, and Refrigeration Mechanics and Installers	49-9021	133.9	49-9021.01	Hea ing and Air Conditioning Mechanics and Installers	84	2	divided bls code 49-9021 (Heating AC refrig mech- 267.8) by 2, es imate should be ok
49	Installation, Maintenance, and Repair Occupations	Heating, Air Conditioning, and Refrigeration Mechanics and Installers	49-9021	133.9	49-9021.02	Refrigeration Mechanics and Installers	94	4	divided bls code 49-9021 (Heating AC refrig mech- 267.8) by 2, es imate should be ok
49	Installation, Maintenance, and Repair Occupations	Millwrights	49-9044	36.5	49-9044.00	Millwrights	77	1	
49	Installation, Maintenance, and Repair Occupations	Electrical Power-Line Installers and Repairers	49-9051	108.4	49-9051.00	Electrical Power-Line Installers and Repairers	95	4	
49	Installation, Maintenance, and Repair Occupations	Telecommunications Line Installers and Repairers	49-9052	160.6	49-9052.00	Telecommunications Line Installers and Repairers	94	4	
49	Installation, Maintenance, and Repair Occupations	Coin, Vending, and Amusement Machine Servicers and Repairers	49-9091	39.1	49-9091.00	Coin, Vending, and Amusement Machine Servicers and Repairers	85	2	
49	Installation, Maintenance, and Repair Occupations	Commercial Divers	49-9092	3.8	49-9092.00	Commercial Divers	92	4	
49	Installation, Maintenance, and Repair Occupations	Locksmiths and Safe Repairers	49-9094	25.7	49-9094.00	Locksmiths and Safe Repairers	87	3	
49	Installation, Maintenance, and Repair Occupations	Manufactured Building and Mobile Home Installers	49-9095	7.8	49-9095.00	Manufactured Building and Mobile Home Installers	86	3	
49	Installation, Maintenance, and Repair Occupations	Riggers	49-9096	15.2	49-9096.00	Riggers	86	3	
49	Installation, Maintenance, and Repair Occupations	Signal and Track Switch Repairers	49-9097	7.1	49-9097.00	Signal and Track Switch Repairers	96	5	
49	Installation, Maintenance,		49-9099	143.6	49-9099.01	Geothermal Technicians	94	4	used employment data from bls code 49-9799

Major BLS SOC code	Major BLS SOC name	BLS SOC name	BLS SOC code	Employed (1000's)	O*NET Code	O*NET Name	Context	Out days/wk	Comment
	and Repair Occupations								(install, repair, other, etc.), possibly an overestimate
51	Production Occupations	Water and Wastewater Treatment Plant and System Operators	51-0000	110.7	51-8031.00	Water and Wastewater Treatment Plant and System Operators	93	4	
51	Production Occupations	Gas Plant Operators	51-8092	13.7	51-8092.00	Gas Plant Operators	92	4	
51	Production Occupations	Petroleum Pump System Operators, Refinery Operators, and Gaugers	51-8093	44.2	51-8093.00	Petroleum Pump System Operators, Refinery Operators, and Gaugers	93	4	
53	Transportation and Material Moving Occupations	Aircraft Cargo Handling Supervisors	53-0000	6.3	53-1011.00	Aircraft Cargo Handling Supervisors	77	1	
53	Transportation and Material Moving Occupations	Commercial Pilots	53-2012	32.7	53-2012.00	Commercial Pilots	92	4	
53	Transportation and Material Moving Occupations	Bus Drivers, Transit and Intercity	53-3021	186.3	53-3021.00	Bus Drivers, Transit and Intercity	83	2	
53	Transportation and Material Moving Occupations	Driver/Sales Workers	53-3031	406.6	53-3031.00	Driver/Sales Workers	76	1	
53	Transportation and Material Moving Occupations	Heavy and Tractor-Trailer Truck Drivers	53-3032	1604.8	53-3032.00	Heavy and Tractor-Trailer Truck Drivers	100	5	
53	Transportation and Material Moving Occupations	Light Truck or Delivery Services Drivers	53-3033	856	53-3033.00	Light Truck or Delivery Services Drivers	100	5	
53	Transportation and Material Moving Occupations	Taxi Drivers and Chauffeurs	53-3041	239.9	53-3041.00	Taxi Drivers and Chauffeurs	78	1	
53	Transportation and Material Moving Occupations	Locomotive Engineers	53-4011	38.7	53-4011.00	Locomotive Engineers	86	3	
53	Transportation and Material Moving Occupations	Locomotive Firers	53-4012	1.1	53-4012.00	Locomotive Firers	86	3	
53	Transportation and Material Moving Occupations	Rail Yard Engineers, Dinkey Operators, and Hostlers	53-4013	5.6	53-4013.00	Rail Yard Engineers, Dinkey Operators, and Hostlers	84	2	
53	Transportation and Material Moving Occupations	Railroad Brake, Signal, and Switch Operators	53-4021	21.7	53-4021.00	Railroad Brake, Signal, and Switch Operators	100	5	
53	Transportation and Material Moving Occupations	Railroad Conductors and Yardmasters	53-4031	40.8	53-4031.00	Railroad Conductors and Yardmasters	87	3	
53	Transportation and Material Moving Occupations	Subway and Streetcar Operators	53-4041	6.5	53-4041.00	Subway and Streetcar Operators	77	1	
53	Transportation and Material Moving Occupations	Sailors and Marine Oilers	53-5011	33.4	53-5011.00	Sailors and Marine Oilers	99	5	
53	Transportation and Material Moving Occupations	Captains, Mates, and Pilots of Water Vessels	53-5021	36.1	53-5021.02	Mates- Ship, Boat, and Barge	95	4	

Major BLS SOC code	Major BLS SOC name	BLS SOC name	BLS SOC code	Employed (1000's)	O*NET Code	O*NET Name	Context	Out days/wk	Comment
53	Transportation and Material Moving Occupations	Motorboat Operators	53-5022	3.1	53-5022.00	Motorboat Operators	85	2	
53	Transportation and Material Moving Occupations	Ship Engineers	53-5031	10.1	53-5031.00	Ship Engineers	77	1	
53	Transportation and Material Moving Occupations	Parking Lot Attendants	53-6021	125.1	53-6021.00	Parking Lot Attendants	76	1	
53	Transportation and Material Moving Occupations	Automotive and Watercraft Service Attendants	53-6031	86.3	53-6031.00	Automotive and Watercraft Service Attendants	94	4	
53	Transportation and Material Moving Occupations	Transportation Inspectors	53-6051	27.4	53-6051.08	Freight and Cargo Inspectors	80	1	
53	Transportation and Material Moving Occupations	Transportation Attendants, Except Flight Attendants	53-6061	24.8	53-6061.00	Transportation Attendants, Except Flight Attendants	77	1	
53	Transportation and Material Moving Occupations	Dredge Operators	53-7031	2.1	53-7031.00	Dredge Operators	86	3	
53	Transportation and Material Moving Occupations	Excavating and Loading Machine and Dragline Operators	53-7032	61.5	53-7032.00	Excavating and Loading Machine and Dragline Operators	88	3	
53	Transportation and Material Moving Occupations	Gas Compressor and Gas Pumping Station Operators	53-7071	4.5	53-7071.00	Gas Compressor and Gas Pumping Station Operators	91	4	
53	Transportation and Material Moving Occupations	Pump Operators, Except Wellhead Pumpers	53-7072	10.8	53-7072.00	Pump Operators, Except Wellhead Pumpers	99	5	
53	Transportation and Material Moving Occupations	Wellhead Pumpers	53-7073	15.1	53-7073.00	Wellhead Pumpers	93	4	
53	Transportation and Material Moving Occupations	Refuse and Recyclable Material Collectors	53-7081	139.9	53-7081.00	Refuse and Recyclable Material Collectors	100	5	
53	Transportation and Material Moving Occupations	Tank Car, Truck, and Ship Loaders	53-7121	10.4	53-7121.00	Tank Car, Truck, and Ship Loaders	91	4	

Attachment 2. Additional mapping of O*NET occupation codes to CHAD/APEX METs occupation activity codes.

```
if occ_code_onet='13-1032.00' then Occ_name_CHAD='ADMSUP';*Insurance Appraisers, Auto Damage;
if occ_code_onet='13-1041.04' then Occ_name_CHAD='ADMSUP';*Government Property Inspectors and Investigators;
if occ_code_onet='13-1074.00' then Occ_name_CHAD='ADMSUP';*Farm Labor Contractors;
if occ_code_onet='13-2021.02' then Occ_name_CHAD='ADMSUP';*Appraisers, Real Estate;
if occ_code_onet='17-3031.01' then Occ_name_CHAD='TECH';*Surveying Technicians;
if occ_code_onet='19-4093.00' then Occ_name_CHAD='TECH';*Forest and Conservation Technicians;
if occ_code_onet='33-9011.00' then Occ_name_CHAD='PROTECT';*Animal Control Workers;
if occ_code_onet='33-9021.00' then Occ_name_CHAD='PROTECT';*Private Detectives and Investigators;
if occ_code_onet='33-9091.00' then Occ_name_CHAD='PROTECT';*Crossing Guards;
if occ_code_onet='37-3013.00' then Occ_name_CHAD='FARM';*Tree Trimmers and Pruners;
if occ_code_onet='43-5021.00' then Occ_name_CHAD='ADMSUP';*Couriers and Messengers;
if occ_code_onet='45-2021.00' then Occ_name_CHAD='FARM';*Animal Breeders;
if occ_code_onet='45-2093.00' then Occ_name_CHAD='FARM';*Farmworkers, Farm, Ranch, and Aquacultural Animals;
if occ_code_onet='45-3011.00' then Occ_name_CHAD='FARM';*Fishers and Related Fishing Workers;
if occ_code_onet='45-4011.00' then Occ_name_CHAD='FARM';*Forest and Conservation Workers;
if occ_code_onet='45-4023.00' then Occ_name_CHAD='FARM';*Log Graders and Scalers;
if occ_code_onet='47-2061.00' then Occ_name_CHAD='LABOR';*Construction Laborers;
if occ_code_onet='47-2071.00' then Occ_name_CHAD='TRANS';*Paving, Surfacing, and Tamping Equipment Operators;
if occ_code_onet='47-2073.00' then Occ_name_CHAD='TRANS';*Operating Engineers and Other Construction Equipment Operators;
if occ_code_onet='47-2121.00' then Occ_name_CHAD='PREC';*Glaziers;
if occ_code_onet='47-4011.00' then Occ_name_CHAD='ADMIN';*Construction and Building Inspectors;
if occ_code_onet='47-4051.00' then Occ_name_CHAD='LABOR';*Highway Maintenance Workers;
if occ_code_onet='47-4061.00' then Occ_name_CHAD='LABOR';*Rail-Track Laying and Maintenance Equipment Operators;
if occ_code_onet='47-5071.00' then Occ_name_CHAD='LABOR';*Roustabouts, Oil and Gas;
if occ_code_onet='49-3022.00' then Occ_name_CHAD='PREC';*Automotive Glass Installers and Repairers;
if occ_code_onet='49-9091.00' then Occ_name_CHAD='PREC';*Coin, Vending, and Amusement Machine Servicers and Repairers;
if occ_code_onet='49-9094.00' then Occ_name_CHAD='PREC';*Locksmiths and Safe Repairers;
if occ_code_onet='49-9096.00' then Occ_name_CHAD='TRANS';*Riggers;
if occ_code_onet='49-9099.01' then Occ_name_CHAD='TECH';*Geothermal Technicians;
if occ_code_onet='53-1011.00' then Occ_name_CHAD='TRANS';*Aircraft Cargo Handling Supervisors;
if occ_code_onet='53-5011.00' then Occ_name_CHAD='TRANS';*Sailors and Marine Oilers;
if occ_code_onet='53-5031.00' then Occ_name_CHAD='TRANS';*Ship Engineers;
if occ_code_onet='53-6051.08' then Occ_name_CHAD='TRANS';*Freight and Cargo Inspectors;
*this following was mischaracterized as 'SALES';
if occ_code_onet='53-3032.00' then Occ_name_CHAD='TRANS'; *Heavy and Tractor-Trailer Truck Drivers;
```

United States Environmental Protection Agency	Office of Air Quality Planning and Standards Air Quality Strategies and Standards Division Research Triangle Park, NC	Publication No. EPA-452/P-14-004c February 2014

www.ingramcontent.com/pod-product-compliance
Lightning Source LLC
Chambersburg PA
CBHW081432170526
45166CB00008B/2185